BASIC LABORATORY PROCEDURES FOR THE OPERATOR–ANALYST

Sixth Edition

Edited by Sidney Innerebner, PhD, PE, CWP, PO

2022

Water Environment Federation
601 Wythe Street
Alexandria, VA 22314-1994 USA
https://www.wef.org

ISBN 978-1-57278-443-7

About WEF

The Water Environment Federation (WEF) is a not-for-profit technical and educational organization of 30,000 individual members and 75 affiliated Member Associations representing water quality professionals around the world. Since 1928, WEF and its members have protected public health and the environment. As a global water sector leader, our mission is to connect water professionals; enrich the expertise of water professionals; increase the awareness of the impact and value of water; and provide a platform for water sector innovation. To learn more, visit www.wef.org.

Sixth edition edited by Sidney Innerebner, PhD, PE, CWP, PO, Indigo Water Group

With contributions by J. Hunter Adams

Review provided by

Michelle Courtroul	Peter Petersen
Richard Finger	Peter Strimple

Reviewers' efforts were supported by the following organizations:

City of Santa Cruz Environmental Laboratory
City of Wichita Falls Cypress Environmental Laboratory

Text originally prepared by the **Basic Laboratory Procedures for the Operator–Analyst Task Force** of the **Water Environment Federation**

Sidney Innerebner, PhD, PE, CWP, *Chair*

Edward F. Askew
Michael Beattie
Rebecca J. Bodnar
Anne Payne Bullock
Srinivasa Rao Chitikela, PhD, PE, BCEE
Brian Clow, PE, BCEE
Janice Conklin
John K. Cooper
Bob Dabkowski
Gil Dichter
Eric Lynne
Kristin Evans, PhD, PE
Ronald Eyma, PE, DWRE
Richard Finger
Alicia Giddens
Richard Huyck

Samuel S. Jeyanayagam, PhD, PE, BCEE
Hari Kapalavai, PE
Jorj Long
Chris Maher, CWP
Annette McMurray
Gregg Mitchell
Indra N. Mitra, PhD, PE, MBA, BCEE
Devon Morgan
Jerry Morgan
Art Newby, CHMM, CPESC
Jill I. Norton
Trinity J. O'Neal
Peter Petersen
Jaana Pietari, PhD
David Riedel
Kevin Winnett, PhD
Frederic J. Winter
Paula Zeller, CWP

Under the Direction of the **Plant Operations and Maintenance Subcommittee** of the **Technical Practice Committee**

Special Publications of the Water Environment Federation

The WEF Technical Practice Committee (formerly the Committee on Sewage and Industrial Wastes Practice of the Federation of Sewage and Industrial Wastes Associations) was created by the Federation Board of Control on October 11, 1941. The primary function of the Committee is to originate and produce, through appropriate subcommittees, publications dealing with technical aspects of the broad interests of the Federation. These publications are intended to provide background information through a review of technical practices and detailed procedures that research and experience have shown to be functional and practical.

Contents

List of Figures

List of Tables

Preface

This manual includes laboratory procedures most often performed in wastewater laboratories and includes general wet chemistry procedures, activated sludge process control tests, bacteriological testing, and special sections about basic laboratory techniques and proper use of spectrophotometers and colorimetry. Each method (procedure) is broken down into easy-to-follow, step-by-step directions that have been updated to adhere to recommendations found in *Standard Methods for the Examination of Water and Wastewater* (23rd ed.; American Public Health Association, American Water Works Association, & Water Environment Federation, 2017) and approved U.S. Environmental Protection Agency (U.S. EPA) guidelines. Each method (procedure) includes a sample benchsheet. The intention of this publication is to help the operator–analyst produce analytical data that are defensible, precise, and accurate for process control and permit reporting. Special emphasis is placed on quality assurance and quality control.

Laboratory results are meaningless unless they can be put into context. The introduction to each method includes a brief discussion of why the test is performed and how it can be used for process control. Most methods include typical ranges for the parameter associated with different types of treatment processes. This information will help the operator–analyst determine the appropriateness (i.e., when and where) of the collected samples and if results are reasonable (i.e., make sense). Recommended sampling locations and parameters for process control are discussed in Chapter 1.

This publication is intended to serve a special need for operating personnel but not to replace or contradict the contents of *Standard Methods*. *Standard Methods* is recommended for those who need further background or a more complete understanding of the subject matter.

On October 10, 1963, the Board of Control of the Water Environment Federation (then Water Pollution Control Federation) authorized preparation of a publication focusing on analytical procedures of special interest to operators. The initial report was published as *Simplified Laboratory Procedures for Wastewater Examination* in 1968, with reprinting and minor corrections in 1969, 1970, 1971, and 1972. The second edition was published in 1976. The third edition was published in 1985, the fourth edition was published in 2002, and the fifth edition was published in 2012.

This sixth edition of this Special Publication was edited under the direction of Sidney Innerebner, PhD, PE, CWP, PO, Indigo Water Group.

Introduction to the Laboratory

1.0 BACKGROUND

Standard Methods for the Examination of Water and Wastewater is a joint publication of the American Public Health Association (APHA), the American Water Works Association (AWWA), and Water Environment Federation (WEF). The text, revised and updated every 3 to 5 years since the first edition in 1905, is designed and written for use by trained laboratory personnel. It should be followed explicitly because of limitations and possible interferences associated with each test.

Some water resource recovery facility (WRRF) operators may find it difficult to follow *Standard Methods* because of the detailed discussions and procedures within the text. To solve this problem, WEF has prepared this special publication, not to be considered as a substitute for *Standard Methods,* but as a guide for operators and technicians of smaller facilities. These operators wear two hats: operator and analyst. They are operator–analysts.

The methods presented in this publication are the same methods given in *Standard Methods,* but do not necessarily contain the same level of detail. As simplified explanations and directions are mastered, the operator–analyst should read details in *Standard Methods* to be aware of possible pitfalls and interferences. The detailed discussions and procedures for the methods presented here are found in the current edition of *Standard Methods.* Simplified methods are not recommended for examination of wastewaters that contain significant industrial wastes.

For a detailed explanation of basic laboratory techniques, including how to use an analytical balance, proper pipetting technique, preparation of solutions, and much more, the operator–analyst is encouraged to consult another WEF publication, *Water and Wastewater Laboratory Techniques, 2nd Edition* (Smith, 2019). This publication should be regarded as a companion publication.

Laboratory results are valuable as a record of facility operation. These data let the operator–analyst know how efficiently the facility is running and help predict and prevent troubles that may be developing in the processes. Laboratory results are required as a record of performance for regulatory agencies and are of value to operations staff and design engineers for performance optimization, troubleshooting, determining loadings, and determining when facility expansions are necessary. For these reasons, laboratory tests should be conducted as carefully and consistently as possible.

2.0 WASTEWATER CONSTITUENTS

2.1 Components

Raw wastewater is comprised of human wastes, ground-up vegetable matter, meat, seeds, fats and oils, and other organic and inorganic matter from garbage disposals, trash, rags, grit, and other materials. Analytically, the different components can be grouped into organics, inorganics, solids, nutrients, microorganisms, and basic chemical parameters. (For discussions on each type of analyte, refer to the individual method introductions in Chapter 3.)

Table 1.1 lists some of the many components of wastewater and their typical concentrations in untreated domestic wastewater. Domestic wastewater includes contributions from schools, stores, and other light commercial operations, but not flows from significant or categorical industrial dischargers. Concentrations are described as low, medium, or high strength. Whether a particular wastewater is low or high strength reflects the makeup of the service area and how much water is being used by the population contained within that service area. In the United States, the amount of water typically used can vary between 370 and 990 L (96 and 255 gal) per person per day (Metcalf & Eddy, Inc./AECOM, 2014). Flows higher than 454 L/cap·d (120 gpd/cap) should be investigated for contributions from industry, inflow, and infiltration. For communities that use less water, concentrations of organic matter and solids will be higher than for communities that

TABLE 1.1 Typical concentrations of selected parameters in raw domestic wastewater.

Parameter	Low strength	Medium strength	High strength
Total solids, mg/L	537	806	1612
Total dissolved solids, mg/L	374	560	1121
TSS, mg/L	130	195	389
Fixed residue, mg/L	29	43	86
Volatile suspended solids, mg/L	101	152	304
Settleable solids, mL/L	8	12	23
BOD, 5-day, 20 °C, mg/L	133	200	400
TOC, mg/L	109	164	328
COD, mg/L	339	508	1016
Nitrogen (total as N), mg/L	23	35	69
Organic nitrogen, mg/L as N	10	14	29
Free ammonia, mg/L as N	14	20	41
Nitrites, mg/L as N	0	0	0
Nitrates, mg/L as N	0	0	0
Phosphorus (total as P), mg/L	3.7	5.6	11
Oil and grease, mg/L	51	76	153

Source: Metcalf & Eddy, Inc./AECOM (2014).

use more water, assuming both communities receive the same number of kilograms (pounds) of organic matter and solids. Higher water use dilutes the strength. Commercial facilities such as schools, businesses, restaurants, and shopping centers may also increase the strength of the wastewater by discharging more kilograms (pounds) of organics and solids. Wastewater strength may also be affected by stormwater, whether it is coming from a combined system or as a result of significant inflow and infiltration. In this situation, wastewater strength can shift significantly between wet and dry weather.

For clarity, a brief discussion of the difference between biochemical oxygen demand (BOD), carbonaceous BOD (CBOD), and 5-day BOD (BOD_5) is warranted. Biochemical oxygen demand is a measure of the amount of oxygen consumed by microorganisms during the test as they consume or "eat" biodegradable organic material in wastewater. The BOD test is a way to estimate the amount of organic matter present because it is not possible to individually measure each organic compound that might be present. If bacteria in the BOD test begin to nitrify (i.e., convert ammonia to nitrate) during the test, then the measured oxygen demand will include both the oxygen required for organics and the oxygen needed for nitrification. The extra oxygen needed for nitrification is called *nitrogen oxygen demand* (NOD). The oxygen needed just to consume the organics is called *CBOD*. For most wastewater influent samples, BOD and CBOD will be the same. For wastewater effluent samples, especially from WRRFs that nitrify, BOD can be much higher than CBOD because the samples included NOD. One pound of BOD requires one pound of oxygen, but converting one pound of ammonia to nitrate requires 2 kg (4.5 lb) of oxygen. If an operator–analyst is interested in just CBOD, or if the discharge permit specifies CBOD, special inhibitors can be added to the BOD test that prevent nitrifying bacteria from converting ammonia to nitrate. Samples that have inhibitor added are reported as CBOD.

Biochemical oxygen demand and CBOD test results are typically reported as BOD_5 or $CBOD_5$. The "5" simply indicates that the test was run over a 5-day period. Some regulatory agencies use 7-day BOD or BOD ultimate in place of BOD_5. Here, the test is run until the oxygen demand is completely satisfied and may take more than 30 days.

Industrial wastewater tends to be much higher in strength than domestic wastewater for certain components, particularly fats, oils, greases, heavy metals, and solvents. Table 1.2 gives concentration ranges for conventional pollutants (BOD, total suspended solids [TSS], oil and grease, total Kjeldahl nitrogen [TKN], and phosphorus [P]) for a wide variety of industrial wastewaters. If any of these industries discharge to the sewer system in significant

TABLE 1.2 Typical ranges of mean concentrations of conventional pollutants and several classic nonconventional pollutants in wastewaters from selected point-source categories.

Point-source category	BOD$_5$	TSS	Oil and grease	COD	TKN	Phosphorus
Battery[a]		210	14			
Carbon black		38–1800				
Coil coating		84–180	52–340			5.5–43
Food processing						
Beverages	1000–10 000	ND–200	50–100		50–150	
Dairies	1000–2500	1000–2000	300–1000		50–100	
Fruit and vegetable processing	300–1000	200–800				
Grain processing	225–4450	81–3500		473–4900		0.5–98
Meat processing						
First processing	2200–7200	1200–3300	150–670		230–310	35–72
Further processing	1500–5000	360–2400	160–1800		24–72	44–82
Poultry processing						
First processing	1600–2200	760–980	160–670		54–90	12–21
Further processing	3300	1660	790		80	72
Rendering	2000	3200	1600		180	38
Electrical and electronic components	5–7.4	185–1440	3–7			
Electroplating[b]		0.1–10 000		50–7200		0.02–140
Explosives	ND–1300	60–520			3–490	

(continued)

TABLE 1.2 Typical ranges of mean concentrations of conventional pollutants and several classic nonconventional pollutants in wastewaters from selected point-source categories. (*Continued*)

Point-source category	BOD$_5$	TSS	Oil and grease	COD	TKN	Phosphorus
Iron and steel manufacturing[c]		31–5000	13–4100	72–9900	838	
Landfills[b]	1–7600	4–16 500	5–65	35–16 700		0.01–23
Leather tanning and finishing	400–5900	710–8600	86–1600	1800–13 600	46–890	
Metal finishing						
Metal cutting and forming	3000–4000	2000–3000	10 000–20 000	20 000–30 000	100–200	
Metal plating			100–500			
Printed wire board			100–500			
Metal products and machinery[d]	2000	1000	2300	11 300	600	170
Nonferrous metals[b]		4.6–4390				
Organic chemicals, plastics, and synthetic fibers[b]	7–2500	15–6100	Present	270–31 000		
Paint formulating[b]	280–65 500	280–148 000	42–3400			
Paving and roofing materials (tars and asphalt)	8–12	11–13 900	Not detected–50			
Pharmaceutical	220–4500	16–1400		718–10 000		
Porcelain enameling		110–32 500	Not detected–96			0.08–9.3

Pulp and paper	0.2–30	8–1100		0–12 000	
Tire and inner tube	9–420	15–770	0.8–96	0.01–300	
Synthetic rubber	200–1000	200–2000	1–200	50–2800	
Cotton	150–300	150–300			
Wool				300–500	
Timber products	56–4000	400–1100	300	2600–19 300	0.17–4 / 0.3–3
Waste combustors[b]	1–10 000	13–19 000	1–420	0.01–1200	

Source: WEF (2008).

Note. Concentrations in mg/L; values are rounded. Range of mean concentrations for different subcategories or for the entire point source category. From U.S. EPA's effluent limitations development document for each point source category and personal database compiled by Terrence Driscoll.

[a] Average concentrations for the lead subcategory.

[b] Range of detected concentrations.

[c] Range of mean concentrations for all subcategories except the byproduct recovery segment of the coke-making subcategory. The TKN value is for the iron-making subcategory.

[d] Mean concentrations.

amounts and without pretreatment, they can affect the overall makeup of the wastewater that arrives at the WRRF.

2.2   Relationships Between Different Wastewater Components

2.2.1   Per Capita Generation Rates

Domestic wastewater has a similar chemical composition from one municipality to another. Large industrial users within a service area can skew the chemical composition; however, even the effects of significant industrial users can be averaged out over a large service area because their contributions are small relative to the total flow and load received by the WRRF. For smaller service areas, the effects of a single industrial or commercial discharger can be significant. Wastewater modeling and WRRF design both depend on being able to make reasonable assumptions about the composition of the influent wastewater when actual data are not available. The same assumptions can be invaluable to the operator–analyst when determining if the flow and load measured at the facility influent are reasonable for the service area. Typical per capita generation rates in kilograms per person per day (pounds per person per day) are presented in Table 1.3. These ratios are for typical domestic wastewater and should not be applied to industrial wastewaters.

As an example, assume that a small wastewater facility serves a town with 5000 residents. The average daily flow for the hypothetical facility is 2018 m^3/d (0.53 mgd). In July, the influent composite sample results were 450 mg/L for BOD, 320 mg/L for TSS, and 45 mg/L for TKN. Refer to Table 1.4 to verify that the data provided are reasonable given what we know about the service area. To find the per capita generation rate, divide the influent flow or kilograms by the number of persons in the service area.

In this instance, all of the per person generation rates are within expected ranges, except for BOD. The operator–analyst may want to investigate why the BOD load is higher than expected. It may be that the BOD sample was not taken correctly and is not representative, or it may be that the BOD

TABLE 1.3   Per capita generation rates.

Parameter	Per capita generation rate
Flow	188 and 338 L/d (50 and 89 gpd)
BOD	0.05 and 0.12 kg/d (0.11 and 0.26 lb/d)
TSS	0.06 and 0.15 kg/d (0.13 and 0.33 lb/d)
TKN	0.09 and 0.018 kg/d (0.020 and 0.040 lb/d)

Source: Metcalf & Eddy, Inc./AECOM (2014).

TABLE 1.4 Per capita flows and loads example.

Parameter	Influent result	Per capita day rate	Expected range
Flow	2018 m³/d (0.53 mgd)	403.7 L/cap·d (106 gpd/cap)	370 and 990 L/cap·d (96–255 gpcd)
BOD	450 mg/L or 723 kg/d (1595 lb/d)	0.15 kg/cap·d (0.32 lb/d/cap)	0.05 and 0.12 kg/cap·d (0.11–0.26 ppcd)
TSS	320 mg/L or 514 kg/d (1134 lb/d)	0.10 kg/cap·d (0.23 lb/d/cap)	0.06–0.15 kg/cap·d (0.13–0.33 ppcd)
TKN	45 mg/L or 72.6 kg/d (160 lb/d)	0.014 kg/cap·d (0.031 lb/d/cap)	0.09–0.018 kg/cap·d (0.020–0.040 lb/d/cap)

lb = (mg/L)(Flow, mgd)(8.34 lb/gal).

result is biased high. Another possibility is that a discharger in the service area is contributing an excess of soluble BOD. The operator–analyst should verify that the laboratory quality control samples were within range and that appropriate dilutions were used for the analysis. It is important to realize that laboratory results are not always correct and that a little troubleshooting may be required from time to time. In this example, it would not matter if the BOD and TSS results came from samples taken on different days because the two types of data are not being compared to one another. Rather, the per capita generation rates are being examined to see if one parameter or the other is out of range. The per capita generation rates should be fairly constant from day to day, even though they may gradually increase or decrease with population changes over long periods of time.

Per capita generation rates can be useful when estimating the effect of a new subdivision. Assume that the hypothetical town in the previous example is planning two new developments. The first subdivision will have 50 new homes and 130 residents. The second subdivision will have 520 new homes and 1250 residents. What is the expected increase in flow and load at the wastewater facility? If it is assumed that each new resident will generate 321.7 L (85 gal) of flow and 0.09 kg (0.2 lb) of BOD, then the new subdivisions will add 444 028 L (117 300 gal) of influent flow and 125 kg (276 lb) of BOD. Although these are rough estimates, they will help the operator–analyst know what to expect in the WRRF influent in the future.

Consider one more example. An operator–analyst at a particular facility is used to seeing influent BOD and TSS concentrations in the 280 to 320 mg/L range. This month, the influent concentrations are much lower, specifically, 178 mg/L for BOD and 213 mg/L for TSS. Is the laboratory analysis skewed or did something else happen to affect the BOD and TSS concentrations? This hypothetical town has 8300 residents. Flows are typically close to 2953 m³/d (0.78 mgd), but this month the average daily flow for the day that the influent samples were collected was 4240 m³/d (1.12 mgd). Refer to Table 1.5 to verify that the data provided are reasonable given what we know about the service area.

In this example, the per capita generation rates for BOD and TSS are within the expected range, but the per capita generation rate for flow is higher than expected. The BOD and TSS results are reasonable. Higher influent flows have diluted the BOD and TSS load, resulting in lower measured concentrations. The operator–analyst may want to review rainfall records to see if there was a significant amount of rain on the day or days before the samples were collected. A sudden increase in influent flows may indicate an inflow and infiltration problem in the collection system.

Another way to look at this would be to multiply the influent BOD and TSS concentrations by the ratio of the flows. For example, 280 mg/L [(2953 m³/d)/(4240 m³/d)] gives a concentration that correlates well with the observed 178 mg/L. This strongly suggests that the change in concentration is because of dilution.

2.2.2 Ratios Between Wastewater Constituents in Domestic Wastewater

Domestic wastewater tends to adhere to a range of ratios for some constituents like chemical oxygen demand (COD), BOD, TSS, nitrogen, and phosphorus. Figure 1.1 illustrates the relationships between COD, BOD, and TSS. Understanding the relationships between these parameters can help the operator–analyst decide whether laboratory results are reasonable and internally consistent. Typical ratios between parameters are presented in Table 1.6. These ratios can be used to determine whether a set of laboratory data is internally consistent. In other words, does the BOD result

---

TABLE 1.5 Infiltration and inflow example.

Parameter	Influent result	Per capita day rate	Expected range
Flow	4240 m³/d (1.12 mgd)	511 L/cap·d 135 gpcd	284 and 492 L/cap·d (70–130 gpd/cap)
BOD	178 mg/L or 753 kg/d (1660 lb/d)	0.09 kg/cap·d (0.20 lb/d/cap)	0.08 and 0.10 kg/cap·d (0.18–0.22 lb/d/cap)
TSS	213 mg/L or 904 kg/d (1992 lb/d)	0.11 kgpcd (0.24 lb/d/cap)	0.09–0.11 kg/cap·d (0.20–0.25 lb/d/cap)

lb = (mg/L)(Flow, mgd)(8.34 lb/gal).

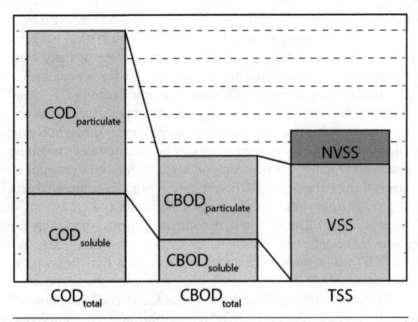

FIGURE 1.1 Relationships between COD, BOD, and TSS. *Courtesy of Indigo Water Group*

TABLE 1.6 Relationships between different wastewater parameters in raw influent.

Parameters	Typical ratio range
Soluble COD/total COD	0.3–0.5
Particulate COD/total COD	0.5–0.7
Particulate COD/volatile suspended solids	1.5–1.7
Nonbiodegradable soluble COD/COD	0.03–0.08
COD/CBOD	1.9–2.2
Soluble BOD/CBOD	0.2–0.4
CBOD/TSS	0.82–1.43
Volatile suspended solids/TSS	0.75–0.85
CBOD/TKN	4.2–7.1
CBOD/total P	20–50
Ammonia-N/TKN	0.5–0.8

Source: EnviroSim Associates (2005); WEF et al. (2018).

agree with the TSS result, and do both of these results agree with the COD result? When laboratory results fall outside of the typical ranges given in Table 1.6, there may be a sampling or analysis error or there may be a significant commercial or industrial discharger in the service area.

Chemical oxygen demand, BOD, total organic carbon (TOC), and TSS are all closely related to one another. The COD test measures all of the substances in wastewater that may consume oxygen under the right conditions. This includes organic compounds that can be eaten by microorganisms in the wastewater facility, organic compounds that cannot be eaten by microorganisms during the 5-day BOD test, and inorganic compounds such as nitrite, ferrous iron, sulfide, manganous manganese, and other compounds (APHA et al., 2017). The BOD test, in contrast, only measures those organic substances that can be consumed by microorganisms within 5 days. Because of this, COD will always be equal to or larger than BOD for a particular wastewater sample.

For domestic influent wastewater, the COD concentration will be between 1.9 and 2.2 times the CBOD concentration (EnviroSim Associates, 2005). The COD-to-CBOD ratio is matrix-specific and will change depending on whether domestic wastewater, septage, or an industrial or commercial waste is being analyzed. The COD-to-CBOD ratio can be used to estimate sample volumes for the BOD test when a new sample is brought into the laboratory. For process control purposes, COD can be used to estimate BOD. The COD test takes about 2 hours to complete, whereas the BOD test takes 5 days to complete.

The ratio of COD to CBOD changes as the wastewater moves through the WRRF. Figure 1.1 and Table 1.6 show that between 30% and 50% of the total COD is soluble; the rest is particulate. Some of the COD is biodegradable and can be measured in the BOD test. The rest of the COD is not biodegradable. Finally, between 3% and 8% of the total COD is both soluble and nonbiodegradable. This means that this fraction of COD will not be consumed by the microorganisms in the facility and it will not settle. This portion of the total COD passes through the entire treatment process essentially unchanged. Particulate COD that cannot be consumed by microorganisms in the treatment process can still be flocculated and settled. Most of this COD ends up as sludge and, eventually, biosolids.

To see how domestic wastewater might look as it moves through the treatment process, the typical ratios shown in Table 1.6 can be applied. As BOD is consumed by the microorganisms and is removed from the wastewater, the COD-to-BOD ratio gradually increases. The final ratio will depend on how much of the influent COD is biodegradable (BOD) and how much of the total COD is both nonbiodegradable and soluble. The example shown in Table 1.7 is a theoretical wastewater. Because of small errors in

TABLE 1.7   Example of changing influent and effluent COD-to-BOD ratios.

	Influent	Effluent
Total COD, mg/L	500	71
Soluble COD, mg/L	150	50
Nonbiodegradable soluble COD, mg/L	40	40
CBOD	250	15
Soluble BOD	100	2
Ratio of total COD to CBOD	2:1	4.7:1

laboratory tests, sampling, and subsampling, actual facility data probably will not match the typical ratios given in Table 1.7 quite this well.

The TOC test uses heat and oxygen, ultraviolet (UV) radiation, chemical oxidants, or a combination of these oxidants to convert organic carbon to carbon dioxide. The carbon dioxide produced is measured directly by a variety of different methods. This test cannot tell the difference between compounds with the same number of carbon atoms and different oxidation states (e.g., oxalic acid [C_2H_2O4] and ethanol [C_2H_6O]). Both of these compounds will give the same TOC result when present in equal molar amounts, even though ethanol will exert 6 times more oxygen demand in the receiving water (Boyles, 1997). Although a relationship exists between TOC, COD, and BOD, the specific relationships or ratios must be established for each sample type such as raw influent, secondary effluent, and final effluent. The ratios should be established at each sampling location where they will be used for process control. Once established, these ratios can be used for process control. In-basin TOC analyzers are available that can be used to continuously monitor wastewater processes. Using TOC or COD for process control suddenly means that process loading and food-to-microorganism ratio (F:M) calculations can be conducted in real time. These ratios can change over time and should be verified on a regular basis by running samples for both or all three parameters.

For domestic wastewater, influent BOD and TSS tend to be roughly equal to one another, with the BOD concentration ranging between 82% and 143% of the TSS concentration (WEF et al., 2018). This is true because most of the solids present in the wastewater are organic in nature. Indeed, most of what goes down the drain (e.g., human waste, vegetable matter, dishwater) is organic. The BOD test is simply a bulk measure of microbially edible organics, which means that most of the solids entering the facility can be measured as BOD. The TSS entering the WRRF is between 75% and 85% organic. Organic materials can be volatilized. In other words, they will burn away at high temperatures. The inert solids such as grit and eggshells do not contribute to the BOD load. These solids are nonvolatile or the inorganic portion of the TSS.

As with COD, not all BOD is particulate. Some BOD is soluble and is not measured by the TSS test. The fraction of soluble BOD depends on the makeup of the service area and the length of the collection system. Breweries, for example, discharge high quantities of soluble BOD. Homes with garbage disposals discharge more particulate BOD. A sprawling collection system can enhance the breakdown of particulate BOD to soluble BOD because of long residence time in the collection system. Generally, the amount of soluble BOD tends to be small relative to the amount of total BOD (in the range of 20% to 40%) (WEF et al., 2018). Because of these variations in particulate versus soluble BOD, the ratio of BOD to TSS is not absolute and varies, with one constituent or the other sometimes being present in greater quantities. Still, the expected ratio of BOD to TSS or TSS to BOD is between 0.37 and 0.64 kg per kg (0.82 and 1.43 lb per 2.2 lb) (WEF et al., 2018).

The ratio of one influent parameter to another (e.g., BOD to TSS) can vary from day to day because of both industrial and behavioral variations. The BOD-to-TSS ratio on the weekend when schools and factories are closed may be different from ratios seen during the week. If results for different parameters are being compared to determine a ratio, they must come from the same sample. One cannot calculate a ratio using last week's influent BOD result and today's TSS result even if the samples were taken at the same location.

So far, we have looked at the relationships between BOD, COD, TOC, and TSS. These relationships hold true for domestic wastewater regardless of whether the WRRF is in South Carolina, California, or Manitoba. The answer to why this is important is best illustrated with another example. Table 1.8 presents some made-up data for a theoretical WRRF.

Look carefully at the influent data. We know that the expected ratio of COD to CBOD is between 1.9 and 2.2. How do the data for this facility compare? Now look at the CBOD to TSS data. These two numbers should be fairly close together (i.e., between 0.82 and 1.43 times CBOD to TSS), but they are much further apart. Which value is more likely to be correct?

TABLE 1.8 Comparing influent and effluent data for consistency.

	Influent	Effluent
Total COD, mg/L	620	45
CBOD, mg/L	344	3
TSS, mg/L	180	5

Because BOD and COD agree, the suspect laboratory result is the TSS value. The operator–analyst can use the ratios in Table 1.4 to start troubleshooting laboratory results and to decide what is reasonable. In this example, the lower TSS number may be attributable to a variety of factors. Perhaps the TSS subsample was not shaken well immediately prior to aliquoting, or maybe there was a hole in the filter paper. It could be that the quality assurance samples for the TSS test were out of limits. Perhaps the BOD and TSS samples were collected on different days, or perhaps they were collected in different sample bottles on the same day. Poor sampling technique can cause variations between different grab samples even when collected at the same time. What if the TSS result had been much higher than the BOD result? This could be caused by a chunk of toilet paper or grease on the filter paper or by poor mixing when aliquoting the BOD sample. Many things can cause inaccurate laboratory results. Using the expected ratios for domestic wastewater can help target which analysis needs to be looked at more closely.

Now look at the effluent data. Here, the COD-to-BOD ratio is 15 to 1. Unfortunately, there is no easy way to tell if the COD and BOD numbers are in agreement or not. The final effluent COD concentration depends on how much of the influent COD was soluble and nonbiodegradable, how much of the biodegradable soluble was consumed, and how much of the particulate COD was settled and removed from the wastewater. The COD-to-BOD ratio for the final effluent will be different from one WRRF to the next and may change from day to day within the same facility. Even if the COD-to-BOD ratio is known for the final effluent, it cannot be used to estimate COD from BOD and vice versa. The BOD and TSS numbers should be fairly close together, but the ratio will vary depending on the type of treatment process. For lagoon facilities, it is possible to have high effluent TSS, but low BOD. This indicates that algae are present. Algae contribute to the solids concentration but contribute little BOD. For mechanical facilities such as trickling filters, rotating biological contactors, activated sludge, and other treatment processes that do not rely on a final polishing or settling pond that can grow algae, effluent TSS and BOD should be closer together.

The ratios cannot be used to justify throwing out or not reporting a particular sample result. If the operator–analyst can document a problem with the test, then this may be reason to rerun the test and replace the previous results. *Poor agreement between one parameter and another is not enough reason by itself.* If sample results are suspect, the operator–analyst should make a note on the benchsheet explaining why results are suspect and keep the data for his or her records. Failed quality control and quality assurance sample results or other irregularity (e.g., broken incubator) may be used to justify not reporting a sample result. It is not always possible for the sample to be reanalyzed because there is not enough sample remaining or because the hold time has expired. In this instance, the operator–analyst should collect another sample. However, it is not always possible to collect another sample because of time constraints. For example, if the discharge permit requires weekly sampling for BOD, then it may be too late in the week to collect another sample by the time the BOD result is obtained. The goal of sampling and analysis is to obtain accurate and representative results, and the operator–analyst should strive to meet this goal at all times.

Organic matter (e.g., plant material, human waste) contains between 6% and 12% nitrogen, by weight, and between 1% and 2% phosphorus. Many cleansers and corrosion control inhibitors also contain phosphorus. In practical terms, nitrogen and phosphorus concentrations for domestic wastewater can be estimated if the CBOD or COD concentration is known. Table 1.6 shows that a typical CBOD-to-TKN ratio for domestic wastewater is between 4.2 and 7.1 (WEF et al., 2018). Flipped around, TKN should be between 14% and 24% of the influent CBOD. Additionally, the ammonia-to-TKN ratio should be between 0.5 and 0.8, with 0.67 being a good assumption for domestic wastewater (EnviroSim Associates, 2005). Typical CBOD-to-total-phosphorus ratios for domestic wastewater are between 20 and 50. Flipped around, influent total phosphorus should be between 2% and 5% of influent CBOD.

Table 1.9 shows some made-up influent and effluent data for a theoretical WRRF. The influent data is in agreement with the expected ratios given in Table 1.5. The CBOD and TSS results are in agreement with TSS being 90.9% of CBOD, which is within the expected range. The TKN result is 18.5% of CBOD, which is also within the expected range. Ammonia is 67% of TKN and total phosphorus is not quite 4% of the influent CBOD. All of the ratios for the influent are within the expected ranges given in Table 1.6 for domestic wastewater. Now look at the final effluent results. Do any of the numbers look odd or out of place?

What kinds of solids are found in the final effluent? For ponds, there may be algae present, but, for mechanical treatment systems, the solids in the final effluent are escaped biological solids from the secondary treatment

TABLE 1.9   Comparing nutrient concentrations to BOD and TSS concentrations.

	Influent	Effluent
CBOD, mg/L	280	12
TSS, mg/L	308	10
TKN, mg/L	52	2.5
Ammonia, mg/L as N	35	<1.0
Total P, mg/L	11	0.05

process. They may be mixed-liquor suspended solids or sloughed material from a trickling filter. Either way, they are biological solids and, because they are organic, they will be between 14% and 24% nitrogen and between 1% and 2% phosphorus. For facilities practicing enhanced biological phosphorus removal, the effluent solids may contain as much as 16% phosphorus. Knowing this, which one of the final effluent numbers is suspect? Is it possible to discharge 10 mg/L of TSS and have the effluent total phosphorus at 0.05 mg/L? No, it is not, because 1% of 10 mg/L is 0.10 mg/L of phosphorus.

As the operator–analyst works through the procedures in this publication and learns what is typical for his or her WRRF, it is important to keep in mind that a laboratory result is not always accurate. Errors in sampling and subsampling and in the analyses themselves can bias results high or low. Ask yourself if the results are reasonable for your service area and if they are typical for your WRRF. Then, ask yourself if the results are internally consistent. Does the BOD result make sense when it is compared to the COD or TSS result? Finally, remember that the ratios given in Table 1.4 cannot be used to justify throwing out or not reporting a particular sample result. They can be used to effectively troubleshoot sampling and analytical errors.

2.3   Mass Balance

Sir Isaac Newton said, "Matter is neither created nor destroyed." Knowing this helps to manage the WRRF. Mass balance can be used for any constituent in the wastewater stream. Typical constituents of interest include solids, nitrogen, and phosphorus. The mass balance tracks the constituent entering a treatment process, constituents removed from the process, and the constituent leaving the process. Ideally, the values should match. In the real world, however, there will be some differences; a discrepancy of less than approximately 10% to 15% is considered acceptable. Discrepancies greater than this range indicate that further investigation is needed to determine the cause of the inconsistency. Mass balances can be done across individual unit processes and married together to create a facility-wide mass balance.

An example of a solids mass balance for a gravity thickener follows. A schematic of the facility (Figure 1.2) shows the data.

Given

Primary sludge, L/s (gpd)	4.53 L/s (103 000 gpd)
Thickened sludge, L/s (gpd)	1.75 L/s (60 800 gpd)
Primary sludge solids, %	2.1
Sludge concentration, %	4.4
Overflow TSS, mg/L	148
Blanket, m (ft)	0.61 m (2 ft)
Gravity thickener diameter, m (ft)	9.1 m (30 ft)

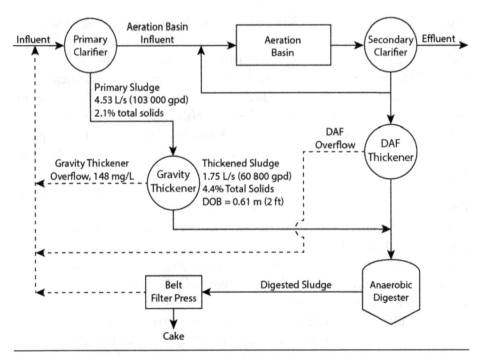

FIGURE 1.2 Operating data for an example of a gravity thickener mass balance (ft × 0.3048 = m; gpd × 0.003785). *Reprinted with permission by Brian Clow*

The basis of the analysis is straightforward, with all calculations determining the kilograms (pounds) of solids in each flow that enter or leave the unit. In this example, 8183 kg (18 039 lb) of primary sludge is discharged to the gravity thickener; solids leave the unit as overflow or thickened solids. Additionally, solids are held in the gravity thickener blanket. The operator–analyst should note that this example is a performance "snapshot" using daily average values. As such, the blanket level does not change. Putting this information into equation form, as follows, the tracking process begins:

Solids in = solids out

$$\frac{\text{Primary sludge, kg/d (lb/d)} = (\text{Overflow, kg/d [lb/d]} +}{\text{Sludge, kg/d [lb/d]} + \text{Blanket, kg/d [lb/d]})} \tag{1.1}$$

The resulting calculations of the individual flows are as follows (see Figure 1.3):

Primary sludge = 8183 kg (18 039 lb)
Gravity thickened sludge (GTS) = 10 120 kg (22 311 lb)
Gravity thickener overflow (GTO) = 58 kg (127 lb)
Gravity thickener blanket solids = 880 kg (1939 lb)

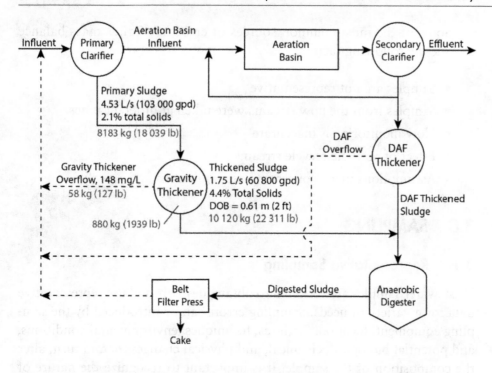

FIGURE 1.3 Example of gravity thickener mass balance (ft × 0.3048 = m). *Reprinted with permission by Brian Clow*

In calculating the blanket solids, it was assumed that the solids concentration was one-half that of the gravity-thickened sludge.

Filling in the equation with calculated values, this becomes

8183 kg (18 039 lb) primary sludge = 58 kg (127 lb) GTO +

10 120 kg (22 311 lb) GTS + 880 kg (1939 lb) in gravity thickener blanket

8183 kg (18 039 lb) in = 11 057 kg (24 378 lb) out.

By inspection, one can see that the solids in the thickener are more than the solids that are discharged from the primary clarifiers. Calculating the difference between the two values shows that the value is more than 10% to 15%.

$$\frac{(\text{Solids In} - \text{Solids Out}) \times 100}{\text{Solids In}} = \frac{(8183 \text{ kg} - 11\,057 \text{ kg}) \times 100}{8183 \text{ kg}} \quad (1.2)$$

$$= -0.35 \text{ or} -35\%$$

Some of the most common sources of error in a solids mass balance include the following factors:

- Samples are not representative
- Samples from the flow streams were taken at different times
- Flow monitoring is inaccurate
- Effect of periodic recycle streams
- Assumptions made concerning accumulations

3.0   SAMPLING

3.1   Representative Sampling

Wastewater samples are collected to obtain, develop, and use representative data for a variety of needs. Sampling errors can be introduced by the sampling equipment, location, methods, techniques, environmental conditions, and potential biological, chemical, and physical changes that, in turn, alter the composition of the sample. It is important to recognize the nature of these changes and know the steps that must be taken to minimize changes and prevent contamination resulting from external influences. Complete preservation of samples, regardless of the source, is impossible. One's best efforts can achieve only a deceleration of the biological, chemical, and physical changes that inevitably will continue after sample collection. Although the complexity of these changes in a sample is beyond the scope of this publication, it is important to be aware of the general nature of the changes so that efforts can be made to control them.

Because the analytical results describing a system are only as accurate as the sample being analyzed, the key link in the analytical process is proper collection and handling of the sample. Field personnel are responsible for the integrity of the sample from the time it is collected until it is in the custody of the laboratory. Poorly collected, improperly identified, damaged, or contaminated samples lead to incorrect data and can result in WRRF malfunction, poor operational decisions, pollution of the environment, poor water quality, and a violation of treatment standards. In the case of an already malfunctioning facility, corrective actions may be impeded or rendered ineffective if based on improper data. Discovery of a sample collection or handling problem is embarrassing if the errors are the result of poor planning or poor technique. The following section discusses some common field sampling problems and offers suggestions on how to eliminate them.

For example, suppose a sample taken from an aeration tank in an activated sludge process is sampled at an incorrect location, unrepresentative of the actual process, and tested for suspended solids. The operator–analyst's results indicate a higher-than-normal suspended solids concentration. These test results are then used to determine the amount of waste sludge to be removed from the system. Wasting too much sludge based on erroneous samples will ultimately have a severe negative effect on the way the process performs. The sample must represent the actual process or true flow stream as closely as possible to provide correct information. The primary goal of proper sampling is to ensure that the sample collected represents the flow stream being analyzed. To collect representative samples, both proper sample site selection and sampling technique are critical. Several guidelines for representative sampling are listed here and should always be followed:

1. Collect samples in containers appropriate to the analysis. See Table 1.10 for a list of sampling containers and preservation methods.

2. Collect samples for nonvolatile constituents at points where the sample stream or tank is well mixed. However, samples should not be collected at the point of maximum turbulence or at the edge of tanks or channels because neither of these areas yields representative samples. Typically, samples in channels should be collected at a point one-third the liquid depth from the channel bottom and midway across the channel between the point of maximum turbulence and the edge.

3. Samples collected for analysis of volatile organic compounds (VOCs) or inorganic hydrogen sulfide (H_2S) compounds should be taken from areas of low turbulence to reduce the amount of entrapped air in the sample. Volatile organic compounds could be driven off to the atmosphere (as offgas) in turbulent sections of the flow stream. Also, if air is entrained in the water, the samples collected cannot be used for VOC analysis because bubbles will develop in the sample container. The "no head space" requirement for VOC sample containers (meaning no air space in the container) is important to ensure that all VOCs are kept in solution for proper analysis.

4. Avoid taking samples at points where solids settling occurs or floating debris is present. These situations typically occur in quiescent areas, where the velocity of the flow has decreased.

5. Avoid sampling nonrepresentative deposits or solids that have accumulated on channel or tank walls.

6. Collect wastewater influent samples at a point upstream of any location at which recycled process streams such as digester supernatant,

TABLE 1.10 Summary of special sampling or handling requirements.

Maximum Parameter number/name	Container[a]	Preservation[b,c]	Maximum Parameter holding time[d]
Table 1A—bacterial tests:			
1–5. Coliform, total, fecal, and Escherichia coli	PA, G	Cool, 10 °C, 0.0008% Na$_2$S$_2$O[e]	8 hours[v,w]
Fecal streptococci	PA, G	Cool, 10 °C, Na$_2$S$_2$O[e]	8 hours[v] 0.0008%
Enterococci	PA, G	Cool, 10 °C,	8 hours[v]
Table 1A—protozoan tests:			
Cryptosporidium Giardia	LDPE; field filtration	1–10 °C	96 hours[u]
	LDPE; field filtration	1–10 °C	96 hours[u]
Table 1A—aquatic toxicity tests:			
10–13. Toxicity, acute and chronic	P, FP, G	Cool, ≤6 °C[p]	36 hours
Table 1B—inorganic tests:			
1. Acidity	P, FP, G	Cool, ≤6 °C[r]	14 days
2. Alkalinity	P, FP, G	Cool, ≤6 °C[r]	14 days
4. Ammonia	P, FP, G	Cool, ≤6 °C[r], H$_2$SO$_4$ to pH 2	28 days
9. Biochemical oxygen demand	P, FP, G	Cool, ≤6 °C[r]	48 hours
10. Boron	P, FP, or quartz	HNO$_3$ to pH < 2	6 months
11. Bromide	P, FP, G	None required	28 days
14. Biochemical oxygen demand, carbonaceous	P, FP, G	Cool, ≤6 °C[r]	48 hours
15. Chemical oxygen demand	P, FP, G	Cool, ≤6 °C[r], H$_2$SO$_4$ to pH < 2	28 days
16. Chloride	P, FP, G	None required	28 days
17. Chlorine, total residual	P, G	None required	Analyze within 15 minutes
21. Color	P, FP, G	Cool, ≤6 °C[r]	48 hours

Parameter number/name	Container[a]	Preservation[b,c]	Maximum holding time[d]
23–24. Cyanide, total or available (or cyanide amenable to chlorination)	P, FP, G	Cool, ≤6 °C[r], NaOH to pH >10[e,f], reducing agent if oxidizer present[e]	14 days
25. Fluoride	P	None required	28 days
27. Hardness	P, FP, G	HNO_3 or H_2SO_4 to pH < 2	6 months
28. Hydrogen ion (pH)	P, FP, G	None required	Analyze within 15 minutes
31, 43. Kjeldahl and organic N	P, FP, G	Cool, ≤6 °C[r], H_2SO_4 to pH < 2	28 days

Table IB—metals:[g]

Parameter number/name	Container[a]	Preservation[b,c]	Maximum holding time[d]
18. Chromium VI	P, FP, G	Cool, ≤6 °C[r], pH = 9.3–9.7[t]	28 days
35. Mercury (EPA Method 1631)	P, FP, G	HNO_3 to pH < 2	28 days
35. Mercury (EPA Method 245.7)	FP, G; and FP-lined cap[q]	5 mL/L 12 N HCl or 5 mL/L $BrCl$[q]	90 days[q]
3, 5–8, 12, 13, 19, 20, 22, 26, 29, 30, 32–34, 36, 37, 45, 47, 51, 52, 58–60, 62, 63, 70–72, 74, 75. Metals, except boron, chromium VI, and mercury	P, FP, G	HNO_3 to pH < 2, or at least 24 hours prior to analysis[s]	6 months
38. Nitrate	P, FP, G	Cool, ≤6 °C[r]	48 hours
39. Nitrate-nitrite	P, FP, G	Cool, ≤6 °C[r], H_2SO_4 to pH < 2	28 days
40. Nitrite	P, FP, G	Cool, ≤6 °C[r]	48 hours
41. Oil and grease	G	Cool to ≤6 °C[r], HCl or H_2SO_4 to pH < 2	28 days
42. Organic carbon	P, FP, G	Cool to ≤6 °C[r], HCl, H_2SO_4, or H_3PO_4 to pH < 2	28 days

(continued)

TABLE 1.10 Summary of special sampling or handling requirements.

Maximum Parameter number/name	Container[a]	Preservation[b,c]	Maximum Parameter holding time[d]
44. Orthophosphate	P, FP, G	Cool, ≤ 6 °C[r,x]	Filter within 15 minutes; analyze within 48 hours
46. Oxygen, dissolved probe	G, bottle and top	None required	Analyze within 15 minutes
47. Winkler	G, bottle and top	Fix on-site and store in dark	8 hours
48. Phenols	G	Cool, ≤ 6 °C[r]; H_2SO_4 to pH < 2	28 days
49. Phosphorus (elemental)	G	Cool, ≤ 6 °C[r]	48 hours
50. Phosphorus, total	P, FP, G	Cool, ≤ 6 °C[r]; H_2SO_4 to pH < 2	28 days
53. Residue, total	P, FP, G	Cool, ≤ 6 °C[r]	7 days
54. Residue, filterable	P, FP, G	Cool, ≤ 6 °C[r]	7 days
55. Residue, nonfilterable (TSS)	P, FP, G	Cool, ≤ 6 °C[r]	7 days
56. Residue, settleable	P, FP, G	Cool, $\leq$ °C[r]	48 hours
57. Residue, volatile	P, FP, G	Cool, $\leq$ °C[r]	7 days
61. Silica	P or quartz	Cool, $\leq$ °C[r]	28 days
64. Specific conductance	P, FP, G	Cool, $\leq$ °C[r]	28 days
65. Sulfate	P, FP, G	Cool, $\leq$ °C[r]	28 days
66. Sulfide	P, FP, G	Cool, ≤ 6 °C[r], add zinc acetate plus sodium hydroxide to pH > 9	7 days
67. Sulfite	P, FP, G	None required	Analyze within 15 minutes

TABLE 1.10 Summary of special sampling or handling requirements. *(Continued)*

Parameter number/name	Container[a]	Preservation[b,c]	Maximum holding time[d]
68. Surfactants	P, FP, G	Cool, ≤ 6 °C[r]	48 hours
69. Temperature	P, FP, G	None required	Analyze
73. Turbidity	P, FP, G	Cool, ≤ 6 °C[r]	48 hours
Table IC—organic tests:[h]			
13, 18–20, 22, 24, 25, 27, 28, 34–37, 39–43, 45–47, 56, 76, 104, 105, 108–111, 113. Purgeable halocarbons	G, FP-lined septum	Cool, ≤ 6 °C[r], 0.008% Na_2S_2O[e]	14 days[i]
6, 57, 106. Purgeable aromatic hydrocarbons	G, FP-lined septum	Cool, <6 °C[r], 0.008% Na2S2O[e], HCl to pH 2[i]	14 days[i]
3, 4. Acrolein and acrylonitrile	G, FP-lined septum	Cool, <6 °C[r], 0.008% Na_2S_2O[e], pH to 4–5[j]	14 days[i]
23, 30, 44, 49, 53, 77, 80, 81, 98, extraction, 40 days after extraction	G, FP-lined cap $Na_2S_2O_3$	Cool, <6 °C[r], 0.008%	7 days until 100, 112. Phenols[k]
7, 38. Benzidines[k,l]	G, FP-lined cap	Cool, <6 °C[r], 0.008% Na_2S_2O[e]	7 days until extraction
14, 17, 48, 50–52. Phthalate esters[k]	G, FP-lined cap	Cool, <6 °C[r], 0.008%	7 days until extraction 40 days after extraction
82–84. Nitrosamines[k,n]	G, FP-lined cap	Cool, ≤ 6 °C[r], store in dark, 0.008% Na_2S_2O	7 days until extraction 40 days after extraction
88–94. Polychlorinated biphenyl (PCBs)[k]	G, FP-lined cap	Cool, ≤ 6 °C[r]	1 year until extraction, 1 year after extraction
54, 55, 75, 79. Nitroaromatics and isophorone[k]	G, FP-lined cap	Cool, ≤ 6 °C[r], 0.008% Na_2S_2O[e]	7 days until extraction 40 days after extraction
1, 2, 5, 8–12, 32, 33, 58, 59, 74, 78, 99, 101. Polynuclear aromatic hydrocarbons[k]	G, FP-lined cap	Cool, ≤ 6 °C[r], store in dark, 0.008% $Na_2S_2O_3$	7 days until extraction 40 days after extraction

(continued)

TABLE 1.10 Summary of special sampling or handling requirements (*Continued*).

Parameter number/name	Container[a]	Preservation[b,c]	Maximum holding time[d]
15, 16, 21, 31, 87. Haloethers[k]	G, FP-lined cap	Cool, ≤6 °C[r], 0.008%	7 days until extraction 40 days after extraction
29, 35-37, 63-65, 107. Chlorinated hydrocarbons[k]	G, FP-lined cap	Cool, ≤6 °C[r]	7 days until extraction 40 days after extraction
60-62, 66-72, 85, 86, 95–97, 102, 103. CDDs/CDFs[k]	G	See footnote k	See footnote k
Aqueous samples: field and laboratory preservation	G	Cool, ≤6 °C[r], 0.008% $Na_2S_2O_3$[e], pH < 9	1 year
Solids and mixed-phase samples: field preservation	G	Cool, ≤6 °C[r]	7 days
Tissue samples: field preservation	G	Cool, ≤6 °C[r]	24 hours
Solids, mixed-phase, and tissue samples: laboratory preservation	G	Freeze, < –10 °C	1 year
Table ID—pesticides tests:			
1-70. Pesticides[k]	G, FP-lined cap	Cool, ≤6 °C[r], pH 5–9[o]	7 days until extraction, 40 days after extraction
Table IE—radiological tests:			
1-5. Alpha, beta, and radium	P, FP, G	HNO_3 to pH < 2	6 months

Source: 40 CFR Part 136 (U.S. EPA 2017).

[a]"P" is polyethylene; "FP" is fluoropolymer (polytetrafluoroethylene (PTFE; Teflon®), or other fluoropolymer, unless stated otherwise in this table; "G" is glass; "PA" is any plastic that is made of a sterilizable material (polypropylene or other autoclavable plastic); and "LDPE" is low-density polyethylene.

[b]Except where noted in Table II and the method for the parameter, preserve each grab sample within 15 minutes of collection. For a composite sample collected with an automated sampler (e.g., using a 24-hour composite sampler; see 40 CFR 122.21[g][7][i] or 40 CFR Part 403, Appendix E), refrigerate the sample at ≤6 °C during collection unless specified otherwise in 40 CFR Part 403, Table II or in the method(s). For a composite sample to be split into separate aliquots for preservation and/or analysis, maintain the sample at ≤6 °C, unless specified otherwise in this 40 CFR Part 403, Table II or in the method(s), until collection, splitting, and preservation is completed. Add the preservative to the sample container prior to sample collection when the preservative will not compromise the integrity of a grab sample, a composite sample, or an aliquot split from a composite sample; otherwise, preserve the grab sample, composite sample, or aliquot split from a composite sample within 15 minutes of collection. If a composite measurement is required but a composite sample would compromise sample integrity, individual grab samples must be collected at prescribed time intervals (e.g., four samples over the course of a day, at 6-hour intervals). Grab samples must be analyzed separately and the concentrations averaged. Alternatively, grab samples may be collected in the field and composited in the laboratory if the compositing procedure produces results equivalent to results produced by arithmetic averaging of the results of analysis of individual grab samples. For examples of laboratory compositing procedures, see U.S. EPA Method 1664A (O&G) and the procedures at 40 CFR 141.34(f)(14)(iv) and (v) (volatile organics).

[c]When any sample is to be shipped by common carrier or sent via the U.S. Postal Service, it must comply with the U.S. Department of Transportation Hazardous Materials Regulations (49 CFR Part 172). The person offering such material for transportation is responsible for ensuring such compliance. For the preservation requirements of 40 CFR Part 403, Table II, the Office of Hazardous Materials, Materials Transportation Bureau, Department of Transportation, has determined that the Hazardous Materials Regulations do not apply to the following materials: hydrochloric acid (HCl) in water solutions at concentrations of 0.04%, by weight, or less (pH about 1.96 or greater); nitric acid (HNO_3) in water solutions at concentrations of 0.15%, by weight, or less (pH about 1.62 or greater); sulfuric acid (H_2SO_4) in water solutions at concentrations of 0.35%, by weight, or less (pH about 1.15 or greater); and sodium hydroxide (NaOH) in water solutions at concentrations of 0.080%, by weight, or less (pH about 12.30 or less).

[d]Samples should be analyzed as soon as possible after collection. The times listed are the maximum times that samples may be held before the start of analysis and still be considered valid . Samples may be held for longer periods of time only if the permittee or monitoring laboratory has data on file to show that, for the specific types of samples under study, the analytes are stable for the longer time, and has received a variance from the regional administrator under § 136.3€. For a grab sample, the holding time begins at the time of collection. For a composite sample collected with an automated sampler (e.g., using a 24-hour composite sampler; see 40 CFR 122.21[g][7][i] or 40 CFR Part 403, Appendix E), the holding time begins at the time of the end of collection of the composite sample. For a set of grab samples composited in the field or laboratory, the holding time begins at the time of collection of the last grab sample in the set. Some samples may not be stable for the maximum time period given in 40 CFR Part 403, Table II. A permittee or monitoring laboratory is obligated to hold the sample for a shorter time if it knows that a shorter time is necessary to maintain sample stability (see § 136.3[e] for details). The date and time of collection of an individual grab sample is the date and time at which the sample is collected. For a set of grab samples to be composited and that are all collected on the same calendar date, the date of collection is the date on which the samples are collected. For a set of grab samples to be composited and that are collected across two calendar dates, the date of collection is the dates of the 2 days (e.g., November 14 to 15). For a composite sample collected automatically on a given date, the date of collection is the date on which the sample is collected. For a composite sample collected automatically and across two calendar dates, the date of collection is the dates of the 2 days (e.g., November 14 to 15).

[e]ASTM D7365-09a specifies treatment options for samples containing oxidants (e.g., chlorine) for cyanide analysis. Also, Section 9060A of *Standard Methods for the Examination of Wastewater* (23rd edition) addresses dichlorination procedures for microbiological analysis.

[f]Sampling, preservation, and mitigating interferences in water samples for analysis of cyanide are described in ASTM D7365-09a. There may be interferences not mitigated by the analytical test methods or D7365-09a. Any technique for the removal or suppression of interference may be employed provided the laboratory demonstrates that it more accurately measures cyanide through quality control measures described in the analytical test method. Any removal or suppression technique not described in D7365-09a or the analytical test method must be documented along with supporting data.

[g]For dissolved metals, filter grab samples within 15 minutes of collection and before adding preservatives. For a composite sample collected with an automated sampler (e.g., using a 24-hour composite sampler; see 40 CFR 122.21[g][7][i] or 40 CFR Part 403, Appendix E), filter the sample within 15 minutes after completion of collection and before adding preservatives. If it is known or suspected that dissolved sample integrity will be compromised during collection of a composite sample collected automatically over time (e.g., by interchange of a metal between dissolved and suspended forms), collect and filter grab samples to be composited (footnote "b") in place of a composite sample collected automatically.

[h]Guidance applies to samples to be analyzed by gas chromatography(GC), liquid chromatography (LC), or gas chromatography/mass spectrometry (GC/MS) for specific compounds.

[i]If the sample is not adjusted to pH 2, then the sample must be analyzed within 7 days of sampling.

[j]The pH adjustment is not required if acrolein will not be measured. Samples for acrolein receiving no pH adjustment must be analyzed within 3 days of sampling.

[k]When the extractable analytes of concern fall within a single chemical category, the specified preservative and maximum holding times should be observed for optimum safeguard of sample integrity (i.e., use all necessary chemical categories, the sample may be preserved by cooling to ≤6 °C, reducing residual chlorine with 0.008% sodium thiosulfate, storing in the dark, and adjusting the pH to 6 to 9; samples preserved in this manner may be held for 7 days before extraction and for 40 days after extraction. Exceptions to this optional preservation and holding time procedure are noted in footnote "e" (regarding the requirement for thiosulfate reduction), and footnotes "l" and "m" (regarding the analysis of benzidine).

[l]If 1,2-diphenylhydrazine is likely to be present, adjust the pH of the sample to 4.0 ± 0.2 to prevent rearrangement to benzidine.

[m]Extracts may be stored up to 30 days at <0 °C.

[n]For the analysis of diphenylnitrosamine, add 0.008% $Na_2S_2O_3$ and adjust pH to 7 to 10 with NaOH within 24 hours of sampling.

[o]The pH adjustment may be performed upon receipt at the laboratory and may be omitted if the samples are extracted within 72 hours of collection. For the analysis of aldrin, add 0.008% $Na_2S_2O_3$.

[p]Sufficient ice should be placed with the samples in the shipping container to ensure that ice is still present when the samples arrive at the laboratory. However, even if ice is present when the samples arrive, it is necessary to immediately measure the temperature of the samples and confirm that the preservation temperature maximum has not been exceeded. In the isolated instances where it can be documented that this holding temperature cannot be met, the permittee can be given the option of on-site testing or can request a variance. The request for a variance should include supportive

data that show that the toxicity of the effluent samples is not reduced because of the increased holding temperature. Aqueous samples must not be frozen. Hand-delivered samples used on the day of collection do not need to be cooled to 0 to 6 °C prior to test initiation.

[q] Samples collected for the determination of trace level mercury (<100 ng/L) using U.S. EPA Method 1631 must be collected in tightly capped fluoropolymer or glass bottles and preserved with BrCl or HCl solution within 48 hours of sample collection. The time to preservation may be extended to 28 days if a sample is oxidized in the sample bottle. A sample collected for dissolved trace level mercury should be filtered in the laboratory within 24 hours of the time of collection. However, if circumstances preclude overnight shipment, the sample should be filtered in a designated clean area in the field in accordance with procedures given in *Standard Method 1669*. If sample integrity will not be maintained by shipment to and filtration in the laboratory, the sample must be filtered in a designated clean area in the field within the time period necessary to maintain sample integrity. A sample that has been collected for determination of total or dissolved trace level mercury must be analyzed within 90 days of sample collection.

[r] Aqueous samples must be preserved at ≤6 °C and should not be frozen unless data demonstrating that sample freezing does not adversely affect sample integrity are maintained on file and accepted as valid by the regulatory authority. Also, for purposes of NPDES monitoring, the specification of "≤ °C" is used in place of the "4 °C" and "<4 °C" sample temperature requirements listed in some methods. It is not necessary to measure the sample temperature to three significant figures (i.e., 1/100th of 1 degree); rather, three significant figures are specified so that rounding down to 6 °C may not be used to meet the ≤6 °C requirement. The preservation temperature does not apply to samples that are analyzed immediately (i.e., less than 15 minutes).

[s] An aqueous sample may be collected and shipped without acid preservation. However, acid must be added at least 24 hours before analysis to dissolve any metals that adsorb to the container walls. If the sample must be analyzed within 24 hours of collection, add the acid immediately (see footnote "b"). Soil and sediment samples do not need to be preserved with acid. The allowances in this footnote supersede the preservation and holding time requirements in the approved metals methods.

[t] To achieve the 28-day holding time, use the ammonium sulfate buffer solution specified in U.S. EPA Method 218.6. The allowance in this footnote supersedes preservation and holding time requirements in the approved hexavalent chromium methods, unless this supersession would compromise the measurement, in which case requirements in the method must be followed.

[u] Holding time is calculated from time of sample collection to elution for samples shipped to the laboratory in bulk and calculated from the time of sample filtration to elution for samples filtered in the field.

[v] Samples analysis should begin as soon as possible after receipt; sample incubation must be started no later than 8 hours from time of collection.

[w] For fecal coliform samples for sewage sludge (biosolids) only, the holding time is extended to 24 hours for the following sample types using either EPA Method 1680 (LTB-EC) or 1681 (A-1): Class A composted, Class B aerobically digested, and Class B anaerobically digested.

[x] The immediate filtration requirement in orthophosphate measurement is to assess the dissolved or bio-available form of orthophosphorus (i.e., that which passes through a 0.45-micron filter), hence the requirement to filter the sample immediately upon collection (i.e., within 15 minutes of collection).

trickling filter recycle, or waste-activated sludge return to the main process flow stream.

7. After selecting a representative sampling location, sample the waste stream consistently at this location. By maintaining this consistency, variations in sample results cannot be attributed to changes in location. The test data can be compared confidently from day to day. The sample site should be permanently marked with paint or a sign to ensure that everyone samples at the same location.

8. Accessibility and safety are also important factors when selecting a sampling site. Do not choose a sample site that is difficult to get to or can result in falls and injuries.

9. For automatic water samplers (auto samplers), sample lines should be flushed or purged for an adequate period of time before taking the sample to ensure that material left in the line from the prior sampling event is not incorporated to the sample. Also, sample lines should be cleaned and/or replaced regularly to avoid the possibility of sediment buildup, which could cause erroneous results. For hand grab samples, the sample wand should be rinsed at least three times prior to collecting the sample.

10. Where samples are to be collected from flowing pipes, sample lines should be kept as short as possible and with a minimum number of bends. Excessively long sample lines pose risks of inadequate flushing before sampling and alterations in the sample resulting from chemical and biological activity within the sample pipe.

11. Sample containers for each location should be clearly identified and marked. This precaution will reduce the possibility of confusing samples from one location with samples from another or mistaking the origin of a sample. Each sample container should be clearly labeled with the date, time, sample location, parameters to be analyzed, and person who collected the sample.

12. A different sampling collection container should be used at each sampling location. The sampling container should be thoroughly cleaned before collecting the sample. The sample from the sampling container should then be transferred to the sample storage container. Sampling equipment should be thoroughly cleaned before sample collection and replaced when needed.

13. To ensure that the sample is representative, the samples should be kept thoroughly mixed throughout the collection, transfer, measurement, and transfer procedure to prevent settling.

14. After collection, samples must be properly preserved and stored. Depending on what constituents are to be analyzed, composite samples may need to be kept refrigerated during collection. Refrigerating and preserving the sample helps ensure that its composition does not change before testing. Depending on the preservation method required, it may be necessary to independently store samples from the same location in different bottles for different analyses. Refrigeration at less than or equal to 6 °C without freezing the sample is prescribed by the U.S. Environmental Protection Agency (U.S. EPA) for permit-required samples and is generally appropriate for samples collected for other purposes.

3.2   Purposes of Sampling

Purposes and functions associated with sampling include compliance with regulatory requirements, process control, determination of process efficiency, support or refutation of capital expenditures, historical data collection, and other purposes such as receiving water studies, collection system studies, and personnel safety programs. All of these purposes of sampling must be considered when establishing a meaningful sampling program.

3.2.1   Regulatory Requirements

Water resource recovery facilities that operate under regulatory permits are generally required to perform specific sampling and analyses on a regular basis. Permits typically specify the sampling location, type (grab or composite), and frequency; required parameters; methods of analysis; and required frequency in reporting the analyses to the regulatory agency. National Pollutant Discharge Elimination System (NPDES) permits also specify that sampling and analysis must comply with either *Methods for Chemical Analysis of Water and Wastes* (U.S. EPA, 1983) or *Standard Methods for the Examination of Water and Wastewater* (APHA et al., 2017) as specified in the *Code of Federal Regulations*, 40 CFR Part 136, Table II, note 4 (see U.S. EPA, 2017).

3.2.2   Process Control

The sampling and analyses required to fulfill most permits will typically reveal how well processes at a WRRF are performing as a whole. However, to properly operate a WRRF and control each unit process, many more testing parameters must be performed. An effective sampling program provides total system information on the loading (influent characteristics), performance (effluent quality), and intermediate conditions of each unit process within a WRRF. The more information known about the process performance and waste stream characteristics as it progresses through the WRRF, the better informed an operator–analyst will be to make proper operational adjustments. Operational staff must be knowledgeable enough to use sampling and test results to make the proper operational adjustments. Too often, WRRF personnel perform tests but do not use the information to make informed decisions. Many of these process control measurements can be made in the field for more accurate and faster results. For example, using a portable dissolved oxygen meter to measure dissolved oxygen in an aeration tank and measuring a sludge blanket level directly in a clarifier are typical field process control tests.

3.2.3   Determine Process and Facility Efficiency

Data from the sampling and testing program can be used to determine the efficiency of an individual process and the facility in general. These data will aid operators in making specific process adjustments. For example, evaluating the secondary clarifier effluent TSS results over an extended period of time may indicate that the secondary clarifier effluent quality is declining. This could indicate that operational changes may be needed to bring the secondary process back to optimal efficiency. It should be noted, however, that operational changes should not be based on secondary effluent TSS results alone but should be coordinated with other activated sludge process controls used at a given facility.

3.2.4   Support or Refute Future Capital Expenditures

A historical sampling and analysis database will show trends in facility loading and performance that can be used to predict when a facility expansion or upgrade is needed. Most operating or discharge permits require periodic loading evaluations to determine when the planning of a facility expansion needs to occur. The database will also supply information that will be useful in designing facility upgrades or expansions. A historical database is simply the retained records of past sampling and testing. Of course, if process control tests are not performed or are rarely performed, little information will be available for the historical database. It should be noted, however, that any historical data must be considered in light of possible changes in WRRF influent characteristics, process adjustments, and/or in-facility system changes that may permanently alter normal conditions. Care should be taken when comparing historical data because detection limits and methods of analysis may differ from current methods and practices used.

3.2.5 Historical Records

A database of historical sampling and analyses is invaluable to operators, analysts, and engineers. For the operator–analyst, a historical database can show seasonal variations at the WRRF, pointing out past abnormal conditions that can help in preparing for treatment process adjustments. A historical database can also point to corrective actions used in the past for recurring abnormal conditions; in other words, it can reveal which corrective actions have worked and which have not.

3.3   Sample and Measurement Collection

3.3.1 Continuous Measurements

There are several types of continuous measurements that are collected at most WRRFs. The most common type of continuous measurement is flow. Other types of continuous measurements include online pH meters, dissolved oxygen meters, conductivity meters, oxygen reduction potential meters, chlorine residual meters, turbidity meters, and other similar devices. Depending on the level of sophistication at the facility, the output from these online devices may be recorded in chart form or stored in the facility's supervisory control and data acquisition system or other computer-based operating and recordkeeping system. It is important that online analyzers and meters be routinely checked for accuracy precision as well as being calibrated and maintained per the manufacturer's written directions.

3.3.2 Grab Samples

A grab sample is one that is taken to represent one moment in time and is not mixed with any other samples. A grab sample is sometimes called an *individual* or *discrete* sample and will only represent sample conditions at the exact moment it is collected. Grab samples are often useful in certain situations in which composite samples would not be adequate. Examples of these situations include the following:

- The characteristics of "slug" discharges must be determined by grab samples to help identify the source and assess potential effects on treatment processes. These discharges are often noticed by a facility operator–analyst performing routine duties through color changes or odor, or by on-line instrumentation monitoring pH or conductivity; the duration typically is unknown.

- Numerous grab samples are used to study variations and extremes in a waste stream during a period of time. Composite samples do not reveal waste variations over time because of the nature of the samples. Composites tend to average both short-duration, high-strength discharges and long-duration, low-strength discharges. The significance of this depends on the volume of flow at the time of collection. Composite samples tend to dilute short, high- or low-strength discharges that could affect a WRRF's performance but go unnoticed. Taking both composite and sequential discrete samples can be advantageous when periodically testing for waste variations over a specific timeframe.

- Grab samples can be used if the flow to be sampled occurs intermittently for short durations.

- Grab samples can be used if the flow composition to be sampled is reasonably constant. Of course, this assumption must be verified with multiple samples over an adequate time period to determine if there really are variations in flow composition.

- Grab samples must be used if the constituents to be analyzed are unstable or cannot be preserved and, therefore, must be analyzed immediately or stored under special conditions. Examples of these parameters include oil and grease, pH, chlorine residual, dissolved oxygen, bacteriological tests, purgeable organic compounds, and phenols.

3.3.3 Composite Samples

A composite sample is prepared by combining a series of grab samples over known time or flow intervals. A composite sample shows the average composition of a flow stream over a set time or flow period if the sample is collected proportional to flow. These samples can be collected manually and mixed together, or they can be collected by automatic sampling equipment. The samples taken by automatic composite sampling equipment may be composited as they are collected into one large receptacle. At most WRRFs, composite samples are required under regulatory permit requirements for most constituents that do not require immediate analyses. Typical composite sampling is required for parameters such as BOD, TSS, ammonia-nitrogen, and total phosphorus. Results from analyses used to calculate facility and process loadings (such as organic loading or F:M) should always be made from composite samples. This practice is important in ensuring that data obtained from a slug or spike flow, using a single grab sample, do not bias the information or provide misleading data. Two different types of composite samples are generally used. These are known as *fixed-volume* or *flow-proportional composites*.

3.3.3.1   Fixed-Volume Composite Samples

The fixed-volume composite, also called a *time-proportional composite,* is the simpler type of composite sample. In fixed-volume composite sampling, a series of individual grab samples, all having the same volume, are collected at equally spaced time periods. Fixed-volume composite samples will only give an accurate representation of average composition of flow if the flow does not vary during the sampling period. This is not often the case in typical WRRFs, even when systems are equipped with flow-equalization tanks to dampen flow variations. The fixed-volume composite sample is more appropriate for sampling activated sludge aeration basins, sludge solids in digesters, constant-flow streams, and sludge cakes from dewatering equipment. The total volume of the composite sample required depends on the types of analyses that must be performed on the sample. The number of individual grab samples required to make up the composite sample depends on the timeframe of the sampling event and other factors such as regulatory requirements and the degree of accuracy. U.S. EPA allows time-proportional sampling, and in its standards requires 15-minute intervals (96 samples per day). Generally, the more individual samples collected, the better the composite will represent the flow stream. For example, 24 samples of 500 mL, each collected to form a 12 L composite, will better represent the flow stream than 12 samples of 1 L each.

To calculate the volume of each individual sample that must be collected and combined into one composite sample, the time interval and total composite sample size required must first be determined. For example, if a 1 L composite sample must be collected during a 24-hour period, with sampling intervals every 2 hours, the calculation is as follows:

First, determine the sampling frequency.

$$Number\ of\ Samples\ Collected = \frac{Total\ Hours}{Sampling\ Frequency} \qquad (1.3)$$

For example: Samples will be collected over a 24-hour period. One sample will be collected every 2 hours. How many sample bottles are required?

$$Number\ of\ Samples\ Collected = \frac{Total\ Hours}{Sampling\ Frequency}$$

$$Number\ of\ Samples\ Collected = \frac{24\ hours}{2\ hours}$$

$$Number\ of\ Samples\ Collected = 12$$

Next, determine the volume required per sample.

Number of Samples Collected = 12

For example: A final sample volume of 1000 mL is required. Twelve samples will be collected over 24 hours. How many milliliters are needed per sample?

$$Volume \ of \ Each \ Sample = \frac{Total \ Sample \ Volume}{Number \ of \ Samples}$$

$$Volume \ of \ Each \ Sample = \frac{1000 \ mL}{12 \ Samples} \qquad (1.4)$$

$$Volume \ of \ Each \ Sample = 84 \ mL \ / \ sample$$

Allowing for surplus sample, the volume should be rounded up to 100 mL. When rounding up, the container size and accuracy of collection volume must be considered. When individual grab samples are combined as a composite, they must be transferred quickly, first from the collection point to the sample measuring device and then to the composite container, while continuously being mixed. Agitating the sample in this manner prevents settling and reduces sampling error. This process is best accomplished using automatic composite-sampling equipment, which eliminates the measure–transfer step.

3.3.3.2 Flow-Proportional Composite Samples

In flow-proportional composite sampling, the sample volume collected varies based on the flowrate of the waste stream being sampled. Either the volume of each individual grab sample or the sample frequency will be varied in direct proportion to flowrate. To be correct, a flow-proportional composite sample must be based on accurate measurements of the waste stream flowrate.

A flow-proportional composite sample is more representative of the waste stream than a fixed-volume sample because it takes into account variations in wastewater characteristics that result from fluctuations in flow. Many NPDES permits require that flow-proportional samples of facility effluent be collected.

Typical parameters that are often analyzed in flow-proportional composite samples include suspended solids, BOD, ammonia-nitrogen, and total phosphorus.

3.3.3.2.1 VARIABLE-VOLUME TECHNIQUE

If manual sampling is used to create a flow-proportional composite sample, the variable-volume technique is typically the easier and more practical way to collect the sample. With this technique, the volume of each sample collected is based on the flowrate of the waste stream at the instant the sample is collected. An example of one procedure used to manually collect a variable-volume composite sample is given here. The procedure estimates the volume of sample that should be collected during each sampling event, as follows:

$$\frac{\text{Composite sample}}{\text{single portion}} = \frac{\text{(Instantaneous flow)(Total sample volume)}}{\text{(Number of portions)(Average flow)}} \quad (1.5)$$

1. Determine the minimum composite sample volume, in milliliters, needed to perform all of the desired analyses. Refer to *Standard Methods* or other references to determine the required sample volumes for each analysis. Keep in mind that the minimum sample volumes recommended in *Standard Methods* may not always be large enough. When the effluent is clean, for example, larger sample volumes will be needed for the TSS test than when the effluent contains higher quantities of solids. Additional volumes of sample should be collected to allow for duplicate quality control analysis or spillage of the sample.

2. Ensure that the composite sampling container is large enough to hold the desired final sample volume. If the sampling container overflows during compositing, discard the entire contents and start over the sampling.

3. Determine how many samples will be collected during the sampling period. Many discharge permits define a composite sample as a total of four samples taken 2 hours apart. In this instance, samples may be collected at 8:00 a.m., 10:00 a.m., 12:00 p.m., and 2:00 p.m. Many discharge permits define a 24-hour composite sample as a minimum of eight samples evenly spaced over a 24-hour period. In this instance, one sample could be collected every hour or once every 3 hours (or with some other frequency) as long as the samples are taken at even intervals during a 24-hour period. In either instance, the minimum sample volume that should be collected at each sampling time is 100 mL. Smaller sample volumes may not be representative. Before collecting samples for permit compliance, check the discharge permit to ensure that all permit requirements are met.

4. Determine the average daily flow for the facility. Flow varies daily and may be very different on weekends compared to flow during the week. An estimated average daily flow number should be chosen that is representative of flows observed over the previous week or two. For example, influent flows for a particular WRRF were 3409 m³/d (0.75

mgd), 3182 m³/d (0.70 mgd), 3546 m³/d (0.78 mgd), and 3319 m³/d (0.73 mgd) in the 4 days before a scheduled sampling event. The average flow received was 3182 m³/d (0.74 mgd), which is a reasonable number to use for calculating the sample sizes needed.

5. The last piece of information needed to calculate sample volumes is instantaneous flow at the time of sampling. This number will not be known until the sample is taken.

6. Prepare a log sheet to track the time of sample collection, the instantaneous flowrate at the time of sampling, the volume of sample collected, the average daily flow used for the sampling calculation, the actual average daily flow, and the initials of the person collecting each sample. An example using the aforementioned procedure is presented here. For the example, an 8-hour, flow-proportional composite sample is to be taken at a sampling interval of once every 2 hours (four samples all together). The parameters to be analyzed are BOD, TSS, and ammonia-nitrogen (NH_3-N). The daily average flow at the WRRF is 5677 m³/d (1.5 mgd), and the minimum measurable reading is 378.5 m³/d (0.1 mgd).

The final minimum sample volume required is calculated based on the volume required to perform each laboratory test, as follows:

$$BOD + TSS + NH_3\text{-}N \qquad (1.6)$$
$$1000 \text{ mL} + 1000 \text{ mL} + 400 \text{ mL} = 2400 \text{ mL}$$

1. Use Equation 1.5 along with instantaneous flow measurement to determine the volume of sample that should be collected each time.

$$\text{Composite sample single portion} = \text{(Instantaneous flow)}$$
$$(2400 \text{ mL})(4)(378.5 \text{ m}^3/\text{d})$$

2. The operator–analyst takes a flow measurement reading immediately before collecting a sample at 12:00 p.m. and finds that the instantaneous flow is 757 m³/d (3.0 mgd).

$$\text{Composite sample single portion} = (757 \text{ m}^3/\text{d})(2400 \text{ mL})(4)(378.5 \text{ m}^3/\text{d})$$
$$\text{Composite sample single portion} = 1200 \text{ mL}$$

The sample collected at 12:00 p.m. should be 1200 mL. Samples collected at other times of day may be bigger or smaller depending on instantaneous flow.

If the facility flowrate varies significantly on a seasonal or other basis, it may be necessary to recalculate the aforementioned steps. If the flow is not known or is intermittent, as is commonly the case with industrial waste streams, equal volumes of samples can be collected and stored. After hourly flowrates are obtained, a composite sample can be prepared by mixing the proportion of each sample that corresponds to the flow at the time of collection. For example, 1-L samples can be collected at hourly increments, and the flow can be recorded each time a sample is taken. After the sampling period has ended, the samples can be proportioned according to the recorded flow each time a liter sample was taken. This procedure requires several clean containers and a large storage area. Also, this procedure must be carefully performed to make sure all of the samples are well mixed before subsampling. This method is better suited to use with automatic sampling equipment, which can collect hourly discrete samples in separate bottles for later composite sampling.

3.3.3.2.2 VARIABLE-FREQUENCY TECHNIQUE

In variable-frequency, flow-proportional composite sampling, the volume of sample collected stays constant, but the time interval between samples or the frequency of sample collection varies. The sample intervals are proportional to the measured flows. Variable-frequency, flow-proportional composite samples are rarely used with manual sampling but are often used with automatic sampling equipment. To collect these types of samples, an automatic sampler is coupled to a flow meter that collects each sample after a preprogrammed amount of wastewater flows past the sampling point. To determine the frequency of each individual sample, the average facility flow, individual sample size, and total composite sample required must be known or accurately estimated. A procedure for determining the wastewater flow volume for each sample interval is described here.

1. Determine the final composite sample volume required. This is determined based on the required volume needed to test all of the parameters to be analyzed.

2. Divide the total composite sample volume by a realistic and convenient volume selected for each sample. This value represents the number of samples to be collected each day.

3. Divide the average daily flowrate of the waste stream by the number of samples to be collected. This number should be adjusted upward to greater than 100 to be statistically significant for a 24-hour period. This number equals the volume of wastewater that must pass through the sampling-point measuring device before a sample is collected. It can be seen from this procedure that the faster a specific quantity of wastewater passes the sampling point, the more frequently a set-volume sample must be collected.

3.4   Use of Automatic Samplers

Use of auto samplers has virtually eliminated human error associated with manual sampling activities and significantly reduced personnel costs associated with collection of representative samples. As with any instrument, auto samplers should be checked routinely and be calibrated, cleaned, and maintained (including sample lines and bottles) according to the manufacturer's specifications.

3.4.1   Programming

An auto sampler can be programmed to collect a variety of samples including time-composite or flow-proportional composite samples. Depending on specific equipment, the samplers can be outfitted with a single sample collection bottle or 24 individual sample bottles. Generally, automatic samplers have a purge cycle programmed into the operating system. The purge cycle is used to flush the sample collection hose before each individual sample collection. It is important that the function of the purge cycle be periodically checked to confirm that it is completely purging the collection system.

3.4.2   Cleanliness

Auto samplers can introduce contaminants to the sample. These contaminants can result from the atmosphere, lubricants, poor cleaning technique, or the materials from which the sampler is constructed. Sampler units should be checked by cycling distilled or deionized water through the sampler and analyzing the water for various constituents. If ambient air is used for purging cycles, the air source should be filtered to remove contaminants. All parts of the sampler, including purging chambers, pumps, and hoses, should be cleaned regularly. The frequency of cleaning depends on what is being sampled and how rapidly the unit gets dirty. Hoses should be as clean and free of collected material as possible. The manufacturer's recommended operation and maintenance procedures should be followed.

When a sampler is relocated from one sampling point to another, it should be, at a minimum, cleaned thoroughly. Ideally, all hoses or parts coming in contact with the sample should be replaced. Setting up dual samplers in the same location can help identify a problem with sampling equipment.

3.4.3   Materials of Construction

Auto samplers used in WRRFs should be made from a durable, watertight casing to protect internal components from submergence and high-humidity conditions. A vandal-proof handle with a locking device is suggested. Depending on the actual location of the sampler in the WRRF, the unit may need to be manufactured to meet explosion-proof requirements. Materials of construction can vary, but it is suggested that corrosion-resistant materials such as plastics, fiberglass, and stainless steel be used.

3.5   Sample Hold Time

Regardless of the hold time, samples should be analyzed as soon as possible. The fresher the sample, the less chance of it changing chemically, biologically, or physically, resulting in a more accurate analysis. The sample hold times listed in Table 1.10 represent maximum times that a sample may be held before the start of analysis. For a grab sample, the holding time begins at the time of collection. For a 24-hour composite sample collected with an auto sampler, the hold time begins when the last sample has been collected. For a set of grab samples composited in the field or laboratory, the hold time begins at the time of collection of the last grab sample in the set (see 40 CFR, Part 136, Table II, note 4 [U.S. EPA, 2017]). the samples must be refrigerated until sampling is complete.

3.6   Sample Containers

The choice of sample container type, size, and material depends on several considerations, including the volume of sample required, interference problems anticipated, type of testing to be performed, cost and availability, and resistance to breakage. The container requirements should be specified as part of the sampling program. To prevent contamination, sample containers must be cleaned before being filled. New containers are not necessarily clean and must also be washed using prescribed procedures. At a minimum, all containers and caps must be washed in a nonionic and nonphosphate detergent, rinsed well with hot tap water, and then rinsed with distilled water. Preferably, after being rinsed in distilled water, sample containers and caps should be soaked in an acid solution at a temperature of approximately 70 °C for approximately 24 hours and then rinsed with distilled water for a second time.

Quartz, polytetrafluoroethylene, or glass containers can be soaked in 1:1 nitric acid, 1:1 hydrochloric acid, or aqua regia (four parts hydrochloric acid to 1 part nitric acid) solutions. Plastic containers should be soaked in 1:1 nitric acid or 1:1 hydrochloric acid solutions.

Solvent rinses are required for oil and grease or pesticide sample containers. These should be rinsed with hexane, followed by acetone and, finally, distilled water. Solvent used for rinsing should be collected for proper disposal. Some container manufacturers certify that their containers are "clean" and therefore to be used without washing or sterilizing once a new case is purchased. Because oil and grease in wastewater easily adhere to a number of materials, the recommended sample container is a wide-mouth jar with a polytetrafluoroethylene-lined screw lid. The wide mouth allows the technician to rinse the inside walls of the jar with the solvent used in the analysis.

The jar should be precalibrated because the initial steps of the procedure are performed in the container. *Precalibrated* means that the jar at whatever volume is premeasured before collecting the sample. Because the initial steps are happening in the sample container, the volume must be accurately measured and determined before adding any solvents. Even if the jar has graduations, the graduations must be verified as accurate because graduations on sample containers (or even non-class-A glassware) are typically "approximate."

Container caps or lids, which can potentially contaminate the sample, can be lined with aluminum foil or polytetrafluoroethylene before being placed on the container.

Sampling containers should be prelabeled with the appropriate information and any special information the technician may need regarding the collected sample, such as unusual odors, appearance, or conditions. Sampling containers that contain a preservative are not typically used to collect the sample. Care should be taken not to overfill the sample container when transferring the sample to the container to avoid flushing out the preservative.

A separate collection device should be used at each sampling location. With the exception of containers to which a preservative has been added and those used for oil and grease, containers should be rinsed with the sample material before collecting the sample. Oil and grease sample containers cannot be rinsed with the sample because of the container preparation and the adherence of oil and grease to the container walls. Sample containers or collection devices that are permanently stained should be discarded.

3.7   Preservatives

Guidelines have been established for preservation of water and wastewater samples. The techniques discussed here are intended to provide an overview of options available for the preservation of samples. Additional information can be obtained from specific procedures described in *Standard Methods*. Legal requirements for sample preservation are contained in 40 CFR 136 (U.S. EPA, 2017). Refrigeration, pH adjustments, and chemical additions are the primary methods recommended for sample preservation. The methods listed in Table 1.10 are presented as guidelines and may differ somewhat from those in other publications. Again, refer to the aforementioned sources for more specific information and legal requirements.

The most frequently used means of sample preservation is refrigeration at temperatures near freezing (≤ 6 °C) or the use of wet ice. Biological activity is decreased because respiration rates are greatly reduced at low temperatures. Chemical reaction rates and the loss of dissolved gases also are reduced. It is often necessary to combine chemical addition or pH adjustment with refrigeration to ensure effective preservation. Refrigeration at temperatures at or below freezing (0 °C) is an effective long-term preservation method for only some parameters. In addition to killing or slowing the metabolism of microorganisms, freezing of samples has other limitations. To avoid breakage, sample containers must have sufficient head space to allow for the expansion of freezing liquid. Particulate material will solubilize readily while samples are thawing because cell structures rupture when frozen. Samples to be analyzed for suspended or dissolved solids should never be frozen.

Acid addition is a common method for decreasing both biological and chemical activity. Sulfuric acid (H_2SO_4) is added to a sample to stop bacterial action; preserving samples for COD and organic carbon analyses are examples of this application. Sulfuric acid addition combined with refrigeration both preserves and pretreats oil and grease. Acids dissolve particulate matter and, therefore, must be avoided if suspended solids are to be determined.

Alkali, typically sodium hydroxide (NaOH), is added to samples to prevent the loss of volatile compounds through the formation of a salt. Organic acids and cyanide are examples of salt-complexing compounds. Chemicals, including acids and bases, are added to samples to stabilize compounds or stop biological activity. Copper sulfate ($CuSO_4$) and mercuric chloride ($HgCl_2$) are two biological inhibitory chemicals commonly used. Zinc acetate [$Zn(C_2H_3O_2)_2$], phosphoric acid (H_3PO_4), and sodium hydroxide (NaOH) are frequently used as complexing agents. The actual addition of chemicals alters the original composition of the sample. It is important to avoid adding chemicals that contain elements for which the sample will be analyzed.

For example, samples being collected for nitrogen analyses should not be preserved with nitric acid; samples for sulfate should not contain sulfuric acid. In many instances, the volume of preservative must be taken into account to properly establish the concentration of contaminants. The combinations of possible cross-contaminations are great; however, they can be avoided by using common sense and carefully planning the sample program.

Pre-cleaned sample bottles are often shipped from contract laboratories with their preservatives. Operator–analysts should exercise caution and wear gloves when handling containers that may contain preservatives because strong acids and bases can cause chemical burns.

Sample preservation can be replaced with an alternative procedure with the proper regulatory approval as outlined in 40 CFR 136 (U.S. EPA, 2021). Currently, biomonitoring is an important tool for monitoring facility operational performance. Preservation for this testing is not covered in 40 CFR 136. Containers should be made of plastic or glass. Samples should be cooled to 6 °C, and the holding time is 36 hours.

3.8   Samples Requiring Special Consideration

Some samples need special consideration during sample collection. For example, microbiological samples need to be collected in sterile containers. Use caution when collecting and analyzing samples for phosphorus. Phosphorus is everywhere and can easily contaminate samples. It is important to never touch the inside of a sample bottle or lid with bare hands. Gloves should be worn and extra caution should be exercised when handling sample bottles, scoops, and so on. The containers and sample scoop should be prewashed with 1 : 1 hydrochloric acid and rinsed with distilled water. Additionally, if oxidants such as chlorine are used for disinfection, reagents capable of neutralizing the disinfectant must be added to the sample container.

3.9   Recommended Sampling Locations for Process Control

Factors that dictate where and how many samples should be collected include regulatory requirements, the size of the facility, facility performance, the intended use of the sampling data, and facility laboratory capabilities. Of course, the type and quantity of samples required on a routine basis will influence what analyses the facility laboratory should be capable of performing. The number of samples that must be taken for regulatory reporting requirements is outlined in the regulatory discharge permit. However, the specific parameter limitations listed in the permit will also influence the type and number of process control tests that need to be performed. For example, if the requirements include removal of ammonia-nitrogen, nitrogen, or phosphorus, more analyses are required at different stages of treatment to ensure that the parameter is being properly reduced. In the same manner, if the WRRF must maintain a high degree of contaminant removal, more process control sampling and testing is required on a regular basis to maintain tighter or stricter control of the facility processes. The size of the facility and availability of laboratory staff are also factors that influence sampling program procedures. The larger and more sophisticated the WRRF, the greater the quantity of sampling typically required and, therefore, the greater the number of laboratory staff needed to perform these duties. It is important to remember that all samples taken and analyzed according to proper protocol must be reported even if they are in excess of the required testing frequency.

How well a facility operates without constant operational adjustment is also a factor in determining the quantity of sampling and analyses needed. A treatment facility that operates well on a continuous basis and with little operator input will require less process sampling and testing than a facility in which operational adjustments must be made constantly or one that historically does not perform well.

Examples of suggested sampling locations and frequencies for common WRRF processes are listed in Tables 1.11 through 1.17. Many of the analyses listed in the tables may actually be tested more frequently based on individual facility conditions. As previously stated, the types of wastewater parameters that must be treated or removed in the facility processes will dictate what process control tests need to be performed. For example, if effluent limitations require total phosphorus removal, total phosphorus should be measured at various stages throughout the facility treatment processes. Each WRRF is slightly different; as such, there is no one sampling and testing program that can be adapted from one facility to another without some type of modification. The person who establishes the sampling program must take into account all factors to produce a useful and meaningful sampling program.

TABLE 1.11   Waste stabilization lagoon.

Sampling location	Analysis[a]	Frequency[b]	Sample type
Pond influent	BOD	Weekly	Composite
	TSS	Weekly	Composite
	pH	Daily	Grab
Pond	DO	Daily	Grab
	pH	Daily	Grab
	Temperature	Daily	Grab
Pond effluent	BOD	Weekly	Composite
	TSS	Weekly	Composite
	pH	Daily	Grab
	DO	Daily	Grab
	Fecal coliform[c]	Daily	Grab
	Chlorine residual[c]	Daily	Grab

[a]May not show all analyses required.

[b]Sampling frequencies may need to be increased for process control or permit requirements.

[c]May not be required if disinfection is not required; check with discharge permit.

TABLE 1.12   Influent.

	Analysis[a]	Frequency[b]	Sample type
Influent	BOD	Daily	Composite
	TSS	Daily	Composite
	pH	Daily	Grab
	TKN[c]	Daily	Composite
	Ammonia[c]	Daily	Composite
	Total phosphorus[c]	Daily	Composite

[a]May not show all analyses required.
[b]Sampling frequencies may need to be increased for process control or permit requirements.
[c]May not be required if disinfection is not required; check with discharge permit.

TABLE 1.13   Primary treatment.

Sampling location	Analysis[a]	Frequency[b]	Sample type
Primary influent	BOD	Weekly	Composite
	TSS	Weekly	Composite
	pH	Daily	Grab
	TKN	Daily	Grab
	Ammonia	Daily	Grab
	Total phosphorus	Daily	Grab
Primary effluent	BOD	Weekly	Composite
	TSS	Weekly	Composite
	pH	Weekly	Grab
	Dissolved oxygen	Weekly	Grab
	TKN[c]	Weekly	Composite
	Total phosphorus[c]	Weekly	Composite

[a]May not show all analyses required.
[b]Sampling frequencies may need to be increased for process control or permit requirements.
[c]May not be required if nutrients (ammonia, phosphorous, etc.) are not a concern.

TABLE 1.14   Trickling filter or rotating biological contactor facility.

Sampling location	Analysis[a]	Frequency[b]	Sample type
Influent	BOD	Daily	Composite
	TSS	Daily	Composite
	pH	Daily	Grab
	Dissolved oxygen	Daily	Grab
Effluent	Dissolved oxygen	Daily	Grab
	pH	Daily	Grab
	Temperature	Daily	Grab
	Ammonia[c]	Weekly	Composite
	Nitrate[c]	Weekly	Grab
Secondary clarifier	BOD	Daily	Composite
Effluent	TSS	Daily	Composite
Facility effluent	pH	Daily	Grab
	Dissolved oxygen	Daily	Grab
	Ammonia[c]	Weekly	Composite
	Nitrate[c]	Weekly	Grab
	Total phosphorus[c]	Weekly	Composite
	Fecal coliform[d]	Daily	Grab
	Chlorine residual[d]	Daily	Grab

[a]May not show all analyses required; check with permit requirements.
[b]Sampling frequencies may need to be increased for process control or permit requirements.
[c]May not be required if nutrients (ammonia, phosphorous, etc.) are not a concern.
[d]May not be required if disinfection is not required; check with discharge permit.

It must be emphasized that Tables 1.11 to 1.17 are just examples of suggested locations and frequencies of various samples. The operator–analyst must read and understand the discharge permit and meet the dictated sampling requirements. Experience and a good knowledge of the biological, physical, and chemical characteristics of a specific WRRF will aid in developing a well-thought-out and meaningful sampling program that will provide the facility what it needs to control the process and meet discharge permit requirements.

TABLE 1.15 Activated sludge.

Sampling location	Analysis[a]	Frequency[b]	Sample type
Aeration basin influent	pH	Daily	Grab
	BOD	Weekly	Composite
	TSS	Weekly	Composite
	TKN[c]	Monthly	Grab
	Ammonia	Monthly	Grab
	Alkalinity	Monthly	Grab
Aeration basin	Dissolved oxygen	Daily (continuous)	In situ
	Temperature	Daily (continuous)	In situ
Aeration basin effluent	TSS (mixed liquor suspended solids)	Daily	Grab
	Settleability	Daily	Grab
	pH	Weekly	Grab
	Microscopic	Weekly	Grab
Return activated sludge	TSS	Daily	Grab
	Flow	Daily	Totalizer
Waste activated sludge	TSS	Daily	Grab
	Flow	Daily	Totalizer
Secondary clarifier	Sludge blanket	Daily	In situ
Secondary clarifier effluent	BOD	Weekly	Composite
	TSS	Weekly	Composite
	Ammonia[c]	Monthly	Composite
	Nitrate[c]	Monthly	Grab
	Nitrite[c]	Monthly	Grab
	Total phosphorus[c]	Monthly	Composite
	pH	Daily	Grab
Final effluent	Turbidity	Daily	Grab
	Fecal coliform[d]	Daily	Grab
	Chlorine residual[d]	Daily	Grab

[a]May not show all analyses required; check with permit requirements.
[b]Sampling frequencies may need to be increased for process control or permit requirements.
[c]May not be required if nutrients (ammonia, phosphorous, etc.) are not a concern.
[d]May not be required if disinfection is not required; check with discharge permit.

TABLE 1.16 Anaerobic digestion.

Sampling location	Analysis[a]	Frequency[b]	Sample type
Digester feed	Total solids	Daily	Composite
	Volatile solids	Daily	Composite
	pH	Daily	Grab
	Alkalinity	2/Week	Grab
Digester contents	Temperature	Daily	Grab
	pH	Daily	Grab
	Volatile acids	2/Week	Grab
	Alkalinity	2/Week	Grab
Digested sludge	Total solids	Daily	Grab
	Volatile solids	Daily	Grab
	Volatile acids	Weekly	Grab
	TKN	Weekly	Grab
Supernatant	TSS	Daily	Composite
	BOD	Daily	Composite
	Ammonia or nitrate	Weekly	Composite
Digester gas	Methane	Daily	Grab

Note. Digester gas sampling is not included in this example.
[a]May not show all analysis required; check with permit requirements.
[b]Sampling frequencies may need to be increased for process control or permit requirements.

TABLE 1.17 Aerobic digestion.

Sampling location	Analysis[a]	Frequency[b]	Sample type
Digester feed	Total solids	Daily	Composite
	Volatile solids	Daily	Composite
	pH	Daily	Grab
	Alkalinity	Weekly	Grab
	Ammonia or nitrate	Weekly	Grab
Digester contents	Temperature	Daily	Grab
	pH	Daily	Grab
	Dissolved oxygen	Daily	Grab
	Total solids	Daily	Composite
	Volatile solids	Daily	Composite
	Ammonia or nitrate	2/week	Grab
	Alkalinity	2/week	Grab

(*continued*)

TABLE 1.17   Aerobic digestion. (*Continued*)

Sampling location	Analysis[a]	Frequency[b]	Sample type
Settled digested sludge	Total solids	Daily	Composite
	Volatile solids	Daily	Composite
	pH	Daily	Grab
	Volatile acids	Weekly	Grab
	Ammonia or nitrate	Weekly	Grab
Supernatant	TSS	Daily	Composite
	BOD	Daily	Composite
	Ammonia or nitrate	Weekly	Composite

[a]May not show all analyses required; check with permit requirements.
[b]Sampling frequencies may need to be increased for process control or permit requirements.

4.0   REFERENCES

American Public Health Association, American Water Works Association, & Water Environment Federation. (2017). *Standard methods for the examination of water and wastewater* (23rd ed.). American Public Health Association.

Boyles, W. (1997). *Chemical oxygen demand: Technical Information Series* (Booklet No. 9). Hach Company.

EnviroSim Associates, Ltd. (2005). *Influent specifier (raw) 2_2.xls*. Spreadsheet for fractionating wastewater into components required for Biowin modeling.

Metcalf & Eddy, Inc./AECOM. (2014). *Wastewater engineering treatment and resource recovery* (5th ed.). McGraw-Hill.

Smith, R.-K. (2019). *Water and wastewater laboratory techniques* (2nd ed.). Water Environment Federation.

U.S. Environmental Protection Agency. (1983). *Methods for chemical analysis of water and wastes.*

U.S. Environmental Protection Agency. (2017, 2021). *Federal register*, Part III, 40 CFR Parts 136 and 503. https://www.ecfr.gov/current/title-40/chapter-I/subchapter-D/part-136

Water Environment Federation. (2008). *Industrial wastewater management, treatment, and disposal* (3rd ed.). McGraw-Hill.

Water Environment Federation, American Society of Civil Engineers, & Environmental and Water Resources Institute. (2018). *Design of water resource recovery facilities* (6th ed.; WEF Manual of Practice No. 8; ASCE Manuals of Reports on Engineering Practice No. 76). Water Environment Federation.

2

Laboratory Basics

1.0   LABORATORY SAFETY

Like most accidents, laboratory accidents happen when work is rushed and a worker's attention is distracted. Therefore, it is important for the operator–analyst to work steadily and methodically and to minimize diversions. The operator–analyst should conduct him- or herself responsibly at all times in the laboratory. Indeed, safety procedures cannot be overemphasized. Before starting, the operator–analyst should be familiar with all procedures and the hazards involved. The operator–analyst is responsible for his or her own safety and the safety of others in the laboratory.

The operator–analyst should develop a chemical hygiene plan that includes a list of all chemicals stored in the laboratory in addition to chemical hazards, emergency procedures, and spill cleanup protocol. The specifics of an operator–analyst's laboratory should be included in the plan. The plan should also be reviewed with all personnel using the laboratory and updated annually.

The following subsections contain general safety guidelines for a wastewater laboratory.

1.1   Personal Protective Equipment

Use of personal protective equipment (PPE) in the laboratory is required. Eye, hand, ear, and clothing protection, such as the following, are the minimum requirements (the operator–analyst should review the chapter in *Standard Methods* on personal protective equipment [1090 E] for more information):

- Safety goggles offer a high degree of protection. Safety glasses are acceptable but should be designed to protect from splashing. Face shields are recommended on occasion (i.e., when sampling ports are at eye level and when working with concentrated reagents or hot materials).

- Disposable latex gloves, preferably powder-free, are the most common type of hand protection. Nitrile gloves should be substituted if latex allergies are an issue. Reusable gauntlet-style gloves should also be available. (For glove compatibility, check Table 1090:IV in *Standard Methods*.)

- It is important to dress appropriately for laboratory work. Closed-toe shoes (preferably chemical-resistant) and long pants should be worn, long hair should be tied back, and shirttails should be tucked in. In addition, a laboratory coat and rubber apron should be worn when needed.

Note that there are many editions of *Standard Methods*. Each edition contains hundreds of test methods, but most of the methods are not updated with each new edition of the print manual. Instead, *Standard Methods* designates the publication or revision date for each test method in both the online and hard-copy versions. The most current edition of any method is available online at https://www.standardmethods.org. The most current print version of *Standard Methods* is the 23rd edition (APHA et al., 2017).

It should also be noted that not every version of every test method is approved for use in 40 CFR Part 136 (U.S. EPA, 2017). 40 CFR lists the approved version of each test method, which may or may not be the most current version. This can sometimes require a laboratory to keep more than one version of *Standard Methods* on hand for reference. For example, the test method for total suspended solids is listed in 40 CFR Part 136 as "Residue, Total, mg/L, 2540 B-2015." The designation refers to the method number (2540), submethod (B), and approved version publication date (2015).

1.2   Chemical Storage

Labeling is the primary rule of chemical storage. All chemicals and reagents should be clearly labeled with the date received, date opened, and initials of the individual responsible. If not prelabeled by a manufacturer, the substance name, concentration, and expiration date should be labeled and a National Fire Protection Association symbol should be affixed to the containment bottle. Chemicals should be used on a "first in, first out" basis so that they do not expire and are not wasted before use.

Chemicals should be stored in cabinets or where there is no risk of being accidentally knocked over. Caution must be used when storing chemicals of different hazard classes. The following items should not be stored together: strong oxidizers and reducing chemicals; strong acids and bases; or oxidizers, flammables, and combustibles. Spills in the storage area can be dangerous when hazard classes mix. If separate storage areas are not available, the smallest amounts of chemical available should be stored. The operator–analyst should plan for storage and use of chemicals to minimize the amount of chemical stored or used. This will achieve safe operation of the laboratory.

The National Fire Protection Association (NFPA) recognizes six classifications of liquids as either flammable or combustible based on their flashpoint and boiling point. Flammable liquids have a flashpoint less than or equal to 22.7 °C (73 °F). NFPA Standard 45: Standard on Fire Protection for Laboratories Using Chemicals limits the volume of flammable and combustible liquids that may be in use or be stored within the enclosed space of a laboratory and sets requirements for their handling and storage. Solvents and other flammable and combustible liquids should be stored within Factory Mutual (FM)–approved, Underwriters Laboratory (UL)–approved, or equivalent listed safety cans or flammable storage cabinets. Some fume hoods are equipped with flammable cabinets that are vented through the back and into the hood.

Compressed gas cylinders should be stored upright and chained in separate, ventilated areas. In addition, oxygen cylinders should be stored away from other cylinders. Every chemical stored in the laboratory must be accompanied by a safety data sheet (SDS), which is readily available. The SDS should be read for all chemicals that will be handled. The operator–analyst should follow recommendations for safe use and disposal of materials.

1.3   Chemical Handling

1. Chemicals should be handled with a spatula, spoon, tongs, or pipet fitted with a suction bulb. Chemicals should never be handled with bare hands and should never be pipetted by mouth. In addition,

chemicals should never be returned to reagent bottles because this can contaminate the stock chemical.

2. Every sample should be handled as if it is pathogenic. Similarly, every chemical should be handled as if it is hazardous. When working with hazardous materials such as volatile solvents, bases, or acids, the operator–analyst should work under a ventilated fume hood.

3. Each chemical's SDS sheet should be consulted for specific disposal instructions. When proper disposal involves disposing of chemicals down the sink, the sink should be flushed with plenty of water because the plumbing system can hold vapor pockets or be damaged by strong acids and bases.

1.4   Chemical Spills

Chemical spills, whether a solid chemical, acid, base, or mixed reagent, are inevitable in the course of laboratory work. The operator–analyst should be prepared to handle spills by having a chemical safety plan and readily available safety equipment. Laboratory-specific spill containment kits are commercially available and should be standard safety equipment. The following are some important instructions regarding handling of chemical spills:

• Spills at a laboratory should be cleaned up immediately and all work areas should be cleaned before leaving the laboratory.

• Access to shut-off valves should never be blocked.

• The operator–analyst should be able to identify and know how to use safety equipment, including the emergency shower station and eyewash. A routine schedule should be established for flushing eyewashes and safety showers and to verify they are functioning properly. Access to the emergency station should never be blocked.

• If a person is splashed with an acid or base, large volumes of water are required immediately to prevent serious burns. Clothing, belts, and shoes that might trap acid against the skin should be removed immediately upon entering the emergency shower because confining heat generated by the reaction could increase the severity of skin burns.

• If there is a chemical exposure to the eye or eyes, both eyes should be flushed immediately with large volumes of water. Flushing should continue for at least 15 minutes. The eyes should be held open throughout flushing and eyeballs rolled around to ensure the water reaches all surfaces and underneath the eyelids. Remove contacts if present.

• A first aid kit should be readily available.

1.5   Fire

All state, local, and other regulatory agency criteria regarding placement and use of fire extinguishers and equipment should always be followed. An ABC-rated fire extinguisher must be wall-mounted in a conspicuous and accessible area of the laboratory. In addition, a fire blanket should be available to wrap and extinguish a person who has caught fire. Personnel must comply with fire regulations concerning storage quantities, types of approved containers and cabinets, proper labeling, and so on. Chemical storage areas must always be labeled. The reaction of water with many chemicals is exothermic, producing heat. Operator–analysts should work closely with local fire departments so they are aware of the unique hazards a wastewater laboratory and a water resource recovery facility (WRRF) present.

1.6   Ingestion Hazards

Care must be taken to avoid ingestion of chemicals used in analyses and the infectious materials found in wastewater and sludge samples. These can be ingested through the mouth, absorbed through the skin, or inhaled to the lungs. The following rules should be followed:

- Never mouth-pipet anything in the laboratory. Use a suction device such as a pipet bulb.
- Never eat, drink, or smoke while working in the laboratory.
- Never store food in laboratory refrigerators (personal food should be stored in a separate area).
- Remove laboratory gloves and coats before leaving the laboratory. Wash your hands before leaving the laboratory and before eating.
- Ensure the laboratory is well-ventilated to avoid buildup of dangerous vapors.

2.0   LABORATORY ANALYSIS

The purpose of analytical chemistry is to determine the concentration of a specific constituent, or analyte, in an unknown sample. Some results are found by direct manipulation of a sample (e.g., a solids analysis). A measured volume of sample is dried to evaporate water, and the remaining solid matter is weighed. Other properties are found using specially designed instruments, the most basic example being the use of a thermometer to determine temperature. The more complex analyses involve subjecting a sample to chemical reactions to producing a characteristic specific to the analyte of interest. Most commonly, a colored compound of the analyte is produced. This is then compared to the same characteristic in a series of solutions of known concentration prepared by the operator–analyst. These solutions are called *standard solutions.* A graph of the standard concentration versus the indicating characteristic is known as a *standard curve* or *calibration curve* (see Figure 2.1).

2.1   Volumetric Analyses (Titrations)

In laboratory procedures classified as *volumetric analyses,* the operator–analyst measures the volume of a solution of known concentration that reacts with a particular substance in a specified volume of unknown sample. The concentration of analyte in the sample is found indirectly from the amount of the known (standard) solution that is required. As standard solution is added, a property of the solution, such as pH or millivolt potential, is monitored to indicate the completion of the reaction. Detecting the completion or "endpoint" of the volumetric reaction can also be achieved with the addition of an organic dye called an *indicator* that exhibits a color change at the desired endpoint pH. Volumetric analyses are commonly called *titrations.*

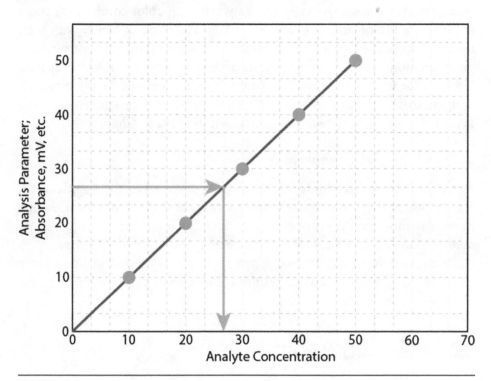

FIGURE 2.1   Basic calibration curve. Trace the unknown sample parameter result over to the line and down to determine analyte concentration. *Courtesy of Indigo Water Group*

2.2 Gravimetric Analysis

In laboratory procedures classified as *gravimetric analyses,* the operator–analyst measures the wastewater or sludge sample and then isolates and weighs an element or one of its compounds. Examples of gravimetric analyses are total solids (residue on evaporation), volatile solids, and suspended matter determinations.

2.3 Colorimetric Analysis

Colorimetric methods of analysis have been developed for several parameters to provide faster, more economical, and convenient ways of obtaining quantitative laboratory data. For a colorimetric method to be quantitative, it must form a compound with definite color characteristics that are directly proportional to the concentration of the substance being measured. Colorimetric measurements may be made with a wide range of equipment. The WRRF operator–analyst may use standard color-comparison tubes, photoelectric colorimeters, or spectrophotometers; each has its place and particular application in wastewater analyses. Color-comparison tubes, sometimes referred to as *Nessler tubes,* have been standard equipment for making colorimetric measurements for many years. Precise work with color-comparison tubes requires the use of optically matched tubes. The main difficulty with their use is that standard color solutions are often unstable, and every time a determination has to be made, a series of fresh standards must be prepared. Use of color tubes and standards is rapidly being replaced by photoelectric and spectrophotometric methods, largely because of their convenience and accuracy. Spectrophotometers are discussed in detail in Chapter 4.

2.4 Electrometric Analysis

These instrumental methods rely on the measurement of the electrical current or potential generated in the sample to indicate the concentration of a particular substance. A probe that is manufactured to select for a particular analyte (ion-selective electrode) is connected to a meter that measures the current across the probe. The meter and probe are calibrated against a range of standard solutions, and the resulting calibration curve is stored in the meter software.

3.0 LABORATORY EQUIPMENT

3.1 Glassware

Laboratory glassware can effectively be divided into two categories: volumetric and other.

3.1.1   Volumetric Glassware

There are five pieces of volumetric glassware that are used only to accurately measure volumes of reagents. These are graduated cylinders, graduated pipets, volumetric pipets, volumetric flasks, and burets. Volumetric glassware is depicted in Figure 2.2. Volumetric glassware can further be categorized as "Class A," which is the most accurate, or "Class B." Class A glass will have an evident "A" marked on it near the top. Class B glass may be marked with a "B," but the absence of either mark indicates Class B. The glass should be marked with a tolerance given in milliliters, shown as "TOL ± 2.4," for example. This is a measure of how accurately the manufacturer places lines on the glass and has nothing to do with how an operator–analyst uses the glass. Class B generally has twice the tolerance of Class A. Proper use of calibrated volumetric glassware requires the user to discern if the ware is calibrated "to contain" (TC) or "to deliver" (TD); the glassware is marked with a "TC" or "TD" accordingly. "To contain" indicates the vessel's calibration includes the amount of water required to wet the inner surface of the vessel in contact with the water. These vessels

FIGURE 2.2   Volumetric glassware (left to right): volumetric flask, buret, volumetric pipet, graduated pipet, and graduated cylinder. *Courtesy of Kelsy Heck, City of Defiance, Ohio Water Pollution Control*

contain the exact volume. "To deliver" indicates the exact volume delivered or dispensed from the vessel. The two are designed for different purposes. For example, a "to contain" volumetric flask is used to dilute an original sample to a known volume; as such, it is critical that it contains the exact volume. A "to deliver" pipet may be used to transfer a solution; as such, it is critical that it delivers a known volume.

3.1.2 Using Volumetric Glassware

The sole use of volumetric glassware is for measuring and mixing reagents. Volumetric glassware should not be heated and reactions should not be performed in it; in addition, reagents should not be stored in volumetric flasks.

Volumetric glass that measures as close to the volume needed as possible should be selected. Because the tolerance applies to every mark on the glass, the percent accuracy will decrease as less than the full volume is measured. For example, the tolerance of a Class A, 50-mL graduated cylinder is ±0.2 mL. Dividing the tolerance by the volume gives a percent accuracy of 0.4%, as follows:

$$Percent\ Accuracy = \frac{tolerance}{volume\ measured} \times 100 \tag{2.1}$$

$$0.2\ \text{mL}/50\ \text{mL} \times 100 = 0.4\%$$

Measuring 25 mL in the 50-mL cylinder gives 0.8% accuracy, or twice as inaccurate. Using a Class A, 25-mL graduated cylinder to measure the same 25 mL is 0.6% accurate. Regardless of the type of volumetric ware, the operator–analyst should try to match the volume needed to the total volume of the glass as closely as possible. Volumetric glassware is most accurately read in a perfectly vertical position, with the mark at eye level and the bottom of the meniscus just touching the top of the line (see Figure 2.3).

The procedure to be followed for a "to deliver" cylinder is to fill to the desired volume and pour out. For "to contain" cylinders, the procedure to be followed is to fill to the desired mark, pour out, and then flush several times with distilled water to remove all samples.

Graduated pipets may be serological or Mohr style (see Figure 2.4). All pipets are calibrated to deliver their marked capacity after they have been wetted down. To wet a pipet, fill it past the measurement mark and allow it to drain completely before filling it for use. The marks on a serological pipet extend to the tip, and the milliliter numbering stops one short of the full volume. A serological pipet should be drained entirely to deliver the measured volume. If it is calibrated "to contain," the operator–analyst

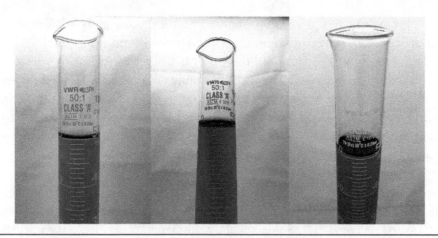

FIGURE 2.3 Reading volumetric glassware (left to right): proper position for reading the 100-mL mark, glass held plumb and read at eye level, glass tilted backwards or 100-mL mark above eye level, and glass tilted forward or 100-mL mark below eye level. *Courtesy of Indigo Water Group*

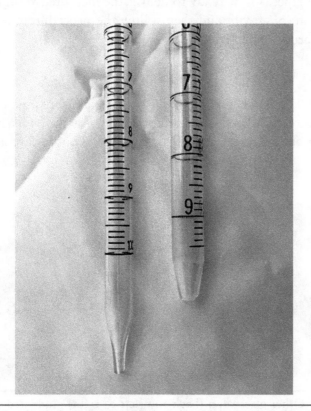

FIGURE 2.4 Serological- (right) and Mohr-style (left) pipets. Drain the serological pipet entirely. Drain the Mohr pipet so that the meniscus is on the 10-mL mark and the tip remains full of sample. *Courtesy of Indigo Water Group*

should blow out the pipet tip to deliver the total volume. If it is calibrated "to deliver," the operator–analyst should let the liquid drain out, hold the glassware at an angle, and hold the tip of the pipet to the inside surface of the glassware for 2 seconds to most accurately deliver the volume of reagent.

The marks on a Mohr pipet end before the taper, with the full volume clearly marked. The operator–analyst should control and stop the discharge when the meniscus reaches the top of the full volume mark. A Mohr pipet should be used to add volumes of nondetermining reagents.

Volumetric pipets have a bubble in the shaft of the pipet and usually have a single volume calibration mark. An example of a volumetric pipet is shown in Figure 2.2. Volumetric pipets may be marked as either to contain (TC) or to deliver (TD). Some special volumetric pipets have two calibration marks, one for TC and one for TD. Volumetric pipets deliver their full volume in one delivery. After wetting the pipet, fill it with liquid past the volume calibration mark. Then, use a low-lint laboratory wipe to wipe off the tip of the pipet. Next, release enough liquid to bring the meniscus down to the volume calibration mark while holding the tip against a clean surface. The pipet is now ready to deliver the measured volume. Allow the pipet to drain by gravity and then follow the procedure for a TC or TD pipet described in the previous paragraphs.

3.1.3 Other Glassware

Beakers and flasks make up the bulk of other glassware used in the laboratory (see Figure 2.5). These are used for a variety of purposes, but primarily for carrying out reactions and analyses. Storage bottles are also common glassware.

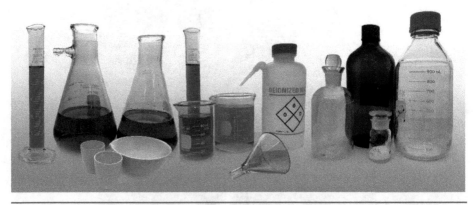

FIGURE 2.5   Other glassware: (back row, left to right) graduated cylinder, filter flask, Erlenmeyer flask, graduated cylinder, tall form beaker, beaker, wash bottle, BOD bottle, amber glass storage bottle, and glass storage bottle; (front row, left to right) Gooch crucibles, evaporating dish, liquid funnel, and weighing/chemical storage bottle. *Courtesy of Indigo Water Group*

The markings on this glassware are approximate, generally good only to ±5%, and are meant only to give the operator–analyst a rough idea of the volume in the glass. Many of the procedures in this manual are carried out using specialized glassware. This includes items such as biochemical oxygen demand (BOD) bottles, filter flasks, filter supports, funnels, and ceramic crucibles. Specialized other glassware should be used only for the intended purpose and never as an accurate volumetric tool.

3.2   Miscellaneous Laboratory Equipment

Laboratory appliances are used to manipulate a sample, while laboratory instruments are used to read an analytical parameter. For example, in a solids analysis, a drying oven is an appliance used to evaporate water from a sample while an analytical balance is an instrument used to determine the mass or weight of solids. There is a wide range of miscellaneous laboratory equipment, from the smallest stir bar to the most expensive spectrophotometer. These are the tools of the operator–analyst. Their functionality has a direct effect on the quality of results obtained. As such, they should be used only for their intended purpose and should be kept clean and well-maintained.

4.0   HOUSEKEEPING

4.1   Laboratory Environment

The most important factor affecting the contamination of samples is the sample preparation environment. The ability to control the environment within the laboratory is extremely important to ensure sample integrity. The operator–analyst should maintain a separate, identifiable area for laboratory procedures. Preferably, the laboratory site should be located away from sources of mechanical vibration, shock, and electrical interference. The noise level should not exceed that of a quiet office.

4.1.1   Climate

The work area of a laboratory should be free from excessive drafts and the temperature should be kept reasonably stable, with a standard temperature of 20 ± 2 °C. Generally, the relative humidity should not exceed 55%. The laboratory should be sensibly, but not clinically, dust-free. Walls, floors, and ceilings should be of a finish that will not make or collect dust. For example, untreated concrete floors or ceilings with excessive loose particles are unsuitable. Lastly, proper lighting is crucial for proper sample handling.

4.1.2 General Cleanliness

The operator–analyst is responsible for keeping the laboratory clean. The bench top or work area should always be cleaned after use and equipment should be washed and dried and returned to its proper storage place. Used paper towels and Kimwipes® should be thrown away, and chemicals and laboratory materials should be put away immediately when finished. Cabinet doors should be kept closed to prevent dust from settling on stored laboratory equipment. The operator–analyst should try to isolate the laboratory from the facility by keeping doors closed. It is important to never eat, drink, smoke, chew gum, or apply cosmetics in sample preparation areas or areas where laboratory chemicals are present. The operator–analyst should always wash hands before and after performing any laboratory procedure. Keeping the work area clean and uncluttered will significantly minimize the potential for contamination.

4.2   Washing Dishes

Washing laboratory dishes (laboratory equipment) may be the most mundane task in a WRRF; however, improper handling of sample containers, glassware, and other apparatuses can lead to faulty and imprecise results. Development of and adherence to a standard operating procedure for cleaning laboratory equipment may be the single most important method of reducing sample contamination. Sources of contamination to be aware of include contaminated sample containers, unclean glassware and filters, use of cleaning products inappropriate for the proposed analysis, and inadequate rinsing of glassware during the cleaning process.

The longer chemicals and reagents stand in glassware, the more chance they have to adhere to the walls of the glass and the more difficult it becomes to get the glass thoroughly clean. The operator–analyst should adopt the concept of "clean as you go" as opposed to "clean before you go." Indeed, a wash station should be set up daily so that it is convenient to wash glassware immediately after use.

The following are general instructions on how to wash common WRRF laboratory equipment. Special consideration should be given to each test performed. For example, nitrogen- or phosphorus-free laboratory-grade detergents are available and should be used with the corresponding containers and glassware. Many analyses have specific instructions for sample bottle preparation. These instructions can be found in *Standard Methods for the Examination of Water and Wastewater* (APHA et al., 2017) in the "Sampling and Sample Preservation" section for each analysis. Additionally, if a contract laboratory is being used, sample bottles are provided in accordance with *Standard Methods* guidelines.

If a laboratory-specific dishwasher is being used, any tape or numbering on the glassware should be removed using a solvent such as acetone. Laboratory glassware and dishwasher-safe laboratory equipment should be placed in a dishwasher immediately after use. The washer is designed to thoroughly clean all laboratory glassware and accessories and is preprogrammed for glassware washing. The operator–analyst should check the owner's manual for the appropriate glassware setting.

Laboratory equipment that is too large for an automatic dishwasher or that cannot be feasibly washed in an automatic dishwasher (i.e., volumetric flasks) requires hand washing. Again, any tape, numbers, and so on from sample bottles should be removed. Laboratory equipment should be immersed in a solution of laboratory-appropriate synthetic detergent such as Liquinox® or Alconox® (Alconox, Inc., White Plains, New York; available through laboratory supply companies). Glassware should be scrubbed with a soft-bristle brush to loosen any adhered particles, triple-rinsed with tap water to remove visible detergent, triple-rinsed with distilled or deionized water (if available), and then allowed to drip dry inverted. Laboratory equipment should be returned to the appropriate cabinet or drawer.

For microbiology sample containers (i.e., *Escherichia coli* and fecals), after the aforementioned washing procedure, designated bottles are sterilized by autoclaving. An autoclave manufacturer's specifications should be followed for sterilization time. Filtering apparatuses should be wrapped in autoclavable paper and taped with heat-sensitive tape prior to autoclaving. Autoclavable bottles should be autoclaved with lids loosely in place and taped with heat-sensitive tape. The heat-sensitive tape ensures that the proper temperature was reached for sterilization.

4.2.1 Acid Washes

For glassware requiring acid washing (i.e., phosphorus sample collection bottles, BOD bottles, and trace metals), the operator–analyst should follow the general washing procedure and then immerse bottles in a 10% hydrochloric acid solution or 10% nitric acid solution for trace metals (1 part acid to 9 parts water). The acid wash should be prepared by slowly pouring and stirring the acid into the water. Water should NEVER be poured into acid. The bottles should be soaked a minimum of 1 hour (preferably overnight), triple-rinsed with distilled or deionized water, and allowed to drip dry inverted. Glassware should be returned to the appropriate cabinet or drawer. It is important to note that extreme care is needed when working with acids. Proper ventilation, goggles, and protective clothing are necessary. The acid bath should be contained with a suitable cover for safety. For acid wash disposal, the acid wash should be neutralized with a saturated solution of sodium carbonate (soda ash) or other basic solution. The operator–analyst should slowly add soda ash or another basic solution into the diluted acid while stirring. The operator–analyst should monitor pH with a pH meter, pH indicator strips, or other pH test method. When the pH is between 6 and 9, the solution should be disposed of down the sink drain and the drain should be flushed with excess water. A neutral pH is preferable to lessen the risk of plumbing damage.

4.2.2 Oxidizers

Specialty cleaning solutions vigorously clean glass by oxidizing the surface. This process is necessary for extremely soiled glassware. Chromic acid is the standard method for cleaning BOD bottles and is available commercially as Chromerge® (Bel-Art Products, Pequannock, New Jersey). Glassware should be cleaned in a similar manner to that described in Section 4.2.1, with extra caution taken to thoroughly rinse the glassware. If chromic acid is retained on any residual grease layers, it will interfere with bacterial growth. Additionally, the operator–analyst should be aware that chromium is highly toxic and requires special disposal following the MSDS protocol. Spent Chromerge® should not be put down the drain. Another commercial product available is NoChromix® (Godax Laboratories, Inc., Cabin John, Maryland). NoChromix® cleans with a similar oxidation process but does not contain chromium, which eases the disposal problem.

5.0 BASIC CHEMISTRY

5.1 Using the Periodic Table of the Elements

The periodic table is the foundation of basic chemistry and provides information about elements that is useful in determining concentrations of solutions. The name, symbol, atomic number, atomic weight, and most common oxidation state, or valence, are generally given. The periodic table should have a legend (see Figure 2.6).

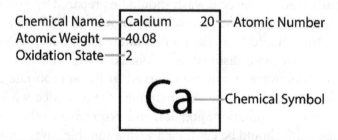

FIGURE 2.6 Periodic table representation of calcium.

5.2   Atoms

Atoms are the basic building blocks of nature. Atoms can't be broken down into smaller pieces without losing their unique properties. Atoms contain protons (positively charged particles) and neutrons (particles without a charge) in their center, and electrons (negatively charged particles) that orbit around the center.

5.3   Elements

An element is a particular type of atom. Elements are defined by the number of protons they contain in their nucleus. For example, all carbon atoms have six protons, all nitrogen atoms have seven protons, and all oxygen atoms have eight protons. The periodic table contains over a hundred boxes. Each box contains a letter or combination of letters and symbolizes a different element. The most common elements in wastewater treatment include aluminum (Al), calcium (Ca), carbon (C), chlorine (Cl), hydrogen (H), iron (Fe), magnesium (Mg), nitrogen (N), oxygen (O), phosphorus (P), and sulfur (S).

5.4   Atomic Numbers

The atomic number is equal to the number of protons in an atom. Atomic numbers are specific to each type of element. If an atom has eight protons, it is oxygen. If an atom has six protons, it is carbon. The atomic numbers are shown in the periodic table along with the chemical symbol for the element.

5.5   Molecules and Compounds

Molecules are combinations of two or more atoms that are chemically bound together. Atoms of some elements, like sulfur, are sometimes found in nature by themselves, unattached chemically to other atoms. A single sulfur atom is not a molecule. The following elements exist in nature as combinations of two of the same type of atom: bromine (Br), iodine (I), nitrogen (N), chlorine (Cl), hydrogen (H), oxygen (O), and fluorine (F). This is why dissolved oxygen (DO) is always shown as O_2 and nitrogen gas is shown as N_2. Putting the chemical symbols for these molecules together makes the word BrINClHOF, which makes it easier to remember which elements form pairs with themselves. The BrINClHOF elements also react with many other elements, not just themselves.

Molecules made from two or more different elements are called compounds. Water (H_2O) is a compound. All compounds are molecules, but not all molecules are compounds. Elements combine to form compounds by sharing electrons to achieve a neutral charge. For example, calcium and chlorine combine to form calcium chloride. The periodic table gives the information needed to predict how these elements will react with each other, as illustrated in Figure 2.7.

Because calcium has a 2^+ charge and chlorine has a single negative charge, the reaction results in a zero charge for the product molecule, as evidenced in the following chemical equation:

$$1(\text{mol})Ca^{2+} + 2(\text{mol})Cl^- \rightarrow 1(\text{mol})CaCl_2$$
$$Ca^{2+} + 2(\text{mol})Cl^- \rightarrow CaCl_2$$

(2.2)

5.6   Moles

It is also important to understand the concept of the mole. A mole is simply a measure of the number of items or particles. Shoes come in a pair, pencils come in a gross, paper comes in a ream, and eggs come in a dozen. Atoms and molecules come in moles. One mole contains 6.022×10^{23} atoms or molecules. That's 602 200 000 000 000 000 000 000! The number assigned to the mole, 6.022×10^{23}, is known as *Avogadro's number.*

5.7   Atomic Weight

The atomic weight of an element is the how much one mole of atoms weighs in grams. For different elements, the atomic weight is equal to the number of protons plus the average number of neutrons. Calcium atoms contain 20 protons and typically have 20 neutrons. The atomic weight of carbon is 40.08 g/mol (see Figure 2.6). Some calcium atoms have more than 20 neutrons. These are known as isotopes. The existence of isotopes is why the atomic weight isn't exactly 40 g/mol. This means that if someone patiently counted out 602 200 000 000 000 000 000 000 carbon atoms and weighed them on a balance, they would weigh 40.08 grams.

5.8   Formula Weight

The formula weight, also called the *molar mass* of a chemical compound, is the sum of the atomic weights, as illustrated in Table 2.1. It is important

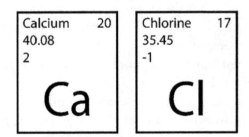

FIGURE 2.7   Using the periodic table to predict chemical compounds.

TABLE 2.1 Calculating formula weight.

Element	Atomic weight, g/mol		Quantity, mol		
Ca	40.08	×	1	=	40.08
Cl	35.45	×	2	=	70.90
Formula weight g/mol			1	=	110.98

to note that the number of moles do not add together; rather, 1 mol of calcium chloride contains 1 mol of calcium and 2 mol of chloride. The lowest number of moles on the left side of the chemical equation limits the number of moles on the right; as such, if one were to react 2 mol of calcium with 4 mol of chloride, 2 mol of calcium chloride would be produced.

Other molecules may be more complicated than calcium chloride. They can contain multiple types of elements and more than one atom of the same element. An example is ferric chloride, $FeCl_3$. Look at the periodic table and find the atomic weight for iron. To get the formula weight for $FeCl_3$, add together the weight of one iron atom and three chlorine atoms.

$$Fe + Cl + Cl + Cl = FeCl_3$$
$$56 \text{ g/mol} + 35.5 \text{ g/mol} + 35.5 \text{ g/mol} + 35.5 \text{ g/mol} = FeCl_3$$
$$162.5 \text{ g/mol} = FeCl_3$$

5.9   Ions

Ions are atoms or molecules that have a charge. Cations are positively charged, and anions are negatively charged. Ions are created when atoms and molecules gain or lose electrons. A chloride ion has one more electron than protons, which results in a -1 charge. Hydrogen ion has one proton and no electrons, which results in a $+1$ charge. Many of the inorganic compounds that are important in wastewater treatment will be present as ions in the water (Table 2.2).

Knowing the charges on common ions can help predict what molecules may be formed in a chemical reaction. For example, magnesium has a $+2$ charge while chloride has a -1 charge. The final compound must be neutral (zero charge). When magnesium and chloride are combined, the final compound will be $MgCl_2$ because two chlorine atoms are needed to neutralize the $+2$ charge on magnesium. When ferrous iron ions ($+2$ charge) are combined with phosphate ions (-3 charge), the quantity of each ion has to be adjusted so the final compound is neutral. In this case, three ferrous iron atoms will have a combined charge of $+6$ and two phosphate ions will have a combined charge of -6. The final compound is ferrous phosphate,

TABLE 2.2 Important ions in wastewater treatment.

Cations	Anions
Aluminum ion (Al^{13})	Bicarbonate ion (HCO_3^-)
Ammonium ion (NH_4^1)	Chloride ion (Cl^-)
Calcium ion (Ca^{12})	Carbonate ion (CO_3^{-2})
Hydrogen ion (H^1)	Hydroxide ion (OH^-)
Ferric ion (Fe^{13})	Nitrite ion (NO_2^-)
Ferrous ion (Fe^{12})	Nitrate ion (NO_3^-)
Magnesium ion (Mg^{12})	Phosphate ion (PO_4^{-3})
Sodium ion (Na^1)	Sulfate ion (SO_4^{-2})

or $Fe_3(PO_4)_2$. The phosphate ion (PO_4^{3-}) is shown in parentheses because the outside subscript 2 applies to the entire ion.

5.10 Molar Solutions

The concentration of a chemical in a solution is often given in molarity (M). *Molarity* refers to the number of moles of chemical per liter of solution (mol/L). The formula weight can be used as a conversion factor to make a molar solution. For 1 M calcium chloride ($CaCl_2$),

$$1 \text{ mol } CaCl_2/L \times 110.98 \text{ g/mol} = 110.98 \text{ g/L} \qquad (2.3)$$

To make this solution, 110.98 g of calcium chloride should be weighed and dissolved in 1 L of water.

Three molar potassium iodide (KI) is a common electrode-filling solution. To make 500 mL of 3 M potassium iodide, the formula weight should first be determined. The atomic weight of potassium is 39.1 g/mol, and 126.9 g/mol for iodide. The formula weight is, therefore, 166.0 g/mol. The following conversion factors can be used to find the correct amount of potassium iodide to use:

$$3 \text{ mol/L} \times 166 \text{ g/mol} \times 0.5 \text{ L} = 249 \text{ g} \qquad (2.4)$$

5.11 Normal Solutions

The term *normality* can be used to refer to strengths of acids and bases. It indicates the amount of hydrogen ion (H^+) or hydroxide ion (OH^-) in a solution. In monoacidic or monobasic chemicals such as hydrogen chloride

(HCl) or sodium hydroxide (NaOH), normality is equal to molarity because 1 mol of HCl contains 1 mol of hydrogen ion. In diacidic or dibasic compounds like sulfuric acid (H_2SO_4) or calcium hydroxide [$Ca(OH)_2$], 1 mol of the chemical contains 2 mol of hydrogen or hydroxide ions so a 1 mol solution is 2 N. A 2-N solution will, therefore, neutralize twice the volume of a 1-N solution.

To prepare a 1-N solution of sulfuric acid, the equivalent mass should first be calculated by dividing the gram-formula weight by the number of acid hydrogens in the compound. Sulfuric acid formula weight is 98.0 g and the number of hydrogens is two; therefore, the mass is 49.0. The amount of sulfuric acid needed for 1 L of 1-N solution can be calculated as follows:

$$\text{(N desired)(equivalent mass)(volume in liters desired)} =$$
$$\text{(1 N)(49.0 g)(1 L)} = 49.0 \text{ g needed}$$

A 1-N solution requires 49.0 g of pure sulfuric acid powder diluted to 1000 mL. However, the acid is a liquid and not pure. Sulfuric acid is commonly supplied at concentrations of 78%, 93%, or 98%, with 98% considered to be concentrated. Concentrated sulfuric acid has a specific gravity of 1.84. The volume of concentrated acid that contains 49.0 g of pure sulfuric acid can be calculated as follows:

$$Volume\ Conc.\ Acid\ Needed = \frac{Grams\ of\ Acid\ Needed}{(Decimal\ Percent\ Concentration \times Specific\ Gravity)}$$

$$Volume\ Conc.\ Acid\ Needed = \frac{49.0\ g}{0.98 \times 1.84}$$

$$Volume\ Conc.\ Acid\ Needed = 27.2\ mL \tag{2.6}$$

Diluting 27.2 mL of concentrated sulfuric acid to a final volume of 1000 mL will produce a 1-N solution.

5.12 Milligrams per Liter Solutions

Concentration is also commonly expressed as milligrams per liter, which is the same as parts per million (ppm) because 1 mL of water weighs 1 g, as follows:

$$\frac{1\ mg}{L} \left| \frac{1\ g}{1000\ mg} \right| \left| \frac{1\ L}{1000\ mL} \right| = \frac{1\ g}{1\ 000\ 000\ mL} \tag{2.7}$$

In the 1-M calcium chloride example, simply convert grams to milligrams as follows:

$$\frac{110.98 \text{ g}}{\text{L}} \left| \frac{1000 \text{ mg}}{1 \text{ g}} \right| = \frac{110\ 980 \text{ mg}}{1 \text{ L}} \qquad (2.8)$$

5.13 Determining Percent Weight of Atoms in Molecules

Often, the operator–analyst is only interested in the concentration of one element or constituent of a chemical compound. To determine this concentration, find the percent weight of the element in the compound by dividing the atomic weight by the formula weight. For calcium and chlorine in calcium chloride,

$$\frac{40.08 \text{ g Ca}}{\text{mol Ca}} \left| \frac{1 \text{ mol Ca}}{1 \text{ mol CaCl}_2} \right| \left| \frac{1 \text{ mol CaCl}_2}{110.98 \text{ g CaCl}_2} \right| = 0.3611 \times 100 = 36.11\% \quad (2.9)$$

$$\frac{35.45 \text{ g Cl}}{\text{mol Cl}} \left| \frac{2 \text{ mol Cl}}{1 \text{ mol CaCl}_2} \right| \left| \frac{1 \text{ mol CaCl}_2}{110.98 \text{ g CaCl}_2} \right| = 0.6389 \times 100 = 63.89\% \quad (2.10)$$

This concept becomes important in making stock and standard solutions and in how the analyte is reported. For example, phosphorus standard can be made by dissolving potassium dihydrogen phosphate (KH_2PO_4) in deionized water. If a 100-ppm phosphate (PO_4) solution is required, the percent weight of phosphorus in potassium dihydrogen phosphate must be determined to find the weight of potassium dihydrogen phosphate to use (see Table 2.3). Then, find the percent weight of phosphorus (P), as follows:

$$P = 30.97/136.09 = 22.75\% \qquad (2.11)$$

TABLE 2.3 Formula weight of potassium dihydrogen phosphate.

Element	Atomic weight, g/mol		Quantity, mol		
K	39.10	×	1	=	39.10
H	1.01	×	2	=	2.02
P	30.97	×	1	=	30.97
O	16.00	×	4	=	64.00
Formula weight g/mol			1	=	136.09

One hundred milligrams of phosphorus is needed to make 1 L of solution, but only 22.75% of potassium dihydrogen phosphate is phosphorus. As an equation,

$$100 \text{ mg P} = (X) \text{ mg KH}_2\text{PO}_4 \times .2275 \qquad (2.12)$$

Solving for (X),

$$100 \text{ mg P}/.2275 = 439.6 \text{ mg KH}_2\text{PO}_4 \qquad (2.13)$$

Analytes such as phosphorus can be reported as phosphorus (indicated as PO_4-P) or as phosphate (indicated as PO_4^{3-}), but these labels do not indicate the same thing. Similarly, ammonia (NH_3) is typically reported as ammonia-nitrogen (NH_3-N), which is different from ammonia. The mass ratio of phosphorus to phosphate can be used to find the concentration of phosphate in the 100-ppm phosphorus standard, as follows:

$$P, \text{ ppm} = 100 \text{ ppm P} \times \frac{94.97 \frac{g}{mol} PO_4}{30.97 \frac{g}{mol} P} \qquad (2.14)$$

6.0   WEIGHING

6.1   Analytical Balances and Weighing Conditions

The ability to accurately weigh small amounts of chemicals and gravimetric dishes and filter papers for solids tests is critical to the success of the operator–analyst. A 200-g capacity analytical balance is probably the most useful weighing instrument for a small-to-medium wastewater laboratory. Analytical balances feature a draft shield over the weighing pan to minimize the effects of air currents and to guard against dust. These balances are extremely sensitive to environmental conditions and, as such, the following guidelines should be observed:

- Vibration—balances should be placed on a specially designed weighing table that features a heavy slab placed on vibration dampeners. Ideally, the table is located on the ground floor. The operator–analyst should be careful not to let the table contact a wall or laboratory bench that could cause vibration. Additionally, the operator–analyst should be aware of the proximity of equipment and machinery.

- Level—the balance must be kept perfectly level. Most balances have adjustable feet and a "bull's-eye" level on one foot.

- Temperature and humidity—temperature and humidity affect the internal components of the balance and should be maintained at a constant level in the laboratory. If the item being weighed is not at room temperature, it will produce air currents sufficient to cause error. Many chemicals absorb water from the air and will be difficult to accurately weigh in a highly humid environment.

- Static electricity—special anti-static brushes are made for cleaning the weighing pan. Static electricity can make it difficult to transfer light chemicals and chemical flakes; it can also affect the weight of these particles.

- Air currents—even light air movements, unnoticeable to the operator–analyst, can affect the modern analytical balance. In addition to keeping the draft shield closed during weighing, the operator–analyst should keep laboratory doors and windows closed and be aware of internal currents caused by heating and ventilation systems and the fume hood.

- Dust—dust accumulation in and on the balance will cause friction on moving components and, therefore, reduce accuracy. As such, the manufacturer's maintenance recommendations regarding cleaning should be followed. The operator–analyst should try to keep the laboratory isolated by keeping doors and windows closed.

- As with all laboratory testing, verifying accurate and consistent readings is crucial. Therefore, it is important to regularly check the balance's calibration. The actual calibration procedure depends on the balance brand and model. The operator–analyst should check the balance's operation manual for details (see Section 6.4).

6.2   Weighing Containers and Accessories

Chemicals or samples should never be weighed directly on the pan. There are a variety of containers designed for weighing. Disposable aluminum or plastic boats and weighing paper are the most common. Reusable ceramic boats, glass weighing bottles, and reagent funnels are also available. The operator–analyst should keep a variety of stainless steel scoops and spatulas near the balance to transfer chemicals from the bottle to the dish and to keep them clean. An anti-static balance brush is useful for brushing away spills. Oil and dust from fingerprints will make accurate use difficult. It is important to never directly touch anything that goes on the balance. Latex gloves, tongs, forceps, and tweezers should be used to handle dishes and papers.

6.3   Drying Dishes and Reagents

Dishes and papers used in gravimetric tests must be predried in the oven and/ or muffle furnace. Some reagents must be dried in the oven before weighing to drive off water that may have been absorbed by the chemical. These laboratory equipment and chemicals must then be kept in a desiccator until they are cooled to ambient temperature and ready for use. A desiccator is an airtight appliance with a pan of desiccant material that adsorbs humidity preferentially to the dishes or chemical. The desiccant must occasionally be regenerated by heating in an oven to the manufacturer's specifications. To dry a chemical, the following procedures generally apply:

1. Place a ceramic dish or glass-weighing bottle on the balance and press tare.

2. Weigh an excess amount of reagent into the container. About 20% more than is needed should be sufficient.

3. Dry the chemical in the oven according to the time and temperature specified in the reagent preparation instructions. Double-check the temperature of the oven because certain chemicals can volatilize at high temperatures, even 180 °C. If drying in a container with a lid, either leave the lid off or cracked open to allow water to escape and prevent the lid from getting stuck when cooled.

4. Cool the reagent in the desiccator to room temperature.

6.4   Weighing Procedures for Reagents

The following procedures should be followed to weigh reagents:

1. Turn on the balance and let it warm up for 15 minutes.

2. Check the balance for level using the bull's-eye level.

3. With the draft shield closed and the pan empty, press tare or zero the balance.

4. When the balance reads zero, use forceps to place a calibration weight (1 to 10 g) on the pan. Most balances will give an indication on the display that the reading is stable. When the reading stabilizes, record the weight.

5. Remove the weight and ensure that the reading returns to zero.

6. Select the appropriate weighing dish and place it on the pan. Choose a dish with the lightest weight possible, such as a disposable boat.

7. When the reading stabilizes, press tare. If the balance does not have a tare feature, document the beginning weight of the dish and subtract this weight from the final weight of the dish and chemical.

8. Carefully transfer the chemical to the dish with a spatula or scoop. Add a bit at a time until you approach the desired weight. Tapping the spatula with the index finger can help add very small amounts of reagent. If you add too much and have to remove reagent, be extra careful not to bump the dish or pan, and do not return the reagent to the bottle. Dispose of this reagent properly.

9. When the desired weight is displayed, close the draft shield and let the reading stabilize. Add or remove reagent to display the desired weight with the draft shield closed.

10. Remove the dish and transfer the chemical per Section 7.3.

7.0   REAGENTS

7.1   Laboratory-Grade Water

Water used to prepare reagent solutions must obviously be free from contaminants, especially elements or ions for which the sample is being analyzed. Although there are a number of methods of purifying water to various specifications, a compact, wall-mounted deionizing unit is likely the best choice for a small- to-medium-size wastewater laboratory. These units purify water by passing it through a number of ion-exchange columns followed by final filtration through a 0.2-micron filter. The purity is indicated by electrical resistivity given in "megohm-cm" (Mohm-cm), and most commercial units have a display that monitors this parameter. Once the resistivity drops below 18 Mohm-cm, the columns are spent and must be replaced. Deionized water will absorb contaminants from the air and grow microorganisms and algae if left exposed to the atmosphere and sunlight. A good rule to follow is to use fresh reagent-grade water daily.

7.2   Reagent Grades

Chemical suppliers manufacture reagents to varying degrees of purity. For the analyses described in this manual, all reagents should be American Chemical Society (ACS) reagent grade. These reagents are certified to specifications given by ACS and have a guaranteed analysis and traceable lot number on the label. Practical and technical grade chemicals are less expensive but are meant for use in industrial processes and are not suitable for use in an analytical laboratory.

7.3   Mixing Reagents

The accuracy and validity of analytical results begins with quality reagents. When mixing reagents, the operator–analyst must use a proper and consistent technique that eliminates contamination and ensures a complete transfer of reagent from the weighing dish to the volumetric flask and, ultimately, to the storage bottle. This technique is known as *quantitative transfer.*

The operator–analyst should use the following quantitative transfer method for making reagents:

1. Fill the volumetric flask to about 70% of the final volume.
2. If stirring is needed to dissolve the reagent in a reasonable amount of time, slide a magnetic stir bar down the neck and into the flask, and put the flask on a stir plate.
3. Place a powder funnel in the neck of the flask.
4. Weigh the given amount of reagent into the proper weighing dish following the weighing procedure outlined in Section 6.0.
5. Slowly and carefully transfer the reagent into the funnel.
6. Rinse the weighing dish into the funnel with a deionized water wash bottle. Repeat for three rinses.
7. Rinse the powder funnel with two rotations of deionized water. Repeat for three rinses.
8. Rinse the neck of the flask three times.
9. Stopper or cover the flask and invert to mix or stir until all reagent is dissolved.
10. Remove stir bars with a Teflon-coated bar retriever. Rinse the retriever and bar into the flask with deionized water as you remove them.
11. When the solution is at room temperature, dilute to the mark and invert to mix.

8.0   MAKING ACIDS AND BASES

Making acids and bases, whether for reactions, cleaning, or titrations, is one of the most common tasks in a laboratory. Although acids and bases are mixed just like reagents, when diluting a concentrated acid or base it is critical that the acid or base is added to the water (it might be helpful for the operator–analyst to remember the following phrase: "do what you oughtta, add acid to watta"). Adding water to acids and bases produces a violent reaction and is likely to splatter from the mixing glassware. The mixing (dilution) of acids and bases results in an exothermic reaction, meaning that it produces a large amount of heat. Because of this, the operator–analyst should allow plenty of time for the solution to cool before diluting to the mark.

Standard Methods (APHA et al., 2017) contains a table for making various strengths of acids from concentrated reagent. The exact concentration of these reagents is still approximate. Concentrated sulfuric acid, for example, is only 95% to 98% sulfuric acid. The exact concentration of the resulting dilution is, therefore, unknown. For many applications, this does not present a problem; however, if the acid is to be used for titrations, it should be standardized to determine the exact concentration. Standardization procedures should be included in the specific method.

9.0   STOCK, STANDARDS, AND STANDARD DILUTIONS

Most analyses require preparing a solution of known concentration of the analyte and then comparing an unknown sample to the prepared known solution. This known solution is called a *standard*. The solution the standard is prepared from is called a *stock solution*. Diluting the standard to various concentrations (standard dilutions) enables construction of multipoint calibration curves.

9.1   Stock Solutions

Stock solutions are concentrated solutions of chemicals that are diluted to make standard solutions. Stock is used because there is less uncertainty in weighing a larger amount of substance, that is, it is more accurate to weigh 1.000 g ± 1.0 mg (0.1% uncertainty) than it is to weigh 10 mg ± 1.0 mg (10% uncertainty).

9.2   Standard Solutions

Standard solutions are made using volumetric glassware to dilute the stock solution into the upper range of the sample being analyzed.

9.3   Standard Dilutions

To make calibration curves for colorimetric procedures and calibrate electrodes and probes, the standard solution must be further diluted to a range of concentrations that are representative of the concentration of the sample being tested. If samples between 0 and 10 ppm are being measured, the standard dilutions could be prepared at 0.5, 1, 2, 5, and 10 ppm.

To make standard dilutions, one of the simplest and most useful equations in analytical chemistry should be used, as follows:

$$C_1 V_1 = C_2 V_2 \qquad\qquad (2.15)$$

Where C_1 and V_1 = the concentration and volume of the standard and C_2 and V_2 are the desired concentration and volume of the dilution. The operator–analyst should use this equation to find the volume of standard needed. For example, if the operator–analyst is using a 50-ppm standard and would like to make 100 mL of 10-ppm standard dilution,

$$(50 \text{ ppm})(V_1) = (10 \text{ ppm})(100 \text{ mL}) \tag{2.16}$$

$$(5'' \text{ ppm})(V_1) = 1000 \tag{2.17}$$
$$V_1 = 20 \text{ mL}$$

Therefore, to make this standard dilution, 20 mL of standard should be diluted to 100 mL of final volume. It can be helpful to make a chart of standard dilutions needed, as shown in Table 2.4.

10.0 ACCURACY, PRECISION, AND ERROR

10.1 Accuracy and Precision

When discussing measurements or the results of measuring instruments, there are several concepts involved. This section describes four important ideas: accuracy, precision, error, and uncertainty.

Accuracy is the degree of closeness of measurements of a quantity to its actual (true) value. Precision, or repeatability, is the degree to which repeated measurements under unchanged conditions show the same results. For example, if you throw several darts toward a target, how far away the darts land from the bull's-eye is accuracy. The grouping of the dart pattern is precision. The concepts of accuracy and precision are depicted in Figure 2.8.

TABLE 2.4 Standard dilution table.

Standard concentration	Volume of standard needed	For standard dilution concentration of	Dilute to final volume of
C_1 (ppm)	V_1 (mL)	C_2 (ppm)	V_2 (mL)
50	1	.5	100
50	2	1	100
50	4	2	100
50	10	5	100
50	20	10	100

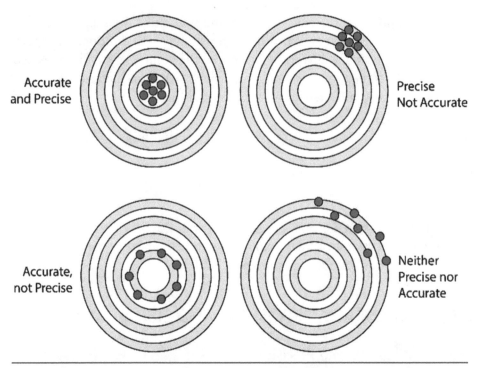

FIGURE 2.8   Accuracy and precision. *Courtesy of Indigo Water Group*

In the first example in Figure 2.8, an accurate but not precise pattern is presented. The darts are not clustered, but their average position is the center of the bull's-eye. The second example in the figure is a precise pattern, but not accurate. The darts are clustered together but did not hit the intended mark. The last example is both accurate and precise. The darts are tightly clustered and their average position is the center of the bull's-eye.

10.2 Error and Uncertainty

Error refers to the disagreement between a measurement and the true or accepted value. In Figure 2.8, this would be the actual distance between the dart and the bull's-eye. Uncertainty of a measured value is an interval around that value such that any repetition of the measurement will produce a new result that lies within this interval. Uncertainty in the laboratory is typically represented by a set value plus or minus a degree of uncertainty.

Errors may fall be classified as determinate or indeterminate. Determinate errors may be discovered, and corrected for or eliminated, while the cause of indeterminate errors are obscure or undetectable and cannot be eliminated.

10.2.1   Determinate Errors

The following are examples of determinate errors:

- Personal errors caused by the operator–analyst such as reading a buret improperly, misidentifying color changes, or preparing volumetric solutions improperly;

- Instrumental errors attributed to instruments such as imperfect weights, volumetric glassware, and balances; and

- Errors in method and sampling including those resulting from such conditions as improper temperature or time of drying of a solids sample or incorrect standard dilution.

10.2.2   Indeterminate Errors

When analyzing unknown samples, the assumption is that the constituent is not present in the sample; the test is done to prove this assumption wrong. Therefore, an operator–analyst can make these two types of indeterminate errors:

- Type I error (alpha error)—determining that a constituent is present when it is actually absent; and

- Type II error (beta error)—not detecting a constituent when it is present.

In general, no laboratory result should be rejected, except in the case of an obvious source of error. Measurements that vary widely from the mean (or average) may be omitted when determining an average if a reasonable explanation can be given.

In the laboratory, determining accuracy, precision, error, and uncertainty validates results and shows possible problem areas. For example, in relation to the set of 5-day BOD (BOD_5) glucose glutamic acid (GGA) quality control averaged values in Table 2.5, the acceptable value range is 198 ± 30.5 mg/L. This value has been predetermined and can be found in *Standard Methods* (APHA et al., 2017; Section 5210: 6, "Quality Control Checks"). The values for week 1 are both precise and accurate; week 2 values are accurate to within the acceptable range, but not precise; and week 3 values are neither accurate nor precise. Evaluating these data is crucial to laboratory operation and sample results validation. For example, BOD sample values from the first week are both accurate and precise and would be considered valid in relation to the quality control check. The second week shows a declining trend, with values that are accurate but not precise, and could be the result of aging GGA quality control. The third week shows a sporadic pattern,

TABLE 2.5 Sample BOD results illustrating accuracy and precision.

BOD5 averaged GGA-control values					
Week 1		Week 2		Week 3	
Date	GGA (mg/L)	Date	GGA (mg/L)	Date	GGA (mg/L)
11/8	196	11/15	220	11/22	135
11/10	210	11/17	211	11/24	105
11/12	194	11/19	170	11/26	160

neither accurate nor precise, with values outside of acceptable uncertainty limits, and could be caused by the presence of a toxic material or poor seed quality. Evaluating data for accuracy, precision, error, and uncertainty allows the operator–analyst to be confident with his or her results.

10.3 Significant Figures

Significant figures are the numbers in any measurement that affect the accuracy or precision of the measurement. There are six rules for determining whether a number is significant:

1. Non-zero digits are always significant. For example, the number 623 has three significant figures because there are no zeroes.

2. Any zeros between two significant digits are significant. For example, in 204, all three numbers are significant.

3. Leading zeroes are not significant. For example, the number 0.0037 only has two significant figures. The 0.00 are only place holders and don't affect the accuracy or precision of the measurement.

4. Trailing zeroes to the right of the decimal are significant. For example, 57.00 contains four significant figures and is more accurate than 57.

5. Trailing zeroes in a whole number without a decimal place shown are not significant. For example, 280 contains only two significant figures. The zero is not significant and could simply be a place holder.

6. Exact numbers, including conversion factors, have an infinite number of significant figures.

There are also some rules for handling significant figures in calculations that are important when calculating laboratory results:

1. When adding and subtracting, the answer should have the same number of decimal places as the term with the fewest decimal places. For example, adding 1.24 and 2.2 together results in an answer of 3.4.

2. When multiplying and dividing, the answer can never have more significant figures than the term with the fewest significant figures. For example, multiplying 2.35 by 2.1 results in an answer of 4.9.

3. In multistep calculations, you may round at each step or only at the end.

The number of significant figures used in reporting results implies the accuracy of the result. A result given as 429.6 mg/L seems much more accurate than a result of 430 mg/L. In general, the last significant figure in a number is considered to be certain to ±1. For example, an operator–analyst running a gravimetric solids analysis must measure an aliquot of 50 mL. On a 50-mL graduated cylinder marked in 1-mL increments, the operator–analyst can reasonably estimate a number between marks to ± 0.1 mL, so that should be the last significant figure. An operator–analyst can pour the aliquot to 50 mL and be confident that it is between 49.9 and 50.1 mL. All these measurements have three significant figures. The operator–analyst then uses an analytical balance to get a tare weight of 24.14297 g and a gross weight of 24.16445 g, both with seven significant figures. The net weight (gross–tare) is 0.02148 g (four significant digits). The operator–analyst then calculates the result as

$$\frac{0.02148 \text{ g}}{50.0 \text{ mL}} \times \frac{1000 \text{ mg}}{1 \text{ g}} \times \frac{1000 \text{ mL}}{1 \text{ L}} = 429.6 \text{ mg/L} \tag{2.18}$$

Although this result has four significant figures, the graduated cylinder limits the result to three significant figures. This result should, therefore, be rounded and reported with three significant figures as 430 mg/L. Considering the implied accuracy assumption of ±1 in the last significant figure, the difference between reporting 429.6 ± 0.1 mg/L (429.5 to 429.7) and 430 ± 1 mg/L (429 to 431) becomes quite significant.

The operator–analyst should be familiar with the rules of significant figures, be aware of the pitfalls of calculators and spreadsheets in introducing extraneous significant figures, and report results that the accuracy and precision of laboratory equipment and instruments can support.

11.0 QUALITY ASSURANCE AND QUALITY CONTROL

11.1 Quality Assurance System

Without quality control results, there can be no confidence in the results of analytical tests. As described in Part 1000 of *Standard Methods* (APHA et al., 2017), essential quality control measures include calibration of methods, standardization of reagents, assessment of each operator–analyst's capabilities, analysis of blind check samples, determination of method sensitivity (method detection level or quantification limit), and daily evaluation of bias, precision, and the presence of laboratory contamination or other analytical interference. Details of these procedures, their performance frequency, and expected ranges of results should be formalized in a written quality assurance manual and standard operating procedures.

Some of the methods an operator–analyst will use include specific quality control procedures, frequencies, and acceptance criteria. These are considered to be the minimum quality controls needed to perform the method successfully. Additional quality control procedures can and should be used. Some regulatory programs may require additional quality control or have alternative acceptance limits.

Each method typically includes acceptance criteria guidance for precision and bias of test results. If not available, the laboratory should determine its own criteria via control-charting techniques. For some procedures, including pH, dissolved oxygen, residual chlorine, and carbon dioxide, the traditional determination of bias (i.e., adding a known amount of analyte to either a sample or a blank) is not possible. This does not, however, relieve operator–analysts of the responsibility for evaluating test bias. Instead, certified ready-made analyte solutions should be obtained for such tests. Precision should be evaluated by analyzing duplicate samples. If one or both results are "nondetect," however, precision cannot be calculated.

To help verify the accuracy of calibration standards and overall method performance, the operator–analyst should participate in an annual or, preferably, semiannual program of analysis of known quality-control quality check samples (QCSs) that are ideally provided by an external agency. Such programs are sometimes called *proficiency testing* or *performance evaluation* studies. These are blind samples, meaning that the outside agency knows the true or certified value while the testing laboratory does not. A QCS is designed to test whether a laboratory is capable of producing accurate and precise data. An unacceptable result on a proficiency testing sample is often the first indication that a test protocol is not being followed successfully. The operator–analyst should fully investigate circumstances to find the cause. In many jurisdictions, participation in proficiency testing studies is a required part of laboratory certification and accreditation.

11.2   Measures of Precision

11.2.1   Mean

When analyzing the precision of a group of results, the operator–analyst should be concerned with how they are distributed around the mean, or average, of the results. The mean is simply the sum of all results divided by the number of results, as follows:

$$\text{Mean} = \frac{x_1 + x_2 \ldots x_n}{n} \tag{2.19}$$

Where x_1, x_2, and x_n = the concentrations of each individual result and n = the number of results.

11.2.2 Relative Percent Difference

To demonstrate the precision of an operator–analyst or method, multiple samples from the same source are analyzed and the difference is expressed mathematically. If only two samples are tested, the relative percent difference (RPD) is used to indicate precision, as follows:

$$\%RPD = \frac{|C_1 - C_2|}{(C_1 + C_2)/2} \times 100 \tag{2.20}$$

where C_1 and C_2 are the determined concentrations of the two samples.

The bars in the numerator are the absolute value signs. The absolute value signs turn any negative number that may result from completing the math inside the absolute value signs into a positive number. Therefore, relative percent difference will always be a positive number. For example, if an operator–analyst determines the concentrations of an analyte to be 15.5 and 14.6 mg/L,

$$\%RPD = \frac{|14.6 - 15.5|}{(14.6 + 15.5)/2} \times 100$$

$$\%RPD = \frac{0.9}{15.1} \times 100 \tag{2.21}$$

$$\%RPD = 6.0\%$$

It is important to note that RPD simply expresses the difference between the results as a percent of the mean, and that the percentage is absolute or never negative. If C_1 is lower than C_2, the absolute difference should be used.

11.2.3 Standard Deviation

If all the errors in a test are random, the results will most likely follow a normal distribution where there are the same number of high and low results and these results have roughly the same RPD (higher or lower) from the mean. In this instance, the standard deviation (SD) can provide the measure of precision, as follows (a low standard deviation indicates high precision):

$$SD = \sqrt{\frac{\Sigma(x - \text{mean})^2}{N}} \qquad (2.22)$$

where x is the determined analyte concentration and N is the number of samples. Equation 2.23 is an example of calculating standard deviation for four determinations of the same sample, with results of 15.5, 14.6, 13.9, and 16.1 mg/L. It is useful to make a table to calculate the mean and the sum of all $(x\text{-mean})^2$ terms, as in Table 2.6.

$$SD = \sqrt{\frac{\Sigma(x - \text{mean})^2}{N}} = \sqrt{\frac{2.83}{4}} = 0.84 \qquad (2.23)$$

Data are typically referred to as being within a number of standard deviations. In this instance, 2 standard deviations equals 1.68 (0.84 × 2). Because the maximum difference between any result and the mean is 1.1, it can be stated that all results are within 2 standard deviations.

11.2.4 Relative Standard Deviation

The relative standard deviation (RSD) allows for precision comparisons between high and low analyte concentrations. It is found by dividing standard deviation by the mean, as in Equation 2.24. When expressed as a percentage (%RSD), it is also known as the *coefficient of variation*.

$$RSD = \frac{SD}{\text{mean}} \qquad (2.24)$$

TABLE 2.6   Calculating standard deviation.

	Result (mg/L)	(x-mean)		(x-mean)^2
	15.5	15.5–15.0 5	0.5	0.25
	14.6	14.6–15.05	−0.4	0.16
	13.9	13.9–15.05	−1.1	1.21
	16.1	16.1–15.05	1.1	1.21
Sum	60.1			2.83
Mean	15.0			

Equation 2.25 is as follows:

$$\%RSD = \frac{SD}{Mean} \times 100 \qquad (2.25)$$

For the example data given in Table 2.6, the RSD is 0.84/15.0 or 0.056. The %RSD is 5.6%.

11.3   Measures of Accuracy

To determine the accuracy of an operator–analyst or a method, the experimentally determined result must be compared to the true value. Obviously, this implies that the true value must be known. This is done by carefully making a solution of known concentration, similar to preparing a standard solution. If this solution is made with deionized water, it is known as a *blank spike*. If the target analyte is spiked into the process control sample, it is known as a *matrix spike*. In the instance of a blank spike, percent recovery (%R) simply compares the determined result to the true value, as in the following equation:

$$\%R = \frac{determined\ result}{true\ value} \times 100 \qquad (2.26)$$

The concentration of the analyte present in the process control sample must be subtracted from the determined result in the case of a matrix spike (see Equation 2.27). In this instance, the "true value" is the final concentration of spike in the sample calculated from $C_1 V_1 = C_2 V_2$.

$$\%R = \frac{determined\ result - background\ concentration}{true\ value} \times 100 \qquad (2.27)$$

A percent recovery of 100% indicates a perfectly accurate test; however, if other specific guidance is not provided in the method, a percent recovery of 80% to 120% is generally considered to be an acceptable result.

12.0   ELEMENTS OF QUALITY CONTROL

A variety of quality control analyses are completed regularly in a laboratory. Characteristically, an acceptable result is obtained for each quality control check before measurement of samples begins. Each check is periodically repeated, validating the analyses of intervening samples. Corrective steps are performed if the check fails, and the intervening samples are remeasured. The frequency of quality control checks varies as determined by the specific methods listed in Table 2.7. A sample set or batch is considered to be 20 samples. More information on quality control elements can be found in Section 1000 of *Standard Methods* (APHA et al., 2017). Table 2.8 provides a specific section reference guide for quality control elements discussed in *Standard Methods*.

12.1 Method Blank

A method blank (also called a *reagent blank*) is a volume of reagent water treated exactly as a sample, including exposure to all equipment, glassware, procedures, reagents, and preservatives. The method blank is used to assess whether analytes or interferences are present in the analytical process. Any constituent(s) recovered must be less than or equal to one-half the reporting level (unless the method specifies otherwise). If any method blank measurements are at or above the reporting level, immediate corrective action should be taken. At least one method blank should be included daily or with each batch of 20 or fewer samples.

12.2 Laboratory-Fortified Blank

A laboratory-fortified blank (LFB) (also called a *blank spike*) is a method blank that has been fortified with a known concentration of analyte from a second source (not the one used to develop working standards, unless the method specifies otherwise). The LFB is used to evaluate accuracy, ongoing laboratory performance, and analyte recovery in a clean matrix.

TABLE 2.7   Frequency of quality control checks.

Quality control element	Frequency
Method blank	One per sample set or 5% basis, whichever is more frequent
Laboratory-fortified blank	One per sample set or 5% basis, whichever is more frequent
Laboratory-fortified matrix	One per sample set or 5% basis, whichever is more frequent
Duplicates	One per sample set or 5% basis, whichever is more frequent
Calibration	One per analysis
Continued calibration verification	One per sample set then 1 every 10 samples following quality control sample

TABLE 2.8 *Standard Methods* quality control reference.

Quality control element	*Standard Methods* section reference
Method blank	1020 B. 4
Laboratory-fortified blank	1020 B. 5
Laboratory-fortified matrix	1020 B. 6
Duplicates	1020 B. 7
Calibration	1020 B. 10
Continued calibration verification	1020 B. 10
Initial demonstration of capability	1020 B. 1
Quality control sample	1020 B. 2
Control charts	1020 B. 12
Method detection level	1020 B. 3; 1030 C.

12.3 Laboratory-Fortified Matrix and Laboratory-Fortified Matrix Duplicate

A laboratory-fortified matrix (LFM) (also called a *matrix spike*) and laboratory-fortified matrix duplicate (LFMD) should be used to evaluate a method's bias and precision, respectively, as influenced by a specific matrix.

To prepare an LFM, a known concentration of analytes from a second source (unless a method specifies otherwise) should be added to a randomly selected routine sample. The addition should roughly double the sample's original concentration without increasing its volume by more than 5%. Ideally, the new concentration should be at or below the midpoint of the calibration curve. If necessary, the sample should be diluted to bring the measurement within the curve. In addition, the range of analyte concentrations should be rotated to verify performance at various levels.

For example, a 50-mL sample is analyzed, and the result is 2 mg/L of analyte. To determine how to spike the sample, the total weight of analyte should first be determined, as follows:

$$2 \text{ mg/L} \times 50 \text{ mL} \times 1 \text{ L/1000 mL} = 0.1 \text{ mg analyte} \qquad (2.28)$$

Calculate 5% of the sample volume, 50 mL $\times$ 0.05 = 2.5 mL. To double the concentration, add another 0.1 mg of analyte to the 2.5 mL. Calculate the concentration of the spiking solution, as follows:

$$0.1 \text{ mg/2.5 mL} \times 1000 \text{ mL/L} = 40 \text{ mg/L} \qquad (2.29)$$

This calculation is meant to give a rough estimate of where to begin. Factors such as available volumetric glassware may make it easier to use 100 mg/L stock spiking solution and to dilute or use a smaller volume of spike. Both the use of an easily prepared 100-mg/L spike and a smaller spike volume are done in the following example of the same 50-mL sample at 2 mg/L spiked with 1 mL of 100 mg/L solution. The determined result was 3.92 mg/L. First, calculate the final concentration of spike in the sample, the "true value," as follows:

$$C_1V_1 = C_2V_2 \tag{2.30}$$
$$100 \text{ mg/L (1 mL)} = C_2 \text{ (51 mL)}$$
$$C_2 = 100 \text{ mg/51 mL} = 1.96 \text{ mg/L}$$

Then, use Equation 2.27, as follows:

$$\%R = \frac{\text{determined result} - \text{background concentration}}{\text{true value}} \times 100$$

$$= \frac{3.92 \text{ mg/L} - 2.00 \text{ mg/L}}{1.96 \text{ mg/L}} \times 100$$

$$\%R = 98.0 \tag{2.31}$$

It is important to notice that the volume of spike used will begin to have a negligible effect on percent recovery as it gets small in relation to the sample volume. Micropipets can be used to deliver spike volumes as low as 0.1 mL.

The operator–analyst should calculate percent recovery and relative percent difference (for duplicates), plot control charts (unless the method specifies acceptance criteria) and determine control limits (Section 1020B of *Standard Methods* [APHA et al., 2017]). The operator–analyst should also ensure that the method's performance criteria are satisfied.

Fortified samples should be processed independently through entire sample preparation and analysis. At least one LFM and LFMD should be included daily or with each batch of 20 or fewer samples.

12.4 Duplicates

Duplicate samples of measurable concentration should be used to measure the precision of the analytical process. Routine samples to be analyzed twice should be randomly selected. Duplicate samples should be processed independently through the entire sample preparation and analysis. At least one duplicate for each matrix type should be included daily or with each batch of 20 or fewer samples. Control limits for duplicates should be calculated when method-specific limits are not provided.

12.5   Calibration

Initial calibration should take place with at least one blank and three calibration standards of the analyte(s) of interest. Calibration standards should be selected that bracket the sample's expected concentration and are within the method's operational range. The number of calibration points depends on the width of the operational range and the shape of the calibration curve. One calibration standard should be at or below the method's reporting limit.

As a general rule, differences among calibration standard concentrations should not be greater than one order of magnitude (i.e., 1, 10, 100, and 1000). However, most methods for inorganic nonmetals do not have wide operational ranges, and therefore the concentrations in their initial calibration standards should be less than one order of magnitude apart. Linear or polynomial curve-fitting statistics should be applied, as appropriate, to analyze the calibration curve. In most instances, a linear regression analysis will be sufficient for the operator–analyst. The appropriate linear or non-linear correlation coefficient ($R2$) for standard concentration to instrument response should be greater than or equal to 0.995.

Initial calibration should be used to quantify analyte concentrations in samples. Calibration verification should only be used to check initial calibration and not to quantify samples. Initial calibration should be repeated daily or when starting a new batch of samples, unless the method permits calibration verification between batches.

12.6   Continuing Calibration Verification

Continuing calibration verification (CCV) is the periodic confirmation that instrument response has not changed significantly since initial calibration. Calibration can be verified by periodically analyzing a calibration standard and calibration blank during a run (typically, after each batch of 10 samples and at the end of a sample run). The CCV standard's analyte concentration should be at the midpoint of the calibration curve or lower. For calibration verification to be valid, standard results must not exceed ±10% of its true value, and calibration blank results must not be greater than one-half the reporting level (unless the method specifies otherwise). If a calibration verification fails, the operator–analyst should immediately cease analyzing samples and initiate corrective action. Then, initial calibration should be repeated and samples run because the last acceptable calibration verification should be reanalyzed.

12.7   Initial Demonstration of Capability

Before a new operator–analyst runs any samples, his or her capability should be verified. A laboratory fortified blank (LFB) should be run at least four times and compared to the limits listed in the method. This process should be repeated after analyzing at least 20 batches of samples to demonstrate proficiency with the method. If no limit is specified, the LFB recovery limits should be set at the mean ± (4.54 × standard deviation). In addition, the operator–analyst should verify that the method is sensitive enough to meet measurement objectives for detection and quantitation by determining the lower limit of the operational range.

12.8 Quality Control Sample

An externally generated, blind QCS (unknown concentration) should be analyzed at least annually (preferably, semiannually or quarterly). This sample should be obtained from a source external to the laboratory and the results compared to that laboratory's acceptance results. If testing results do not pass acceptance criteria, the operator–analyst should investigate why, take corrective action, and analyze a new QCS. This process should be repeated until results meet acceptance criteria.

12.9 Control Charts

Control charts plot the results of quality control analyses versus time, allowing the operator–analyst to see if a method is in control or tending to bias. A simple accuracy control chart plots the percent recovery of an LFB or LFM analysis against the data, as shown in Figure 2.9. A measure of precision, such as RPD, between duplicates can be plotted on a control chart in a similar way.

The data in Figure 2.9 are randomly distributed around 100%, indicating that the error responsible for the distribution is random error. If the control chart reveals that a quality control parameter is consistently high or low, it is evidence of a systematic error in the method or analysis, and the operator–analyst should look to correct this error. Warning limits are also commonly established in addition to the upper and lower control limits shown in Figure 2.9. The operator–analyst must decide where to set all these limits. Control charts can be constructed in many ways and become complex and specific to the individual laboratory. The operator–analyst should refer to *Standard Methods* for guidance on control charts and on setting warning and control limits.

12.10 Limits and Levels

12.10.1 Instrument Detection Limit

Most analytical instruments produce a signal even when a blank is analyzed (i.e., the noise level) S. The instrument detection limit (IDL) is a measure of the relative strength of analytical signal to the average strength of the

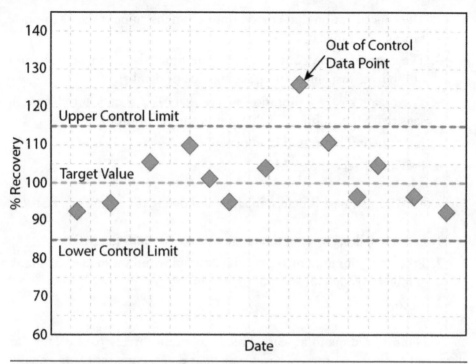

FIGURE 2.9 Simple control chart for percent recovery.

background instrument noise. This ratio is useful for determining the effect of the noise on the relative error of a measurement. The strength-to-noise ratio can be measured many ways. One way to approximate the ratio is by dividing the arithmetic mean (average) of a series of replicates (preferably eight) by the standard deviation of the replicate results.

12.10.2 Method Detection Level

The method detection level (MDL) is the concentration at which it is 99% probable that the sample will produce an instrument signal that is greater than the blank. The MDL is calculated from a minimum of seven spiked samples and seven method blank samples prepared in three batches on three separate calendar dates and analyzed on three separate calendar dates. Because the samples are carried through the entire method, the MDL is specific to method and operator–analyst.

There is much and varied guidance on determining the concentration of spike to use for determining an MDL. First, an estimate of the initial MDL must be made. According to 40 CFR Part 136 B (U.S. EPA, 2017), at least one of the following must be used to estimate the initial MDL:

1. The mean determined concentration plus three times the standard deviation of a set of method blanks;

2. The concentration that corresponds to an instrument signal-to-noise ratio in the range of 3 to 5;

3. The concentration equivalent to three times the standard deviation of replicate instrumental measurements of blanks;

4. The region of the calibration where there is a significant change in sensitivity (i.e., a break in the slop of the calibration);

5. Instrumental limitation; and/or

6. Previously recognized MDL.

Once the MDL estimate is determined, the following procedure can be used:

1. Select a spiking level in the range of 2 to 10 times the MDL;

2. Process a minimum of seven spiked samples and seven method blank samples through all steps of the method; and

3. Record results.

This must be done in at least three batches on three separate calendar dates and analyzed on three separate calendar dates. Preparation and analysis may be on the same day. This will result in a minimum of 21 spiked sample results and 21 method blank results.

Record results, and calculate the mean standard deviation for the spiked sample results as in Section 11.2.3 using a variation of Equation 2.22, as follows:

$$SD = \sqrt{\frac{\Sigma(x - \text{mean})^2}{N - 1}} \tag{2.32}$$

The $N = 1$ term in this equation refers to the number of samples minus one and is used to approximate the behavior of an infinite number of samples by the behavior of the seven replicates. The MDL for the spiked sample results is then calculated as

$$MDL_S = SD(t) \tag{2.33}$$

Where t is the students' t value at the 99% confidence level, a statistical constant found from a table and equal to 3.143 for seven replicates. If more than seven replicates are used, use the appropriate t value from Table 2.9 for the MDL calculation.

Next, compute the MDL_b using the method blank results as follows:

1. If none of the method blanks give numerical results for the analyst of interest, the MDL_b does not apply.

TABLE 2.9 Students' t value at the 99% confidence level.

Number of replicates	Degrees of freedom (N-1)	$t_{(n-1,0.99)}$
7	6	3.143
8	7	2.998
9	8	2.896
10	9	2.821
11	10	2.764
16	15	2.602
21	20	2.528
26	25	2.485
31	30	2.457
32	1	2.453
48	47	2.408
50	49	2.405
61	60	2.390
64	63	2.387
80	79	2.374
96	95	2.366
100	99	2.365

Source: U.S. EPA (2016).

2. If some, but not all, of the method blanks give numerical results, set the MDL_b equal to the highest method blank result. If more than 100 method blanks were analyzed, refer to 40 CFR Part 136, Appendix B (U.S. EPA, 2017) for additional guidance.

3. If all of the method blanks give numerical results, then calculate the MDL_b as:

$$MDL_b = \bar{X} + t_{(n-1,1-\alpha=0.99)} S_b \qquad (2.34)$$

Where: MDL_b = the MDL based on method blanks, $\bar{X}$ = mean of the method blank results (use zero in place of the mean if the mean is negative), $t_{(n-1,1-\alpha=0.99)}$ = the Student's t value, and S_b = sample standard deviation of the replicate method blank sample analyses. Calculate using formula 2.32, then select the greater of the MDL_s or MDL_b as the initial MDL.

The MDLs must be confirmed though ongoing data collection every quarter. Review 40 CFR Part 136, Appendix B (U.S. EPA, 2017) for further details.

12.10.3   Quantitation Limits

A quantitation limit is the lowest concentration of an analyte that can be consistently measured within specific limits of precision, accuracy, representativeness, completeness, and comparability during routine laboratory operating conditions. Factors influencing the quantitation limit include sample size, analytical instrument, method, and analytical uncertainties in the sample matrix. For example, a quantitation limit may be lower for an analyte when analyzing a drinking water sample versus a wastewater sample because of the matrix of each. Quantitation limits come in many varieties, such as minimum quantitation limit, practical quantitation limit, and lower quantitation limit; these are defined differently by different laboratories, but typically fall in the range of 3 to 5 times the MDL.

12.10.4   Reporting Limits

Reporting limits are even more abstract than quantitation limits. Reporting limits can be set by laboratory staff according to what they feel comfortable reporting in a legal framework. They can also be specified by a client when using a contract laboratory. Many states have guidance on reporting limits regarding discharge monitoring reports.

13.0   REFERENCES

American Public Health Association, American Water Works Association, & Water Environment Federation. (2017). *Standard methods for the examination of water and wastewater* (23rd ed.). American Public Health Association.

U.S. Environmental Protection Agency. (2016). *Definition and procedure for the determination of the method detection limit, Revision 2* (EPA 821-R-16-006).

U.S. Environmental Protection Agency. (2017). Definition and procedure for the determination of the method detection limit, Revision 2. *Code of Federal Regulations*, Part 136, Title 40, Appendix B. *Federal Register 5*, 23703.

3

Physical and Chemical Examination

1.0 pH

The following sections comprise an expanded version of Method 4500-H⁺B-2011 from *Standard Methods* Online. Note that there are many editions of *Standard Methods for the Examination of Water and Wastewater*, a joint publication of the American Public Health Association, the American Water Works Association, and the Water Environment Federation. Each edition contains hundreds of test methods, but most of the methods are not updated with each new edition of the print manual. Instead, *Standard Methods* designates the publication or revision date for each test method in both the online and hard-copy versions. The most current edition of any method is available online at https://www.standardmethods.org. The most current print version of *Standard Methods* is the 23rd edition (APHA et al., 2017).

It should also be noted that not every version of every test method is approved for use in 40 CFR Part 136 (U.S. EPA, 2017). 40 CFR lists the approved version of each test method, which may or may not be the most current version. This can sometimes require a laboratory to keep more than one version of *Standard Methods* on hand for reference. For example, the test method for total suspended solids listed in 40 CFR Part 136 as "Residue, Total, mg/L, 2540 B-2015." The designation refers to the method number (2540), submethod (B), and approved version publication date (2015).

1.1 General Description

pH is a numerical expression of the relative intensity of acidity or basicity. A pH of less than 7.0 denotes acidity with the intensity of acidity increasing as the numbers decrease (see Figure 3.1). Numbers between 7.0 and 14.0 denote basicity, with intensity increasing as the numbers increase.

Table 3.1 gives the pH for some common substances. Technically, pH is the inverse log of the hydrogen ion concentration in a solution. The "p" in pH actually stands for negative log. Because pH is a logarithmic scale, a substance with a pH of 4 is 10 times more acidic than a substance with a pH of 5 and 100 times more acidic than a substance with a pH of 6.

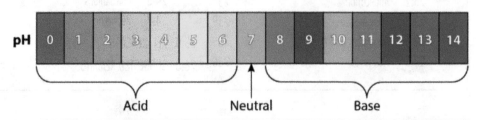

FIGURE 3.1   The acidic range on the pH scale is 0 to 7. The basic range is 7 to 14 (7 is neutral). *Courtesy of Indigo Water Group*

TABLE 3.1   pH of common substances.

Substance	pH
Hydrochloric acid	0.1
Lemon juice	2.3
Milk	6.5
Pure water	7.0
Blood	7.4
Baking soda	8.0
Ammonia	11.0
Sodium hydroxide	14.0

pH can be measured either colorimetrically or electrometrically (with a pH probe and meter). Colorimetric measurements include pH paper. The colorimetric method requires a less-expensive investment in equipment but suffers from severe interference by color, turbidity, high saline content, colloidal matter, free chlorine, and various oxidants and reductants. Indicators are subject to deterioration, as are the color standards with which they are compared. Moreover, no single indicator encompasses the pH range of interest in water and wastewater. In poorly buffered liquids (i.e., low alkalinity), which is a description that applies to some waters, the indicators themselves may alter the pH of the sample that they are expected to measure unless they are adjusted beforehand to nearly the same pH as the sample. For these reasons, the colorimetric method is suitable only for rough estimation and will not be described further. Colorimetric methods are not as accurate as the electrometric method; therefore, the electrometric method is preferred.

Because of the differences between the makes and models of pH meters commercially available, it is impossible to provide detailed instructions for correct operation of every instrument. In each instance, the manufacturer's instructions must be followed. pH measurements are temperature-dependent. Only pH probes equipped with temperature compensation should be used. In low ionic strength solutions such as distilled water, electrometric methods tend to drift and take excessive time to reach equilibrium, if at all.

1.2   Application

pH is commonly measured on liquid streams to identify the presence of corrosives, typically by pH < 2.0 or > 12.5. Sometimes, wastewater streams have pH meters to monitor discharges, and the meters need to be checked for accuracy. Often, when air scrubbers are used to remove hydrogen sulfide and other noxious sulfur compounds for odor control, pH is monitored to maintain removals depending on the type of scrubber. For example, a biofilter with synthetic media coated with activated sludges may need to maintain a pH of 2.0 to 3.0 for good foul air removals, whereas a chemical scrubber using caustic soda would require a pH of 10.0 for good hydrogen sulfide removals. pH ranges and effects for select treatment processes are presented in Table 3.2.

TABLE 3.2   pH ranges and effects for select wastewater processes.

Wastewater process	pH range and effect
Preliminary and primary treatment	Anaerobic conditions lower the pH of a wastewater. Raw wastewater primary treatment typically has a pH near 7. Consistently low pH values indicate either partial decomposition of wastewater in the collection system or regular discharges of partially decomposed or acid waste to the collection system. At a pH less than 7 (acidic), grease and scum may tend to stay in suspension, and some may enter the settled sludge instead of floating to the surface.
Lagoons	pH should be on the alkaline side, that is, 8.0 to 8.4. Decreasing pH followed by a drop in dissolved oxygen is indicative of green algae die-off caused by overloading, long periods of adverse weather, or the presence of higher life forms Daphnia feeding on the algae.
	Odor problems are evident in anaerobic ponds at low pH (<6.5) as acid formers are outpacing methane formers.
Fixed film	Reduced treatment outside of a pH range of 6.5 for carbonaceous; 8 to 8.5 for nitrification.
	Changes in pH may cause excessive sloughing. pH > 10 or pH < 5.
Activated sludge	Optimum growth for nitrifiers is at a pH between 7.0 to 8.3. By the time the pH processes decreases to less than 7.0, inhibition of the nitrifying organisms has already occurred.
	Running at a pH consistently lower than 7 can result in nitrite accumulation because ammonia oxidizers are less sensitive to low pH than nitrite oxidizers.
	Low pH can support filamentous organisms and create bulking. At pH less than 6.5, MLSS settleability may be affected.
Secondary effluent	Effluent from a secondary treatment system pH typically ranges between 6.0 and 9.0 pH (on a 30-day average basis), according to 40 CFR 133.
Disinfection	Adding chlorine to water results in the formation of HOCl and HCl. The reaction is temperature- and pH-dependent. The equilibrium approaches 100% dissociation (H^+ + OCl^-) when the pH is above 8.5 and approaches 100% HOCl when the pH < 6.0. Chlorine exists predominately as HOCl at pH levels between 2.0 and 6.0. Below a pH of 2.0, HOCl reverts to Cl_2 gas. Above a pH of 7.8, hypochlorite ions (OCl^-) dominate.

(continued)

TABLE 3.2 pH ranges and effects for select wastewater processes. (*Continued*)

Wastewater process	pH range and effect
Aerobic digestion	Aerobic digestion occurs best at a pH around 7.0. If pH drops to 6.0 to 6.5, look for causes—low alkalinity, nitrification, or carbon dioxide accumulation in a covered digester.
Anaerobic digestion	Best pH range is 6.8 < pH < 7.2.
	If the pH falls below 6.0, un-ionized volatile acids become toxic to methane-forming microorganisms. At pH values above 8.0, un-ionized aqueous ammonia (dissolved ammonia) becomes toxic to methane-forming microorganisms.
	pH is the last process control parameter to change.
Odor control	At a pH of approximately 9, more than 99% of the sulfide dissolved in water occurs in the form of the nonodorous hydrosulfide ion. Consequently, odorous amounts of hydrogen sulfide gas will not be released if a pH above 8 is maintained. Below this pH value, hydrogen sulfide gas is released from the wastewater.
	During lime stabilization where pH 12 must be maintained, ammonia is released because of high pH. (Odorous ammonia gas is released primarily at a pH greater than 9).

Source: WEF et al. (2018).

1.3 Interferences

Oils, greases, and fats may cause electrodes to become coated, requiring occasional cleaning with solvents or detergents as recommended by the manufacturer.

1.4 Apparatus and Materials

The apparatus and materials required for pH measurement include a pH meter, pH electrode and reference electrode or combination pH electrode, manufacturer's instructions, stir bar, stir plate, and beakers.

1.5 Stock and Standards

The operator–analyst should check with the manufacturer of the pH meter to determine which buffer solutions are needed. For permit reporting, a minimum of three standard buffers are needed. Most pH meters are preprogrammed to calibrate using pH 4, pH 7, and pH 10 buffers. According to *Standard Methods*, an initial buffer solution should be selected, followed by a second buffer that is within 2 pH units of the anticipated sample pH, followed by a third buffer that is below pH 10 and within 3 pH units of the second buffer. The vast majority of wastewater samples will have a pH near 7, so the standard pH 4, pH 7, and pH 10 buffers usually meet these requirements.

Some meters will only allow a single-buffer or two-buffer calibration and should not be used for regulatory compliance reporting. The operator–analyst should refer to the manufacturer's instructions for his or her particular meter.

Buffer solutions are best purchased as premade solutions rather than being prepared on site. Buffers are intended for a single use. Buffer aliquots should be discarded following meter calibration. A pH buffer should never be poured back into its original container.

1.6   Quality Assurance and Quality Control

The pH meter should be calibrated once per day or each time it is used. Although not a requirement of the method, it is a good idea to keep an instrument log and record the calibration slope. The slope should be within the manufacturer's recommendations. Changes to the slope may indicate a problem with the meter or the probe, or the need for maintenance. Calibration may be verified by taking a pH measurement of a standard buffer solution that was not used for to calibrate the meter. A second source standard—one from a different manufacturer or lot number—can also be used to verify meter calibration or to troubleshoot, but is not required by the method.

At least 10% of all samples should be analyzed in duplicate; duplicates should agree within 5%. Buffers are intended for single use. In addition, buffer aliquots should be discarded following meter calibration.

1.7   Sample Collection, Preservation, and Holding Times

A 50-mL grab sample can be collected from the wastewater stream and taken directly to the laboratory for immediate analysis. There can be no holding time or preservation of the sample, although 40 CFR Part 136 requires that pH samples analyzed for compliance reporting be analyzed within 15 minutes.

Anaerobically digested sludge contains high concentrations of dissolved carbon dioxide. Care should be taken to minimize agitation of the sample during collection because this can release carbon dioxide gas and change the sample pH. It may be preferable to analyze these samples in the field rather than transporting them back to the laboratory.

1.8   Procedure

- **Step 1**—Prepare or calibrate electrodes for the test. Electrodes should be wetted thoroughly and prepared for use according to the instructions. The instrument should be standardized against a buffer solution with a pH approaching that of the sample, and then the linearity of electrode response must be checked against at least one additional buffer of a different pH. Both buffers used should bracket the expected sample pH.

- **Step 2**—Clean electrodes. After standardization and between samples, the electrodes should be washed carefully with deionized water from a wash bottle and blotted with a soft tissue or cleaned with solvents or detergents as directed by the manufacturer. Electrodes are delicate instruments that must be handled carefully.

- **Step 3**—Collect a 50-mL grab sample to analyze. Place the sample into a beaker similar to the ones used with the buffers to check the pH reading.

- **Step 4**—Run the test. Place electrodes into the beaker of the sample. Samples and standard buffers should be stirred during measurement. If a stirrer bar is being used, it should be rinsed and dried between measurements. Once the reading is stable, record the pH and temperature as displayed (see Figure 3.2).

- **Step 5**—If the pH meter does not automatically correct for temperature, record the temperature of the buffers and samples. For precise pH measurements, sample results can be corrected for temperature.

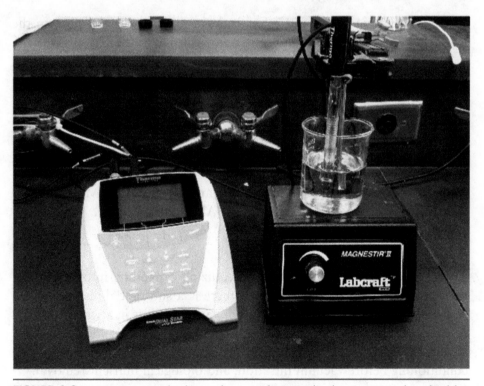

FIGURE 3.2 pH meter with electrodes in solution; display is pH in hundredths. *Courtesy of Indigo Water Group*

For additional information, see Method 4500^+ B-2011 in *Standard Methods* Online.

- **Step 6**—Remove electrodes and rinse with distilled water. Blot dry and store in manufacturer's recommended solution. Some manufacturers recommend storage in a 4-M solution of potassium chloride or pH 4 buffer. Other manufacturers recommend storage in a pH 7 buffer. pH electrodes should not be allowed to dry. For rapid response, electrodes should be stored in a solution recommended in the manufacturer's manual.

1.9   Reporting

If the meter cannot be adjusted to display pH in tenths, round to the nearest tenth. Look at the digit in the hundredths place; if it is 6 or greater, round the number in the tenths place up one number. If it is 4 or less, round the number in the tenths place down one number. If it is a 5, round up when the preceding number is odd and down when the preceding number is even. pH is commonly written by tenths (one decimal point) without specific units, such as pH 9.5. Refer to Chapter 6 for additional guidance on reporting.

1.10   Troubleshooting and Tips

The operator–analyst should refer to the troubleshooting section in the manufacturer's manual for the meter and electrode in use.

1.11   Benchsheet

Table 3.3 shows a sample benchsheet for pH.

2.0   TOTAL ALKALINITY OF WASTEWATER AND SLUDGE

The following sections comprise an expanded version of Method 2320 B-2011 from *Standard Methods* Online (APHA et al., 2017).

2.1   General Description

The alkalinity of water is its acid-neutralizing capability; in other words, how much acid it can absorb before there is a substantial change in pH. The determination of alkalinity levels at various points in a water resource recovery facility (WRRF) aids the understanding and interpretation of the treatment process and management of digesters, sludge elutriation, sludge conditioning before vacuum filtration, and biological nutrient removal.

TABLE 3.3 Sample benchsheet for pH.

Method: _____

Date: _____

Time of Collection: _____

Time of Analysis: _____

Operator-Analyst: _____

Location: _____

Slope: _____

Sample ID	Sample Date	pH Value (SU)	Comments

2.2   Application

Measurement of alkalinity is important for a number of treatment processes. Alkalinity ranges and effects for select treatment processes are presented in Table 3.4.

TABLE 3.4   Alkalinity ranges and effects for select wastewater processes.

Wastewater process	Alkalinity range and effects
Activated sludge system	Every time 1 mg of ammonia-nitrogen is oxidized to nitrate, 7.14 mg CaCO3 alkalinity is consumed. Likewise, every time 1 mg of nitrate is converted to nitrogen gas, 3.57 mg CaCO3 alkalinity is recovered. A residual alkalinity of 50 to 100 mg CaCO3/L is recommended for stable operation in nitrifying facilities. Supplemental alkalinity may be provided through chemical additions of lime, soda ash, or magnesium hydroxide. Metal salts (e.g., alum and ferric chloride) typically consume alkalinity and lower pH if sufficient alkalinity is not naturally present in the water. As a general rule, approximately 0.5 mg/L of alkalinity is required to react with each 1 mg/L of alum, and approximately 1.0 mg/L of alkalinity is required to react with each 1 mg/L of ferric chloride.
Trickling filtration	When trickling filters are nitrifying, maintain alkalinity at 7 times the influent ammonia-nitrogen concentration.
Anaerobic digestion	In a properly operating anaerobic digester, the bicarbonate alkalinity should be maintained at a level no lower than 1000 mg/L as calcium carbonate to ensure adequate pH control. Typical ranges are 1500 to 5000 mg/L. The volatile acid concentration is typically compared to alkalinity in the volatile acid to alkalinity ratio. The VA:ALK ratio is an indicator of the progress of digestion and the balance between the acid fermentation and methane fermentation microorganisms. Because of the need for balance between volatile acids and alkalinity, the VA:ALK ratio is an excellent indicator of digester health. Careful monitoring of the rate of change in this ratio can indicate a problem before a pH change occurs. The VA:ALK ratio should be approximately 0.1 to 0.2. Digester pH depression and inhibition of methane production occur if the ratio exceeds 0.8; however, ratios higher than 0.3 to 0.4 indicate upset conditions and the need for corrective action.
Aerobic digestion	Typically, aerobic digestion is a nitrifying process, especially in the absence of primary sludge. When nitrification occurs, both pH and alkalinity are reduced. See "activated sludge."

(*continued*)

TABLE 3.4   Alkalinity ranges and effects for select wastewater processes. (*Continued*)

Wastewater process	Alkalinity range and effects
Disinfection	When added to wastewater, chlorine gas solution will reduce alkalinity by 1.4 mg/L as calcium carbonate (CaCO3) per mg/L of chlorine.
	Breakpoint chlorination: large amounts of chlorine lower the water's pH, so large amounts of alkalinity must be added to maintain a near-neutral pH of 7, which is also where the breakpoint reaction requires the shortest contact times. About 30 mg/L of alkalinity must be present for every 1 mg/L of ammonia-nitrogen present to maintain the proper pH range. Theoretically, 14.3 mg/L of alkalinity is required to neutralize the hydrochloric acid formed by oxidizing 1 mg/L of ammonia-nitrogen. To keep the pH higher than 6.3, at least twice that amount of alkalinity is needed. Unless the water is highly alkaline or lime treatment precedes chlorination, alkali addition facilities should be provided. If sodium hypochlorite is used instead of chlorine, the alkalinity requirement is reduced by 75% (WEF et al., 2018).

Source: WEF et al. (2018).

2.3   Interferences

Some operator–analysts use color indicators in place of a pH meter when titrating for alkalinity. The bromocresol green–methyl red indicator changes color at pH 4.5. It is blue above pH 4.5 and red below pH 4.5. The color change shows that the desired pH has been reached and that alkalinity has been consumed. If the operator–analyst titrates to pH 4.5, the result is referred to as *total alkalinity.* Color indicators do not always work well for samples that are turbid or colored.

When a pH meter is used instead of a color indicator, soaps, oily matter, suspended solids, and precipitates may coat the glass electrode and cause a sluggish response. In these situations, the operator–analyst should add a small amount of titrant and wait for the pH to stabilize before adding more.

2.4   Apparatus and Materials

- A 50-mL buret
- Assorted beakers, graduates, volumetric flasks, and bottles as needed
- Pipet bulb as needed
- pH meter and electrodes

- Magnetic stirrer
- Stir bar

2.5   Reagents

For expected alkalinities of less than 200 mg/L, 0.02 N of sulfuric acid (H_2SO_4) or hydrochloric acid (HCl) should be used as the titrant. For expected alkalinities of more than 200 mg/L, 0.1 N of sulfuric acid or hydrochloric acid should be used. The following reagents will be needed to perform an alkalinity titration:

- Certified standard: Standards may be purchased commercially from a variety of sources
- Concentrated sulfuric acid, reagent-grade
- Concentrated hydrochloric acid, reagent-grade
- Sulfuric acid, 1.0 N: Take a 1-L volumetric flask filled with approximately 500 mL of deionized water. Carefully add 28.0 mL of concentrated sulfuric acid while stirring. Cool, then dilute to the 1.0-L mark. This solution may be purchased ready-made.
- Hydrochloric acid, 1.0 N: Take a 1-L volumetric flask filled with approximately 800 mL of deionized water. Carefully add 83.0 mL of concentrated hydrochloric acid while stirring. Cool, then dilute to the 1.0-L mark. This solution may be purchased ready-made.
- Sulfuric or hydrochloric acid, 0.1 N: Dilute 100 mL of either the 1.0-N sulfuric acid or hydrochloric acid to 1.0 L in a 1-L volumetric flask. This solution may be purchased ready-made.
- Sulfuric or hydrochloric acid, 0.02 N: Dilute 20.0 mL of 1.0-N sulfuric acid or hydrochloric acid to 1.0 L in a 1-L volumetric flask. This solution may be purchased ready-made.
- Bromocresol green–methyl red indicator solution, pH 4.5 indicator

2.6   Quality Assurance and Quality Control

For permit reporting, a certified standard must be analyzed. The result for the standard should be within ±10% of its true value. Certified standards may be purchased commercially from a variety of sources.

It is important for the operator–analyst to note that when titrants are made in-house from concentrated acids, they must be standardized prior to use. The standardization procedure can be found in the alkalinity procedure (Method 2320 B-2011) in *Standard Methods* Online (APHA et al., 2017). Certified acid titrants purchased commercially do not require standardization.

2.7   Sample Collection, Preservation, and Holding Times

Samples for alkalinity should be collected in glass or plastic containers. If not analyzed immediately, samples should be cooled to $\leq 6\ °C$ without freezing the sample and analyzed within 14 days. *Standard Methods* allows samples to be held for up to 14 days; however, the preferred and recommended hold time is less than 24 hours.

Samples collected from biological treatment processes may contain large numbers of active microorganisms. Nitrification or denitrification occurring in the sample can rapidly change the amount of alkalinity present. For this reason, rapid analysis is recommended whenever practicable.

2.8   Procedure

- **Step 1**—Use a pipet or graduated cylinder to place 50 mL of waste-water or sludge sample in a 100-mL beaker, as shown in Figure 3.3.
- **Step 2**—Fill a buret with 0.1 or 0.02 N sulfuric acid or hydrochloric acid (see Section 2.5). Using a properly calibrated pH meter and constantly mixing the beaker contents (preferably with a magnetic stirrer and stir bar), titrate to pH 4.5 with the sulfuric acid or hydrochloric acid from the buret. The recommended setup is shown in Figure 3.4, with the pH meter placed next to the titration assembly and the stir plate and the sample located in between. (As an alternative, instead of a pH meter, the operator–analyst can add a few drops of bromocresol

FIGURE 3.3   Add measured sample to a 100-mL beaker. *Courtesy of Kelsy Heck, City of Defiance, Ohio Water Pollution Control*

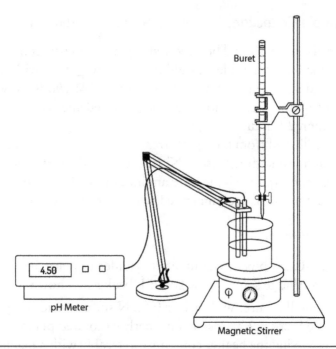

FIGURE 3.4   Titrate to pH 4.5.

green–methyl red indicator solution to the sample. Care should be taken to add just enough to get a good color. If too much is added, it will be difficult to see the color change. If the sample is colored or turbid, a pH meter should be used. While constantly mixing the beaker contents, titrate until the sample turns from blue to purple. Stop and wait a moment; then add titrant, one drop at a time, until the sample turns red.

- **Step 1**—Record the amount of acid added for calculation (milliliter [mL] titrant to reach endpoint pH 4.5). It is important to note that samples for alkalinity cannot be diluted. If alkalinity concentrations are high, the amount of sample volume should be decreased or a higher concentration of titrant should be used.

2.9   Calculations

$$\text{Alkalinity as } CaCO_3, \frac{mg}{L} = \frac{(\text{Volume of Titrant, mL})(\text{Normality of Titrant}) \quad \text{Equivalent Weight of } CaCO_3 \frac{g}{\text{equivalent}}}{(\text{Volume of Sample, mL})}$$

(3.1)

For example, given

$$\text{Volume of titrant (acid added to reach endpoint)} = 46 \text{ mL}$$
$$\text{Normality of acid} = 0.1 \text{ N (equiv/L)}$$
$$\text{Equivalent weight of } CaCO_3 = 50 \text{ g/equiv}$$
$$\text{Volume of sample} = 50 \text{ mL}$$
$$1000 \text{ mg} = 1 \text{ g}$$

$$\text{Alkalinity as } CaCO_3 = \frac{(46 \text{ mL}) \; 0.1 \text{ N} \frac{\text{equiv}}{\text{L}} \; 50 \frac{\text{g}}{\text{equiv}} \; (1000 \text{ mg})}{(50 \text{ mL})(1 \text{ g})}$$

$$= 4600 \text{ mg/L}$$

2.10 Reporting

Alkalinity is reported as calcium carbonate in milligrams per liter (mg/L). This standardized way of reporting means that an operator–analyst would need to add this much calcium carbonate to water to get the same alkalinity as the sample. The method number and titration endpoint should be recorded on the benchsheet and reported with the results.

2.11 Troubleshooting and Tips

An important safety factor for the operator–analyst to remember is to always add acid to water and to never add water to acid. In addition, for continuous mixing, the operator–analyst may need another person's assistance or may need to use a magnetic stirrer to constantly mix the wastewater or sludge sample.

2.12 Benchsheet

Table 3.5 shows a sample benchsheet for total alkalinity of wastewater and sludge.

3.0 VOLATILE ORGANIC ACIDS IN WASTEWATER SLUDGE

The following sections are an expanded version of Method 5560 C-2001 from *Standard Methods* Online (APHA et al., 2017). This procedure is used for process control purposes only and is not listed as an approved reporting method in 40 CFR Part 136 (U.S. EPA, 2017).

TABLE 3.5   Sample benchsheet for total alkalinity of wastewater sludge.

Method: _____

Date: _____

Time: _____

Operator-Analyst: _____

Location: _____

pH meter slope: _____

H_2SO_4 normality (N): _____

Standard concentration: _____

$$\text{*mg/L as } CaCO_3 = \frac{(\text{mL } H_2SO_4 \text{ titrant})(\text{N titrant})(50{,}000)}{\text{Sample volume (mL)}}$$

Sample date	Sample identification	Sample volume (mL)	H_2SO_4 titrated (mL)	Alkalinity (mg/L)*	Comments

3.1   General Description

Method 5560 C-2001 from *Standard Methods* Online (APHA et al., 2017) can be applied to control anaerobic digestion and biological nutrient removal. Because the method is empirical, it should be carried out exactly as described. It is assumed that approximately 70% of the volatile acids will be found in the distillate. The assumption is corrected for in the computation. This factor has been found to vary from 68% to 85%, depending on the nature of the acids and rate of distillation.

This method can be used to recover acids containing up to six carbon atoms in the supernatant of a sludge sample. In this method, an acidified sample is distilled, and the distillate is titrated to an endpoint with sodium hydroxide (NaOH).

3.2   Application

In anaerobic digesters, increasing volatile acids is one of the first signs of poor digester performance. Causes include

- Inhibition of methane-forming bacteria (toxicity or temperature change),

- Insufficient detention times that reduce the digester's buffering capacity (its ability to neutralize volatile acids), and

- Excessive solids loading rate.

The volatile acid concentration is typically compared to alkalinity in the volatile acid to alkalinity ratio (VA:ALK). The VA:ALK is an indicator of the progress of digestion and the balance between the acid fermentation and methane fermentation microorganisms. Because of the need for balance between volatile acids and alkalinity, the VA:ALK is an excellent indicator of digester health. Careful monitoring of the rate of change in this ratio can indicate a problem before a pH change occurs. The VA:ALK should be between 0.1 and 0.2 to ensure adequate buffering capacity. Additionally, volatile acids should be below 500 mg/L in a healthy digester and will be below 100 mg/L most of the time. Digester pH depression and inhibition of methane production occur if the ratio exceeds 0.8; however, ratios higher than 0.3 to 0.4 indicate upset conditions and the need for corrective action.

In addition to monitoring anaerobic digester performance, volatile fatty acids (VFAs) are important for biological phosphorus removal (BPR). Phosphorus accumulating organisms (PAOs) absorb large quantities of VFAs in the anaerobic zone of a biological nutrient removal activated sludge process. The VFAs are stored within the PAOs. When the PAOs move into the aerated zone, they metabolize their supply of stored VFAs and take up large quantities of phosphorus. Given a sufficient supply of VFAs (primarily a mix of acetic and propionic acids), the biology for phosphorus removal is forgiving. Typically, 5 to 10 mg/L of VFA per milligram per liter of total phosphorus in the raw wastewater are needed in aeration basin influent.

3.3   Interferences

Fractional recovery of each acid increases with increasing molecular weight. Because it is empirical, the still-heating rate, presence of sludge solids, and final distillate volume affect recovery.

Hydrogen sulfide and carbon dioxide are liberated during distillation and will be titrated as a positive error. This is why the first 15 mL of distillate is discarded.

3.4   Apparatus and Materials

- Centrifuge and tubes

- Distillation flask, 1000-mL capacity

- Distillation assembly, shown in Figure 3.5, consisting of a minimum of an adapter tube, a condenser column, and an electric heating element

- pH meter and probe or recording titrator

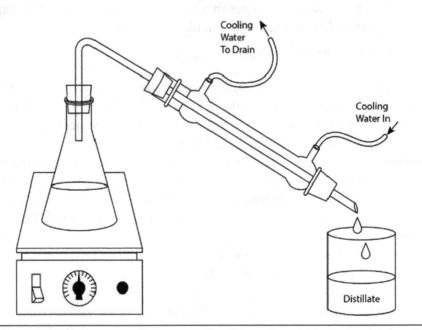

FIGURE 3.5   Distill off 150 mL at the rate of approximately 5 mL/min.

- Titration buret, minimum of 10 mL capacity
- Beakers, 250-mL capacity
- Graduated cylinders
- Pipet, Class A, Class B wide-tipped, and serological
- Pipet bulb
- Boiling beads

3.5   Reagents

- Sulfuric acid (1+1): Put approximately 200 mL of deionized water in a 500-mL volumetric flask. Cautiously add 250 mL of concentrated sulfuric acid. Stir the solution and allow it to cool. Dilute to volume with deionized water. It is important to note that extreme heat can be generated from this addition and the acid may have to be added in small quantities.
- Sodium hydroxide stock, 1 N: Place 4.0 g of sodium hydroxide in a 100-mL volumetric flask and dilute to volume with deionized water. Prepare fresh every 6 months.
- Sodium hydroxide titrant, 0.02-N standardized: Dilute 4.0 mL of 1-N sodium hydroxide stock to 200 mL in a volumetric flask with deionized water. Titrate 100 mL to a pH of 7.0 with standardized 0.50-N

sulfuric acid or follow the procedures in Method 2310B (3c) of *Standard Methods* (APHA et al., 2017). Calculate the actual normality and use it in the calculation. Prepare fresh weekly.

- Acetic acid stock solution (2000 mg/L): Put about 400 mL of deionized water in a 500-mL volumetric flask. Add 0.95 mL of concentrated glacial acetic acid. Dilute to volume with deionized water. Prepare fresh at least every 6 months.

- Acetic acid standard (50 mg/L): Put 2.5 mL of acetic acid stock solution in a 100-mL volumetric flask. Dilute to volume with deionized water. Prepare fresh daily.

3.6   Sample Collection, Preservation, and Handling

Samples should be collected in a 500-mL high-density polyethylene bottle with no preservatives. The sample should be stored at ≤ 6 °C and analyzed as soon as possible.

3.7   Quality Assurance and Quality Control

For each set of 20 samples or for each daily analysis set, a duplicate sample will be run. The relative percent difference (RPD) of the two samples must be no greater than 20%. If RPD is greater than 20%, the cause of the discrepancy should be determined and the samples reanalyzed. Relative percent difference is calculated by taking the difference between two measurements and dividing by the average of the two measurements. This section also includes an example calculation.

3.8   Procedure

- **Step 1**—Centrifuge 200 mL of each sample for 5 minutes or any volume that results in 100 mL of supernate. (This may also be done while some of the samples are being distilled if all of the samples cannot be distilled at one time.)

- **Step 2**—Prepare two acetic acid standards. (One standard will be analyzed undistilled; the other will be analyzed as distilled. The results will be used in the final calculations.)

- **Step 3**—Turn heaters on.

- **Step 4**—Turn on condenser water.

- **Step 5**—Place 100 mL of blank, supernatant liquor or standard to be distilled in a distillation flask.

- **Step 6**—Add 100 mL of deionized water and three boiling beads.

- **Step 7**—Add 5 mL of 1 + 1 sulfuric acid and mix well.

- **Step 8**—Connect the flask to the distillation apparatus and distill at a rate of approximately 5 mL/min.
- **Step 9**—Discard the first 15 mL of distillate and collect 150 mL of distillate in a 250-mL glass graduated cylinder. Do not distill to dryness.
- **Step 10**—Transfer the distillate to a beaker and titrate to a pH of 8.3 with 0.1-N sodium hydroxide.

3.9   Calculations

Find the acetic acid concentration as follows:

$$\left(\frac{A \cdot B \cdot 60\,000 \text{ g/equivalent}}{C}\right) = \text{Acetic acid, mg/L} \tag{3.2}$$

Where

A = mL of NaOH
B = N or NaOH
60 000 = equivalent weight of CH_3COOH, g/equivalent
C = volume of original sample, in mL.

Calculate the recovery factor as follows:

$$\left(\frac{A}{B}\right) = F \tag{3.3}$$

Where

A = distilled acetic acid, mg/L
B = undistilled acetic acid, mg/L
F = recovery factor

Find the VOA concentration of the sample as follows:

$$\left(\frac{A \cdot B \cdot 60\,000}{C \cdot F}\right) = \text{Acetic acid, mg/L} \tag{3.4}$$

Where

A = mL of NaOH
B = N of NaOH
C = 100 mL of original sample
F = recovery factor

For example, given the following:

Volume of titrant   = 12 mL
Normality of titrant   = 0.02 N (equiv/L)
Equivalent weight of CH3COOH = 60 000 mg/equiv
Approximate recovery   = 0.7 (70%)
Volume of sample   = 100 mL

$$\text{Volatile acids as } CH_3COOH = \frac{(12 \text{ mL})\left(0.02 \frac{\text{equiv}}{\text{L}}\right)\left(60\,000 \frac{\text{mg}}{\text{equiv}}\right)}{(100 \text{ mL})(0.7)}$$

Volatile acids as CH_3COOH = 206 mg/L

3.10   Reporting

There are no reporting guidelines. This test is for process control only.

3.11   Benchsheet

Table 3.6 shows a sample benchsheet for volatile acids.

4.0   SOLIDS PROCEDURES

This section is an expanded version of Method 2540 A-G-2015 from *Standard Methods* Online (APHA et al., 2017). Solids include all solid material that is either suspended or dissolved in water or wastewater. This includes salts, oil and grease, vegetable matter, grit, hair, and so on. Solids in water and wastewater are characterized by their size (i.e., can or cannot pass through a 2.0-μm filter) and their volatility (i.e., volatile vs. nonvolatile or organic solids vs. inorganic solids). Figure 3.6 illustrates how solids are broken down. The figure is simplified and does not include every detail of analysis.

Figure 3.6 shows that when a sample is taken, poured into a dish, dried, and the remains weighed, the *total solids* in the sample are being measured. Total solids include both suspended solids (e.g., grit and mixed liquor flocs) and dissolved solids (e.g., salts and colloidal material small enough to pass through the filter disk). There are two different total solids tests, one for samples with solids concentrations up to 20 000 mg/L (2% solids) and one for solids concentrations higher than 20 000 mg/L. The primary differences between these two methods are the drying time required, how the samples are aliquoted, and how the results are expressed. Samples that can

TABLE 3.6   Sample benchsheet for volatile acids.

Method: _____

Date: _____   pH meter slope: _____

Time: _____   NaOH normality (N): _____

Operator-Analyst: _____

Location: _____

$$*VOA\ (mg/L)= \frac{(mL\ NaOH\ titrant)(N\ titrant)(60\ 000)}{Sample\ volume\ (mL)}$$

Sample date	Sample identification	Sample volume (mL)	NaOH titrated (mL)	VOA (mg/L)*	Comments

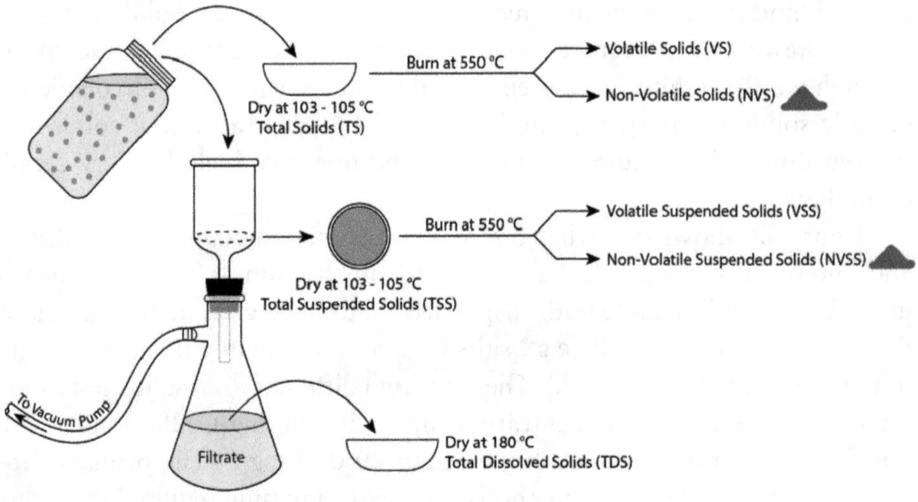

FIGURE 3.6   How types of solids are defined. *Courtesy of Indigo Water Group*

be poured are measured in milligrams per liter (mg/L). Samples that must be weighed to aliquot because they are too thick to pour are expressed as milligrams per gram (mg/g) or as a percent. In the first method, samples are dried for at least 1 hour and results are expressed as mg/L. In the second method, samples are typically dried overnight and results are expressed as percent total solids. The second method should be used for thickened sludges, biosolids, and dewatered cake.

Suspended solids can be separated from dissolved solids by filtering the sample. The solids that remain on the filter disk are *suspended solids*. Solids that were able to pass through the filter disk and ended up in the filtrate in the flask are dissolved solids. *Standard Methods* (APHA et al., 2017) strictly defines dissolved solids as the solids that can pass through a 2-μm (or smaller) filter disk. So far in this text, solids are separated and classified by their size.

Solids can also be classified by whether they are volatile or nonvolatile. This determination is done by taking either the total solids or the total suspended solids (TSS) and placing them into a muffle furnace at 550 °C. Heating the solids at such a high temperature causes the organic portion of the solids to burn away or volatilize. What is left over on the filter disk is the material that cannot be burned away. These are the nonvolatile or *fixed solids*. In the test, it is assumed that the solids that volatilize are organic and that the remaining solids are inorganic.

As an example, let's look at a sample of mixed liquor. The operator-analyst runs the TSS test and finds that the mixed liquor suspended solids (MLSS) concentration is 3000 mg/L. The operator–analyst then takes the same filter disk, heats it to 550 °C, and reweighs the filter. In this example, there are enough solids left on the filter disk to equal 600 mg/L. The solids that stay behind are the nonvolatile suspended solids (NVSS). To find the volatile suspended solids (VSS), subtract NVSS from TSS as follows:

$$\text{MLSS, mg/L} - \text{NVSS, mg/L} = \text{MLVSS, mg/L} \qquad (3.5)$$
$$3000 \text{ mg/L} - 600 \text{ mg/L} = 2400 \text{ mg/L}$$

It is assumed that VSS are 100% organic and that they represent live, active microorganisms in the activated sludge basin. The NVSS are assumed to be 100% inert material such as grit, fibers, and other materials that are not part of the active biomass. For the aforementioned example, MLSS is 80% VSS and (it can be assumed) is 80% live, active biomass. In reality, the separation between organic and inorganic solids is not quite as neat or perfect. Sometimes, inorganic solids like lead carbonate are volatilized and end up being counted as part of the "organic" fraction. Overall, however, the test gives a good estimate of organic versus inorganic fractions of solids in the sample.

One final distinction that can be made for solids is whether or not they are readily settleable. If a sample of raw influent, primary clarifier effluent, or even final effluent is allowed to sit quietly in a container, some of the solids in the sample will eventually settle to the bottom of the container. These solids have settled and are settleable. The remaining solids are either too small, too light, or both to go to the bottom of the container by themselves in a reasonable amount of time. These are the non-settleable solids.

Each of the different tests for solids are necessarily intertwined. As shown in Figure 3.6, filtrate from the TSS test can be collected for the total dissolved solids (TDS) test. Theoretically, if the operator–analyst knows the total solids concentration and the TSS concentration, the dissolved solids concentration can be *estimated* by subtraction. All laboratory tests have some amount of built-in error that comes from sampling, aliquoting, and minor differences in how each sample was handled by the operator–analyst. For this reason, an operator–analyst should not expect to be able to have TSS and TDS numbers add up to exactly the number obtained from the total solids test. Although they should be close, they probably will not be exact.

In summary,

- Total solids are all of the solids in a whole, unseparated sample. This manual includes two methods for analyzing total solids.
- Total volatile solids (TVS) are the solids from a total solids test that are burned away at 550 °C.
- Fixed solids or fixed residue is another way of saying nonvolatile solids or nonvolatile suspended solids.
- Total suspended solids are solids that cannot pass through a 2.0-μm or smaller filter.
- Volatile suspended solids (VSS) are solids from a TSS test that are burned away at 550 °C.
- Nonvolatile suspended solids (NVSS) are solids from a TSS test that remain after VSS have been burned away.
- Total nonvolatile solids (TNVS) are the solids from a total solids test that remain after VS have been burned away at 550 °C.
- Total dissolved solids (TDS) are those solids that will pass through a 2.0-μm or smaller filter.

5.0 SETTLEABLE SOLIDS

The following sections comprise an expanded version of Method 2540-F-2015 from *Standard Methods* Online (APHA et al., 2017).

5.1   General Description

This simple test can be performed to quickly and qualitatively show whether the primary and secondary processes are functioning properly. The volume of settleable solids in raw wastewater and effluent is readily seen and the turbidity removed by secondary treatment processes is immediately observable. The results are not quantifiable but are useful to both the untrained and trained observer.

5.2   Application

The settleable solids test estimates removal efficiency of primary and secondary treatment processes. The Imhoff cone represents ideal settling conditions. If the Imhoff cone removes more material than the primary or secondary treatment process being evaluated, then treatment process efficiency could be improved. If settleable solids in the Imhoff cone remove about the same amount of material as the process being evaluated, then the process is performing as expected.

5.3   Apparatus and Materials

- Imhoff cone (see Figure 3.7)
- Cone support
- Long stirring rod made of glass or other inert material

FIGURE 3.7   After filling the Imhoff cone to the 1-L mark, allow to settle for 45 minutes, stir gently, and allow to settle for 15 minutes. *Courtesy of Indigo Water Group*

5.4 Sample Collection, Preservation, and Holding Times

Fresh grab samples should be collected and analyzed as soon as reasonably practicable. If the sample is allowed to sit for more than 1 hour, the settling characteristics can change. The *Code of Federal Regulations*, Part 136 (U.S. EPA, 2017), allows a hold time of up to 14 days.

5.5 Quality Assurance and Quality Control

There are no required quality assurance and quality control (QA/QC) samples for the settleable solids test. This test is for process control purposes.

5.6 Procedure

- **Step 1**—Fill the Imhoff cone to the 1-L mark with a well-mixed sample. Occasionally, settleable matter in a given wastewater sample may exceed the 40-mL/L graduation of an Imhoff cone. In these instances, a 500-mL volume of the sample should be used and transferred to an Imhoff cone. The operator–analyst should follow this procedure and then multiply the value of the settleable matter measured by a factor of 2 to express the final result in milliliters per liter.
- **Step 2**—Let the sample settle for 45 minutes.
- **Step 3**—Gently stir the upper portion of the sample with the glass rod to dislodge suspended matter clinging to the tapered sides of the cone.
- **Step 4**—Let the sample settle 15 minutes longer.
- **Step 5**—Read the volume of settleable matter in milliliters per liter.

5.7 Calculations

Sample results are expressed in milliliters of solids settled per liter of original sample volume.

5.8 Reporting

The lowest concentrations that can be reasonably measured with this test are in the range of 0.1 to 1.0 mL/L depending on where the lowest gradation mark is located on the Imhoff cone. Most Imhoff cones have the lowest marking at 0.2 mL/L.

5.9 Troubleshooting and Tips

If there is a separation of settleable and floating materials, the floating material should not be included as part of the settleable matter measurement.

5.10   Benchsheet

Table 3.7 shows a sample benchsheet for settleable solids.

TABLE 3.7   Settleable solids.

Method: _____

Date: _____

Time: _____

Operator-Analyst: _____   TSS whole: TSS of untreated sample

Location: _____   TSS super: TSS of supernatant

Sample date	Sample identification	TSS whole (mg/L)	-	TSS super (mg/L)	=	Settleable (mg/L)
			-		=	
			-		=	
			-		=	
			-		=	
			-		=	
			-		=	
			-		=	
			-		=	
			-		=	
			-		=	
			-		=	
			-		=	
			-		=	
			-		=	

6.0 TOTAL FIXED AND VOLATILE SOLIDS IN SOLID AND SEMISOLID SAMPLES

The following sections comprise an expanded version of Method 2540 G-2015 from *Standard Methods* Online (APHA et al., 2017), which is a total solids method for samples with greater than 20 000 mg/L (2%) solids. For lower solids concentrations, the total solids dried at 103 to 105 °C method (APHA et al., 2017) should be used. In practice, the method used should be selected based on whether the sample is liquid enough to be accurately measured with a pipet or graduated cylinder. For thicker samples that do not pour easily, this method should be used.

6.1 General Description

Method 2540-G-2015 is used for measuring total solids in samples with greater than 20 000 mg/L (2%) solids. It works well for thickened sludge, digester sludge, and dewatered biosolids. To find total solids in samples with solids concentrations below 20 000 mg/L, the method described in Section 7.0 of this chapter should be used.

Volatile and fixed solids in water or wastewater are determined by evaporating a weighed sample in a drying oven. Unlike TSS in wastewater, which is expressed as milligrams per liter, solids in sludge are expressed in terms of percent by mass of the total amount of solids, or in milligrams per kilogram (mg/kg). This procedure makes calculations easier.

A well-mixed aliquot of the sample is quantitatively transferred to a pre-weighed evaporating dish and evaporated to dryness at 103 to 105 °C. If TVS are to be determined, the sample is reweighed and then ignited at 550 °C in a muffle furnace. The loss of weight on ignition is reported as milligrams per kilogram (or percent) volatile residue.

6.2 Application

Total solids should be determined on samples with expected solids concentrations above 20 000 mg/L in place of either the TSS test or the total solids dried at 103 to 105 °C test. The total solids test is useful for estimating loadings to solids handling processes such as belt filter presses, centrifuges, and digesters; for calculating the percent volatile solids reduction through a digester; and for complying with monitoring requirements.

Untreated primary clarifier sludge ranges from 5% to 9% solids and is between 60% and 80% volatile (Metcalf & Eddy, Inc./AECOM, 2014). Secondary sludge tends to be much more dilute at less than 1.2% solids.

Calculating volatile solids reduction in an aerobic or anaerobic digester is a way to tell how well the digester is working and how digested or stabilized the solids are. Volatile solids reduction cannot be calculated with the simple percent removal equation because of the way the digester works. For example,

$$\% \text{ Removal} = \left[\frac{\text{In} - \text{Out}}{\text{In}}\right] \cdot 100 \qquad (3.6)$$

A special equation, as follows, is used to calculate volatile solids reduction:

$$\% \text{ Volatile solids reduction} = \left[\frac{(\text{In} - \text{Out})}{\text{In} - (\text{In} \cdot \text{Out})}\right] \cdot 100 \qquad (3.7)$$

The more complicated equation is required because we are only concerned about the volatile solids destroyed, not the total solids destroyed. Most MLSS are about 80% volatile solids. Although there is some variation from facility to facility, generally speaking, this will be true. If the MLSS is 80% volatile, what is the other 20% made up of? The answer is inert material (e.g., sand, grit, eggshells, hair) that cannot be broken down or digested by bacteria. This inert material passes through the digester unchanged. This means that if 100 kg (45.5 lb) of 80% volatile MLSS were put into the digester, 20 kg (9.1 lb) will come out the other side unchanged. If the same digester is achieving 50% volatile solids reduction, then the 80 kg (36.4 lb) of volatile solids that went into the digester will become only 40 kg (18.2 lb) of volatile solids plus some carbon dioxide and water. In this example, 80 kg of volatile solids and 20 kg of inert solids went into the digester (100 kg total) and 40 kg of volatiles and 20 kg of inerts came out (60 kg total). Even though the digester achieved 50% volatile solids reduction, the total solids reduction was only 40% because of the inert material.

6.3   Interferences

Because solid samples must be dried and ignited for longer periods of time, there is a risk of losing material from the sample (ammonium carbonate and volatile organic matter). This loss can cause sample results to be biased low.

Nonrepresentative particles such as leaves, sticks, algae, and lumps of fecal material should not be included in the sample aliquot if the operator–analyst determines that these are not representative. In practice, this requires the operator–analyst to make a judgment call. If the entire sample has leaves mixed throughout, then the leaves are probably representative and should be left in the sample aliquot. If the sample has only one or two leaves floating on the top, then the leaves are not representative and should be excluded.

If the sample contains floating oil and grease, it should be mixed in with a blender and included in the sample.

6.4   Apparatus and Materials

- Drying oven
- Evaporating dishes (porcelain, 90-mm diameter)
- Muffle furnace
- Desiccator
- Balance capable of weighing to the nearest 0.01 g (10 mg)
- Graduated cylinder
- Pipet and pipet bulb (as needed)
- Blender

6.5   Stock and Standards

For permit-required samples including biosolids analyses, a certified standard will be needed. Certified standards can be ordered from a variety of companies.

6.6   Quality Assurance and Quality Control

When analyzing samples for process control, the operator–analyst may not have the time or resources to run all of the recommended quality assurance samples. Certified standards can be expensive. It is acceptable to decrease the analysis frequency for most or all QA/QC samples when analyzing process control samples; however, QA/QC samples should still be analyzed on a regular basis to ensure data quality. At a minimum, a blank and a duplicate should be analyzed with each batch. The operator–analyst must balance the benefit gained by running a particular QA/QC sample against available resources. It is important to remember that process control is only as good as the laboratory data on which it is based.

For permit reporting, *Standard Methods* (APHA et al., 2017) recommends the following minimum quality assurance samples:

- Blanks should always be lower than the method detection limit (MDL). When blanks are higher than this, it is an indication that some or all of the samples in the batch may have been contaminated. A blank should be run once every 20 samples or once per batch, whichever is more frequent.

- The certified standard should agree within 10% of its certified value or within the limits specified on the manufacturer's certificate. A standard should be run once per 20 samples or once per batch, whichever is more frequent.

- Analyze at least 10% of all samples in duplicate. Duplicates should agree within 5% of their average weight.

- It is good laboratory practice to analyze at least one standard and one duplicate for each set or batch of samples even if fewer than 20 samples are analyzed at a time.

- The online version of *Standard Methods* recommends running samples in replicate to increase precision. For highly variable samples like raw wastewater, it may be desirable to analyze two aliquots for each sample and average the results. This is in addition to analyzing duplicates at the recommended 10% frequency.

6.7   Sample Collection, Preservation, and Holding Times

Samples for solids analyses should be collected in plastic or glass containers. A preservative is not used. Samples should be cooled to $\leq$ 6 °C when not analyzed immediately. Samples should be analyzed within 24 hours whenever possible, but may be held up to 7 days.

6.8   Procedure

- **Step 1**—Clean glassware.
  1. Wash evaporating dishes and crucibles with warm water and laboratory detergent; rinse with deionized water and allow to air dry.
  2. Mark glassware with nail polish or ceramic marker and allow them to dry at room temperature. Special heat-resistant marking pens may also be purchased for this purpose. It may be necessary to etch or engrave the crucible before applying a ceramic marker or heat-resistant pen.
  3. Preheat clean porcelain evaporation dishes to 550 °C for 1 hour in a muffle furnace.
  4. Remove dish from muffle furnace using tongs.
  5. If the desiccator is made of plastic, allow the dish to partially cool in air until most of the heat has dissipated. A hot porcelain dish can melt or mar a plastic desiccator. After the dishes are cool enough to handle, place them into a desiccator. Placing a ceramic tile in the bottom of the desiccator can alleviate this problem.
  6. If the desiccator is heat-tolerant, crucibles should be placed directly into the desiccator from the drying oven.
- **Step 2**—Obtain initial dish weight.
  1. Dishes must be completely cooled before weighing.
  2. Weigh immediately before use.

3. Verify balance calibration with a certified check weight. Some balances are equipped with internal check weights that may be used instead.

4. Weigh the dish and record the weight on the benchsheet to the nearest measurable value (e.g., 0.01 g).

- **Step 3—Aliquot sample.**

1. Transfer a well-mixed, measured aliquot of sample to the pre-weighed dish as shown in Figure 3.8.

2. If the sample has enough liquid in it that it can be easily poured, stir it thoroughly. Then, pour between 25 and 50 g of sample into the pre-weighed dish. The operator–analyst does not need to be careful to get exactly 25 g because the final sample result will take the initial weight into account.

3. Record the crucible plus sample weight on the benchsheet.

4. If the sample cannot be poured easily and consists of individual pieces of solid material, *Standard Methods* (APHA et al., 2017) recommends using a no. 7 cork borer to take cores of different parts of the sample. A cork borer is a metal tool typically used for cutting

FIGURE 3.8   Aliquot sludge sample. *Courtesy of Indigo Water Group*

holes in a cork or rubber stopper. A no. 7 cork borer will take plugs that are approximately 21 mm in diameter at the top and 16 mm in diameter at the bottom. Practically speaking, any method or tool may be used as long as enough different parts of the total sample are selected to result in a subsample that is representative of the whole. The operator–analyst should place between 25 and 50 g of material into the pre-weighed dish. Again, the exact weight is not that important. It is much better to get a representative subsample than it is to get exactly 25 or 30 g in the dish. The operator should record the crucible plus sample weight on the benchsheet.

- **Step 4**—Transfer the dish to a drying oven for at least 1 hour or until dry at 103 to 105 °C; make sure to use tongs.

- **Step 5**—Dry the sample until constant weight is achieved. *Constant weight* is defined as less than 4% difference or 0.5 mg (0.0005 g), whichever is less.

1. Place the dish with the sample into a drying oven. Figure 3.9 shows evaporating dishes with samples being transferred to the drying oven with a heat-resistant fiberglass tray.

2. Dishes may be left in the oven overnight if necessary.

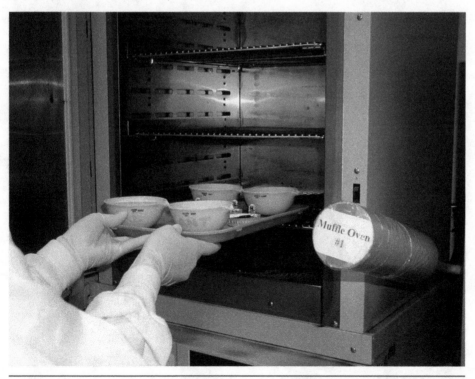

FIGURE 3.9   Load samples into drying oven. *Courtesy of Indigo Water Group*

3. Remove dishes from the drying oven and place in a desiccator to cool.

4. Allow to cool to room temperature. The dried sludge samples will pull away from the dish edges as shown in Figure 3.10.

- **Step 6**—Weigh the dishes. Record the final weight on the benchsheet. It is important to note that *Standard Methods* (APHA et al., 2017) requires that Steps 6 and 7 be repeated until the weight no longer changes. This is defined as no more than 4% or 50 mg difference between weighings. In practice, the operator–analyst will quickly learn how long it takes for samples to dry to a consistent weight. For process control samples, consistency is more important than following the precise details of the method. If the same drying time is used for process control samples each time they are analyzed, the results from one day can easily be compared to the next.

- **Step 7**—Calculate the percent total solids.

- **Step 8**—If VS is needed, place the dish from previous step in a muffle furnace at 550 °C until the sample is completely burned. The length of time for complete burning (ashing) depends on the size of the sample. For thickened sludge, dewatered cake, and biosolids, 1 hour is typically sufficient to completely ash the sample. It is important

FIGURE 3.10 Dried sludge samples. *Courtesy of Indigo Water Group*

to note that *Standard Methods* (APHA et al., 2017) recommends repeated ashing, cooling, and weighing cycles until a constant weight is achieved. This is not necessary for process control samples.

- **Step 9**—Remove the dish from the muffle furnace. Wear heavy, heat-resistant gloves and use tongs to transfer the dishes from the muffle furnace to a heat-resistant surface. A few ceramic tiles on the bench top will protect the laboratory counter from the heat.
- **Step 10**—Allow the dish to partially cool in air until most of the heat has dissipated.
- **Step 11**—Transfer the dish to a desiccator and allow it to cool to room temperature;
- **Step 12**—Weigh the dish and record the weight to the nearest 0.01 g (10 mg). If the operator–analyst is unsure that constant mass has been obtained, record the weight, then reheat and weigh again until the weight no longer changes.
- **Step 13**—Calculate the percent VS.

6.9 Calculations

Percent total solids can be calculated as follows:

$$\% \text{ Total solids} = \frac{(A - B) \cdot 100}{C - B} \tag{3.8}$$

Where
   A = weight of dried solids plus dish (mg),
   B = weight of dish (mg), and
   C = weight of wet sample plus the dish (mg).
   If desired, calculate the percent volatile solids (VS) as follows:

$$\% \text{ Volatile solids} = \frac{(A - D) \cdot 100}{A - B} \tag{3.9}$$

where D is the weight of solids remaining after ignition plus the dish after ignition (milligrams).

6.10 Reporting

Sample results are reported as percent with three significant figures. For sample results that are lower than 10% solids, three significant figures will mean reporting to two decimal places (e.g., 5.62% total solids). For sample results greater than 10% solids, three significant figures will mean reporting to one decimal place (e.g., 18.9% total solids).

6.11 Troubleshooting and Tips

Use nail polish to mark crucibles before firing at 550 °C. This inexpensive shellac will ash and turn black in the muffle furnace but will not flake off during routine testing. Labels are easily removed with a little scrubbing, hot water, and soap. Special heat-resistant marking pens may also be purchased for this purpose. Table 3.8 provides troubleshooting for the total solids test.

7.0 TOTAL SOLIDS DRIED AT 103 TO 105 °C

The following sections comprise an expanded version of Method 2540 B-2015 from *Standard Methods* Online (APHA et al., 2017), which is a total solids method for samples with between 4 and 20 000 mg/L (2%)

TABLE 3.8   Troubleshooting for total solids in solid and semisolid sample test.

Problem	Probable cause
Standard result is lower than certified value.	1. Standard not shaken or mixed enough before aliquoting.
	2. Standard poured too slowly or allowed to sit on bench after mixing.
	3. Oven temperature too high, volatilized standard.
Standard result is higher than certified value.	1. Dirty glassware.
	2. Oven temperature too low. Trapped water.
	3. Standard may have been compromised. If previous uses were out of range low, the standard may have been concentrated in the bottle.
Blank changed by more than 50 mg between initial and final weighing.	1. Dirty glassware.
	2. Balance not calibrated at initial or final weighing or both.
Sample replicates do not agree within 5% of their average weight.	1. Chunky samples. Look for obvious visual differences between sample aliquots.

solids. For higher solids concentrations, the total, fixed, and volatile solids in solid and semisolid samples method should be used (APHA et al., 2017).

In practice, the method used should be selected based on whether the sample is liquid enough to be accurately measured with a pipet or graduated cylinder. Method 2540 B-2015 should be used for thinner samples that pour easily.

7.1   General Description

Volatile and fixed solids in water or wastewater are determined by evaporating a measured sample on a steam bath or in a drying oven. Sample results are expressed in milligrams per liter. A well-mixed aliquot of the sample is quantitatively transferred to a pre-weighed evaporating dish and evaporated to dryness at 103 to 105 °C. If total volatile solids are to be determined, the sample is then ignited at 550 °C in a muffle furnace. The loss of weight on ignition is reported as milligrams per liter (or percent) volatile residue.

7.2   Application

Total solids should be determined on samples with expected solids concentrations between 10 000 and 20 000 mg/L in place of the TSS test. Samples in this concentration range will not pass easily through the filter disk.

7.3   Interferences

Samples that contain high concentrations of calcium, magnesium, chloride, and/or sulfate may need longer drying times to obtain good results. Samples with these characteristics can be hygroscopic, which means they hold on tightly to water. Because of this, these samples must be properly desiccated and weighed as soon as they reach room temperature. If they are allowed to sit, they may reabsorb water and need to be dried again.

Sample aliquots should be adjusted so that no more than 200 mg of solids accumulate in the dish during the test. Higher amounts of solids can form a water trapping crust. Even long dry times will not remove trapped water. In practice, this means that, for a raw wastewater sample with an expected total solids concentration of 720 mg/L, the sample volume should be smaller than about 300 mL.

If the sample contains chunky materials (e.g., leaves, algae, toilet paper bits) that are not representative, they should be removed from the sample before drying. Whether or not something should be removed is up to the professional judgment of an operator–analyst. Samples that contain visible floating oil and grease should be treated with a blender before taking a subsample.

7.4   Apparatus and Materials

- Drying oven
- Steam bath (optional) (see Figure 3.11)
- Evaporating dish (porcelain, 90-mm diameter)
- Muffle furnace
- Desiccator
- Balance capable of weighing to the nearest 0.0001 g (0.1 mg)
- Wide-mouth, "to deliver" (TD) pipet (10- and 25-mL size)
- Pipet bulb
- Class A "to contain" (TC) graduated cylinder

7.5   Reagents

No reagents are required other than deionized water.

7.6   Stock and Standards

For permit-required samples, a certified standard will be needed. Certified standards may be ordered from a variety of companies.

FIGURE 3.11   Place evaporating dish containing 25- to 50-mL sample on steam bath and evaporate to dryness. *Courtesy of Hach Company*

7.7   Quality Assurance and Quality Control

When analyzing samples for process control, the operator–analyst may not have the time or resources to run all of the recommended quality assurance samples. Moreover, certified standards can be expensive. It is acceptable to decrease the analysis frequency for most or all QA/QC samples when analyzing process control samples; however, QA/QC samples should still be analyzed on a regular basis to ensure data quality. At a minimum, a blank and a duplicate should be analyzed with each batch. The operator–analyst must balance the benefit gained by running a particular QA/QC sample against available resources. Process control is only as good as the laboratory data it is based on.

For permit reporting, *Standard Methods* (APHA et al., 2017) recommends the following minimum quality assurance samples:

- Analyze all samples in replicate. Replicate results are averaged together to get a single reportable result for each sample.
- Blanks should always be lower than the MDL. When blanks are higher than this, it is an indication that some or all of the samples in the batch may have been contaminated. A blank should be run once every 20 samples or once per batch, whichever is more frequent.
- The certified standard should agree within 10% of its certified value or within the limits specified on the manufacturer's certificate. A standard should be run once per 20 samples or once per batch, whichever is more frequent.
- Analyze at least 10% of all samples in duplicate. Duplicates should agree within 5% of their average weight.
- It is good laboratory practice to analyze at least one standard and one duplicate for each set or batch of samples, even if fewer than 20 samples are analyzed at a time.

7.8   Sample Collection, Preservation, and Holding Times

Samples for solids analyses should be collected in plastic or glass containers. A preservative is not used. Samples should be cooled to ≤ 6 °C when they are not analyzed immediately. Samples should be analyzed within 24 hours whenever possible, but may be held up to 7 days.

7.9   Procedure

- **Step 1—Clean glassware.**
  1. Wash evaporating dishes and crucibles with warm water and laboratory detergent. Rinse with deionized water and allow to air dry.

2. Mark glassware with nail polish or permanent marker and allow it to dry at room temperature. Special heat-resistant marking pens may also be purchased for this purpose.

3. Preheat clean, porcelain evaporation dishes to 550 °C for 1 hour in a muffle furnace.

4. Remove dish from muffle furnace using tongs.

5. Allow the dish to partially cool in air until most of the heat has dissipated.

6. After dishes are cool enough to handle, place them into a desiccator.

- **Step 2**—Obtain initial dish weight.

1. Dishes must be completely cooled before weighing.

2. Weigh immediately before use.

3. Verify balance calibration with a certified check weight. Some balances are equipped with internal check weights that may be used instead.

4. Weigh the dish and record the weight on the benchsheet to the nearest 0.0001 g.

- **Step 3**—Aliquot sample.

1. Select a sample volume that will leave between 2.5 and 200 mg of solids in the dish at the end of the test.

2. Transfer a well-mixed, measured aliquot of sample to the pre-weighed dish. Mix the sample by stirring or shaking immediately before taking a subsample.

3. For sample volumes smaller than 25 mL, use a class A, TD wide-mouth pipet to measure and transfer the aliquot.

4. For sample volumes larger than 25 mL, use a class A, TC graduated cylinder to measure the sample.

5. Rinse the inside of the graduated cylinder (if used) with deionized water. Add the rinse water to the dish with the sample.

- **Step 4**—Follow either Step 4a or Step 4b.

a) Place the dish on a steam bath and evaporate to dryness. The reason for using the steam bath is that the water in the sample will slowly be steamed away and will not boil. If the liquid were allowed to boil, it might spatter and solids could end up on the bench top instead of in the dish.

b) If a steam bath is not available, the samples can be dried inside a drying oven. First, lower the oven temperature to 2 °C below the boiling point to prevent spattering. At sea level, this will be about

100 °C, so the oven should be set at 98 °C. At higher elevations, the boiling point will be lower. In Denver, Colorado, which is 1-mile high, water boils at 95 °C.

- **Step 5**—Transfer the dish to a drying oven for at least 1 hour at 103 °C. Be sure to use tongs.

- **Step 6**—Dry the sample until constant weight is achieved.

  1. Place dish with the sample into a drying oven.

  2. Dishes may be left in the oven overnight if necessary.

  3. Remove dishes from drying oven and place in a desiccator to cool to room temperature.

- **Step 7**—Weigh the dishes. Record the final weight on the benchsheet. It is important to note that *Standard Methods* (APHA et al., 2017) requires that Steps 6 and 7 be repeated until the weight no longer changes. This is defined as no more than 4% or 50 mg difference between weighings. In practice, the operator–analyst will quickly learn how long it takes for samples to dry to a consistent weight. For process control samples, consistency is more important than following the precise details of the method. If the same drying time is used for process control samples each time they are analyzed, the results from one day can easily be compared to the next.

- **Step 8**—Calculate total solids in milligrams per liter.

- **Step 9**—If VS are needed, place the dish from the previous step in a muffle furnace at 550 °C until the sample is completely burned. The length of time for complete burning (ashing) depends on the size of the sample. For most samples, 15 minutes is typically sufficient to completely ash the sample. It is important to note that *Standard Methods* (APHA et al., 2017) recommends repeated ashing, cooling, and weighing cycles until a constant weight is achieved. This is not necessary for process control samples.

- **Step 10**—Remove dish from muffle furnace. Wear heavy, heat-resistant gloves and use tongs to transfer the dishes from the muffle furnace to a heat-resistant surface. A few ceramic tiles on the bench top will protect the laboratory counter from the heat.

- **Step 11**—Allow the dish to partially cool in air until most of the heat has dissipated.

- **Step 12**—Transfer the dish to a desiccator and allow it to cool to room temperature.

- **Step 13**—Weigh the dish and record the weight to the nearest 0.0001 g (0.1 mg). If the operator–analyst is unsure that constant mass has been obtained, record the weight, then reheat and weigh again until the weight no longer changes.

- **Step 14**—Calculate VS in milligrams per liter.

7.10 Calculations

Calculate milligrams per liter of total solids as follows:

$$\text{Total solids, mg/L} = \frac{(A - B) \cdot 1000 \text{ mL/L}}{\text{Sample volume, mL}} \qquad (3.10)$$

Where
    A = weight of dried solids plus the dish (mg) and
    B = weight of dish (mg).

If desired, the milligrams per liter of volatile solids may be calculated as follows:

$$\text{Volatile solids, mg/L} = \frac{(C - B) \cdot 1000 \text{ mL/L}}{\text{Sample volume, mL}} \qquad (3.11)$$

Where
    C = weight of solids remaining after ignition plus dish after ignition (mg)
       and
    B = weight of dish.

Percent solids and percent volatile solids may be calculated by referring back to Equations 3.8 and 3.9.

7.11 Reporting

Sample results are reported as milligrams per liter with three significant figures. The calculated result of 2432 mg/L is reported as 2430 mg/L and the calculated result of 18.34 mg/L is reported as 18.3 mg/L.

7.12 Troubleshooting and Tips

Nail polish should be used to mark crucibles before firing at 550 °C. This inexpensive shellac will ash and turn black in the muffle furnace but will not flake off during routine testing. Labels are easily removed with a little scrubbing, hot water, and soap. Special heat-resistant marking pens may also be purchased for this purpose. Table 3.9 provides troubleshooting for the total solids test.

TABLE 3.9 Troubleshooting for total solids dried at 103 to 105 °C test.

Problem	Probable cause
Standard result is lower than certified value.	1. Standard not shaken or mixed enough prior to aliquoting. 2. Standard poured too slowly or allowed to sit on bench after mixing. 3. Oven temperature too high, volatilized, or spattered and boiled standard.
Standard result is higher than certified value.	1. Dirty glassware. 2. Oven temperature too low. Trapped water. 3. Standard may have been compromised. If previous uses were out of range low, the standard may have been concentrated in the bottle.
Blank changed by more than 0.5 mg between initial and final weighing.	1. Dirty glassware. 2. Balance not calibrated at initial or final weighing or both.
Sample replicates do not agree within 5% of average weight.	1. Chunky samples. Look for obvious visual their differences between sample aliquots.

7.13 Benchsheet

Table 3.10 shows a sample benchsheet for total solids.

8.0 TOTAL SUSPENDED SOLIDS (TOTAL NONFILTERABLE RESIDUE)

8.1 General Description

Method 2540 D-2015 from *Standard Methods* Online (APHA et. al., 2017) measures suspended solids, also referred to as *nonfilterable residue*. These parameters are determined by filtering a sample through a glass-fiber filter disk.

TABLE 3.10 Sample benchsheet for total solids in water.

Method: _____   Time: _____

Date: _____

Analyst: _____   Location: _____

Oven temperature: _____

		A	B	B	B–A		C	B–C		
Sample date	Sample identification	Dish identification	Sample volume (mL)	Dish weight (g)	Dry weight (g)	TS residue (g)	TS (mg/L)	Ash weight (g)	Ash residue (g)	TVS (mg/L)	QC

8.2   Application

The TSS test is one of the most important tests the operator–analyst can run. The results of the test can be used to estimate process loading to the WRRF as a whole and the efficiency of screening and sedimentation processes; calculate the mean cell residence time (MCRT) or sludge age; and determine the sludge wasting rate, loading to solids handling processes, and removal and capture efficiency of solids handling processes. The TSS test result is also needed to calculate the sludge volume index (SVI). Table 3.11 lists typical ranges of suspended solids concentrations for some wastewater processes.

TABLE 3.11   Total suspended solids concentrations for various wastewater processes.

Process	Total suspended solids, mg/L
Raw domestic wastewater	120 to 400 mg/L
Primary clarifier effluent	Primary clarifiers can remove between 50% and 70% of influent TSS and between 25% and 40% of influent BOD. High-rate clarification can remove up to 95% of influent TSS. Primary sludge is typically between 2% and 8% total solids.
Activated sludge basin	Mixed liquor concentrations vary with the type of process, influent (mixed liquor) wastewater characteristics, and sludge age. Most activated sludge MLSS will be in the range of 1000 to 4000 mg/L. Sequencing batch reactors and extended aeration activated sludge systems may reach 6000 mg/L. Membrane bioreactor facilities can have MLSS concentrations up to 20 000 mg/L.
Return activated sludge and waste activated sludge	RAS and WAS will be thicker than MLSS by a factor of 2 to 5.
Secondary clarifier effluent	Secondary clarifiers are capable of producing treated effluent with solids concentrations well below 30 mg/L when functioning properly. The average solids loading rate (SLR) to a secondary clarifier should not exceed 98 to 146 kg MLSS/m^2·d (20 to 30 lb MLSS/d/sq ft). The maximum SLR should not exceed 195 to 244 kg MLSS/m^2·d (40 to 50 lb MLSS/d/sq ft).
Dissolved air flotation	A typical SLR for a DAFT 10 kg/m^2·h (2 lb/hr/sq ft) when the thickeners solids are preconditioned with polymer. For unconditioned solids, decrease the SLR accordingly.
Belt filter press	SLR for a belt filter press ranges from 150 to 300 kg/m·h (100 to 200 lb/hr/ft).
Mesophilic anaerobic digester	SLR should be held as constant as possible for an anaerobic digester. Keep the volatile solids loading rate between 1.9 and 2.5 kg volatile solids/m^3·d (0.12 to 0.16 lb/d/cu ft).

Source: WEF et al. (2018).

8.3 Interferences

The TSS test suffers from the same interferences as the other solids tests. If too many solids are collected on the filter disk, a water trapping crust may form. For this reason, it is important to limit the amount of material collected to between 2.5 and 200 mg total. If smaller diameter filter disks are used, the amount of sample filtered may need to be decreased because there is less available filter area. As solids accumulate on the available filter area, it will become more difficult to filter additional sample. The operator–analyst should be aware that collecting a large amount of material on a smaller filter area may result in a water-trapping crust even when less than 200 mg of solids are collected.

Samples dried at 103 to 105 °C may contain chemically bound water (water of crystallization) and also some free water trapped under and within dried portions of the sample (mechanically occluded water). This extra water can bias results high. At the same time, some material can be lost during drying by being converted to carbon dioxide.

Nonrepresentative chunks of material may be removed from the sample and/or filter disk at the operator–analyst's discretion. If it is determined that the offending material is not representative (i.e., only a few pieces in the entire 2-L sample), it may be removed.

Samples with visible oil and grease should be mixed well in a blender to disperse the oil and grease prior to taking a sample aliquot.

8.4 Apparatus and Materials

Several different filter apparatus are acceptable for the TSS test. Figure 3.12 shows three different filter types: a magnetic filter funnel, Gooch crucible, and a Buchner funnel. The magnetic filter funnels consist of two easily separated parts, the base and the funnel. The filter disk is supported by a plastic, glass, or ceramic screen. The funnel snaps into place and eliminates the possibility of the sample getting under the filter disk. These funnels may also be purchased as a two-part glass funnel with a large clamp that holds the two pieces together. The magnetic or glass filter funnels are best suited to the TSS test because they take a larger diameter filter disk (4.7 cm) than the older style Gooch crucibles. Larger diameter filter disks can accommodate larger sample volumes without clogging and tend to filter faster than smaller diameter filter disks.

Gooch crucibles may be purchased in a variety of sizes. For the TSS test, *Standard Methods* (APHA et al., 2017) calls for a 25- to 40-mL capacity Gooch crucible. A 25-mL Gooch crucible is 36 mm in diameter at the top and takes a 20-mm diameter filter disk. A 40-mL Gooch crucible is 40 mm

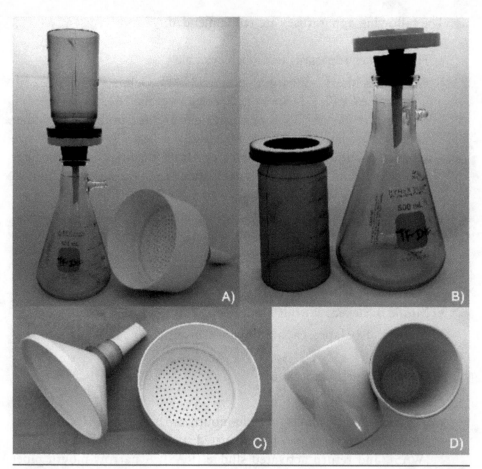

FIGURE 3.12 Filter funnel types. A) filter flask with magnetic funnel, B) disassembled magnetic funnel, C) Buchner funnel, D) Gooch crucible. *Courtesy of Indigo Water Group*

in diameter at the top and takes a 24-mm diameter filter disk. Compared to the 47-mm filter disk commonly used with magnetic or glass filter funnels, the Gooch crucible filter disks provide about 25% of the surface area as the larger filters. Consequently, it is often necessary to reduce the total sample size when using Gooch crucibles. Reducing the sample size can result in nonrepresentative subsamples and can magnify analytical error.

The last filter shown is a Buchner funnel. This type of filter is not appropriate to use for the TSS test. Rather, it is used for rapidly filtering large volumes of sample. It does not work well for the TSS test because the filter disk does not completely cover the bottom of the funnel and some solids may not be captured. It is also difficult to remove filter disks from this type of filter funnel without damaging them. Buchner funnels, however, work well for the TDS test.

The following apparatuses are recommended:

- Filter funnel apparatus or Gooch crucible
- Glass fiber filters—Whatman-grade 934AH, Gelman-type A/E, Millipore- type AP40, E-D Scientific Specialties grade 161; Environmental Express Pro Weigh filters or equivalent are acceptable. Glass fiber filters should have a pore size of 2.0-μm or less. Choose a filter size between 2.2 and 12.5 cm in diameter. Match the filter size to the filter funnel apparatus so the filter covers the entire bottom of the funnel.
- Vacuum flasks, 500 mL or larger
- Vacuum pump or water aspirator
- Drying oven, 103 to 105 °C
- Aluminum weighing pans
- Porcelain evaporating dishes (for VSS only)
- Muffle furnace, controlled at 550 °C ± 50 °C
- Analytical balance capable of measuring to the nearest 0.0001 g (0.1 mg)
- Desiccator with desiccant
- TC graduated cylinders
- TD wide-mouth pipet—a wide-mouth pipet is critical for accurate measurement of samples like activated sludge, waste activated sludge (WAS), and return activated sludge (RAS). A narrow-mouth pipet will be prone to clogging and can shear up the solids in an activated sludge sample
- Pipet bulb

Figure 3.13 shows how two vacuum flasks can be connected in a series to form a water trap between the filter funnel and the vacuum pump. The second flask can be loosely filled with cotton or other absorbent material. The sole purpose of the water trap is to prevent moisture from entering the vacuum pump, where it can cause corrosion and eventual failure of the pump.

8.5   Reagents

No reagents are required for the TSS test other than deionized water.

8.6   Stock and Standards

For permit reporting, a certified standard is required. Standards may be purchased from a variety of sources.

FIGURE 3.13 Figure of filtration setup with water trap for vacuum pump. A) filter funnel, B) primary filter flask, C) secondary filter flask or trap, D) vacuum pump. *Courtesy of Indigo Water Group*

8.7   Quality Assurance and Quality Control

When analyzing samples for process control, the operator–analyst may not have the time or resources to run all of the recommended quality assurance samples. Certified standards can be expensive. It is acceptable to decrease the analysis frequency for most or all QA/QC samples when analyzing process control samples; however, QA/QC samples should still be analyzed on a regular basis to ensure data quality. At a minimum, a blank and a duplicate should be analyzed with each batch. The operator–analyst must balance the benefit gained by running a particular QA/QC sample against the available resources. Process control is only as good as the laboratory data on which it is based

For permit reporting, *Standard Methods* (APHA et al., 2017) recommends the following minimum quality assurance samples:

- Although not required by *Standard Methods*, it is recommended that all samples be analyzed in replicate (two filters per sample). Replicate results are averaged together to get a single reportable result for each sample.

- Blanks should always be lower than the MDL. When blanks are higher than this, it indicates that some or all of the samples in the batch may have been contaminated. A blank should be run once every 20 samples or once per batch, whichever is more frequent. For the TSS test, the filter blank consists of 100 mL of Type III water that is run through a prewashed filter disk and treated as if it were a regular sample. The filter blank initial weight and final weight should not differ by more than 0.5 mg.

- The certified standard should agree within 10% of its certified value or within the limits specified on the manufacturer's certificate. A standard should be run once per 20 samples or once per batch, whichever is more frequent.

- At least 5% of all samples should be analyzed in duplicate, or at least one duplicate sample with each batch of 20 or fewer samples. Duplicates should agree within 10% relative percent difference.

- It is good laboratory practice to analyze at least one standard and one duplicate for each set or batch of samples even if fewer than 20 samples are analyzed at a time.

8.8   Sample Collection, Preservation, and Holding Times

Samples should be collected in glass or plastic bottles. The operator–analyst should ensure that the solids in the sample do not stick to the container walls. In addition, the sample should be refrigerated at ≤ 6 °C until ready for analysis; the samples should be brought to room temperature just before analysis. Ideally, samples should not be held for more than 24 hours before analysis. Samples should never be held longer than 7 days.

8.9   Procedure

This procedure assumes the operator–analyst is using a two-part funnel such as a magnetic funnel. If Gooch crucibles are used, the crucibles and the filters they contain remain together throughout all steps of the procedure. Filters are not removed from Gooch crucibles for weighing.

- **Step 1**—Prewash the filter. The operator–analyst should note the following about Step 1:
  - The prewashing step removes loose filter material. This step is critical for obtaining accurate and precise results. If the filters are not prewashed, the final results may be biased low. Always include a filter blank, even when analyzing process control samples.

- If prewashed glass fiber filter disks are used, eliminate this step.
- Two glass fiber filter disks should be washed for each sample to be analyzed. Consider preparing additional filters in case they are needed.

1. Place a new, clean filter disk on the funnel stand. Use tweezers as shown in Figure 3.14. Touching the filter disk may contaminate it with oils and dirt from your fingers. Magnetic filter funnels or Gooch crucibles may be used for this step. If Gooch crucibles are used, the filter remains in the crucible throughout the entire procedure.

2. Add funnel top and secure.

3. Add 20 mL of laboratory-grade water to the filter funnel to wash the filter. Repeat this step two more times for a total of 60 mL of rinse water. Be sure to allow the filter to drain completely before adding the next aliquot.

4. Remove the funnel and inspect the filter disk visually to verify that it is as dry as possible before removing it. If the disk it too wet, it will stick to the bottom of the drying pan. This may cause part of the filter to tear off after drying.

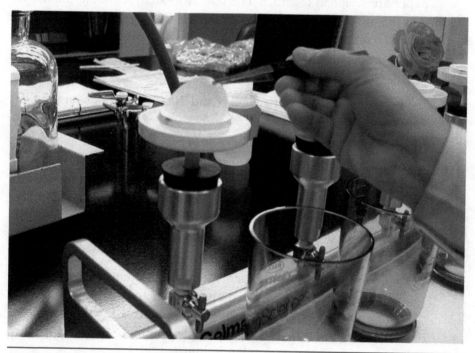

FIGURE 3.14 Remove dry filter disk with tweezers. *Courtesy of Indigo Water Group*

5. Use tweezers to carefully remove the washed filter disk and place it into a numbered aluminum pan.

6. Angle the filter disk so that as little of the filter disk is touching the pan as possible.

- **Step 2**—Dry filters until a constant weight is reached.

  1. Dry the filters in their numbered pans at 103 to 105 °C for at least 1 hour.

  2. Remove filters and pans from the drying oven.

  3. Place filters and pans into the desiccator to cool.

  4. Ensure that the filter disks are completely cool before proceeding to the next step. Many laboratories prewash filters the night before and leave them in the desiccator until morning. That way, they are always ready to use.

- **Step 3**—Record the initial filter weight to the nearest 0.0001 g (0.1 mg) and weigh using an analytical balance. Record the weight in grams on the data sheet and place the filter disk in a numbered aluminum pan. Weigh only the filters, not the support pans or dishes, unless a Gooch crucible is used in which case the crucible and filter paper are weighed together. The operator–analyst should make sure that the number on the data sheet corresponds to the number on the pan for each filter disk.

- **Step 4**—Aliquot samples. The operator–analyst should note the following about Step 4:

  ○ The goal is to have between 2.5 and 200 mg of solids on the filter disk.

  ○ Because excessive residue on the filter may form a water-entrapping crust, limit the sample size to that yielding no more than 200 mg of residue. If this requirement is met, the samples should not take more than a few minutes to filter. For sample volumes of 25 mL or greater, a graduated cylinder should be used. Sample volumes less than 25 mL should be measured using a wide-mouth pipet.

  ○ Suggested sample volumes are listed in Table 3.12. These volumes are only a starting point. Each WRRF is different, and sample volumes may need to be adjusted up or down accordingly. *Remember to always use the largest volume of sample that will easily pass through the filter disk.* Larger sample volumes are more likely to be representative than smaller volumes.

    1. Place each filter disk in consecutive order onto the vacuum filtration apparatus.

TABLE 3.12 Suggested sample volumes for TSS test.

Sample type	Suggested volume
Facility influent	25 to 50 mL
Primary clarifier influent and effluent	25 to 100 mL
Trickling filter influent and effluent	25 to 100 mL
Aeration tank (MLSS)	5 to 10 mL
Return activated sludge	3 to 5 mL
Waste activated sludge	3 to 5 mL
Secondary clarifier effluent	200 to 1000 mL
Tertiary unit processes	200 to 1000 mL
Facility final effluent	250 to 1000 mL

2. Place a magnetic filter funnel over each filter.

3. Add the sample to the filter funnel.

   a) Filter the blanks and the standard first. The next set of filters should be the final effluent samples. Filter all liquid samples before starting on the thick sludges. Moving from clean to dirty samples will help prevent cross-contamination problems.

   b) Thoroughly mix or shake each sample immediately before taking an aliquot. Pour the sample volumes according to the data sheet through the corresponding filter disk and apply suction (see Figure 3.15). If a sample looks as if it has substantially more or less particulates in it than normal, adjust the aliquot size accordingly and note the new aliquot size on the data sheet.

- Step 5—If a graduated cylinder was used to measure the sample, rinse the inside of the graduated cylinder with Type III water as shown in Figure 3.16. Add the rinsate to the filter funnel with the sample.

- Step 6—If a pipet is used to aliquot the sample, care should be taken to avoid touching the pipet to the sides of the filter funnel while it is draining. Be sure to completely cover the filter disk with sample. If dry spots are left on the filter disk, there will not be enough suction to completely dry the filter disk. Air will simply pass through the drier portions of the disk. To deliver pipets should be gently touched to the side of the filter near the filter disk for 2 seconds after they have drained. This will transfer the remaining sample to the filter. Rinse the pipet to transfer any remaining solids to the filter disk.

FIGURE 3.15   Add sample to filter funnel. *Courtesy of Indigo Water Group*

FIGURE 3.16   Rinse inside of graduated cylinder. *Courtesy of Indigo Water Group*

- ○ Note: When filtering MLSS samples for process control, the analyst may prefer to blow out the pipet rather than touch it to the side of the funnel and skip rinsing the filter disk in Step 7. This may be done for process control samples only and not for compliance reporting.

- **Step 7**—Once the sample has completely passed through the filter, rinse the filter disk with three portions of 10 mL of Type III water. Wait between each rinse until the water has completely drained. It is important to note that for high solids process control samples like MLSS that were aliquoted with a pipet, it may not be desirable to wash the funnel and filter disk. For these samples, it is likely that the water will pond on top of the filter disk instead of passing through. Some samples for which washing the filter disk may cause difficulties include MLSS, RAS, and WAS.

- **Step 8**—Carefully remove the top of the filter funnel. Visually inspect the filter disk to ensure that it is dry. It may be necessary to wait for several minutes to get the filter as dry as possible.

- **Step 9**—Remove the filter disk with tweezers and return it to the appropriately numbered aluminum pan. (In Figure 3.17, the laboratory depicted uses metal trays instead of aluminum pans.)

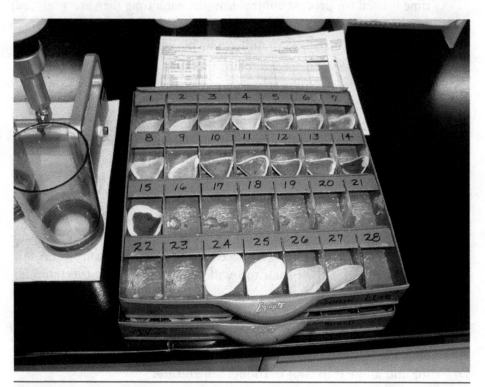

FIGURE 3.17   Filter disks in numbered pans. *Courtesy of Indigo Water Group*

- Step 10—Repeat the aforementioned steps for all remaining samples.
- Step 11—Dry filters until a constant weight is reached:
  1. Dry the filters in their numbered pans at 103 to 105 °C for at least 1 hour.
  2. Remove filters and pans from drying oven.
  3. Place filters and pans into the desiccator to cool.
  4. Ensure that the filter disks are completely cool before proceeding on to the next step.
- Step 12—Record the final filter weight to the nearest 0.0001 g (0.1 mg) and weigh using an analytical balance. Record the weight in grams on the data sheet and place the filter disk in a numbered aluminum pan. Make sure that the number on the data sheet corresponds with the number on the pan for each filter disk. *Standard Methods* (APHA et al., 2017) requires that Steps 11 and 12 be repeated until the weight no longer changes. This is defined as no more than 4% or 0.5 mg difference between weighings. In practice, the operator–analyst will quickly learn how long it takes for samples to dry to a *consistent weight*. For process control samples, *consistency* is more important than following the precise details of the method. If the same drying time is used for process control samples each time they are analyzed, results from one day can easily be compared to the next.
- Step 13—Calculate TSS in milligrams per liter.
- Step 14—If VSS are needed, place the filter and pan (or Gooch crucible) from the previous step in a muffle furnace at 550 °C until the sample is completely burned. The length of time for complete burning (ashing) depends on the size of the sample. For most samples, 15 minutes is typically sufficient to completely ash the sample. *Standard Methods* (APHA et al., 2017) recommends repeated ashing, cooling, and weighing cycles until a constant weight is achieved. This is not necessary for process control samples.
- Step 15—Remove filter and pan (or Gooch crucible) from muffle furnace. Wear heavy, heat-resistant gloves and use tongs to transfer the dishes from the muffle furnace to a heat-resistant surface. A few ceramic tiles on the bench top will protect the laboratory counter from the heat.
- Step 16—Allow the filter and pan (or Gooch crucible) to partially cool in air until most of the heat has dissipated.
- Step 17—Transfer the filter and pan (or Gooch crucible) to a desiccator and allow it to cool to room temperature.
- Step 18—Weigh the filter (or Gooch crucible) and record the weight to the nearest 0.0001 g (0.1 mg). If unsure that constant mass has

been obtained, record the weight, then reheat and weigh again until the weight no longer changes.

- **Step 19**—Calculate VSS in milligrams per liter.

8.10   Calculations

$$\text{Total suspended solids, mg/L} = \frac{(A - B) \cdot 1000}{\text{Sample volume, mL}} \qquad (3.12)$$

Where
  A = weight of filter plus dried residue (mg) and
  B = weight of filter (mg).

$$\text{Volatile suspended, solids mg/L} = \frac{(A - C) \cdot 1000}{\text{Sample volume, mL}} \qquad (3.13)$$

Where
  A = weight of filter plus dried residue (mg) before ashing and
  B = weight of filter plus dried residue (mg) after ashing.

Percent solids and percent volatile solids may be calculated by referring back to Equations 3.8 and 3.9.

8.11   Reporting

Sample results are reported as milligrams per liter with three significant figures. The calculated result of 2432 mg/L is reported as 2430 mg/L and the calculated result of 18.34 mg/L is reported as 18.3 mg/L.

8.12   Troubleshooting and Tips

Table 3.13 contains troubleshooting tips for the TSS test.

8.13   Benchsheet

Table 3.14 shows a sample benchsheet for TSS.

9.0   TOTAL DISSOLVED SOLIDS DRIED AT 180 °C

9.1   General Description

In Method 2540 C-2015 of *Standard Methods* Online (APHA et al., 2017), samples are filtered through a glass fiber filter. The filtrate (i.e., the water that passes through the filter disk) is collected and dried at 180 °C. The solid

TABLE 3.13 Troubleshooting for TSS test.

Problem	Probable cause
Standard result is lower than certified value.	1. Standard not shaken or mixed enough prior to aliquoting.
	2. Standard poured too slowly or allowed to sit on bench after mixing.
	3. Hole in filter disk.
Standard result is higher than certified value.	1. Filter disks not dried long enough.
	2. Dirty glassware.
	3. Standard may have been compromised. If previous uses were out of range low, the standard may have been concentrated in the bottle.
Filter blank changed by more than 0.5 mg between initial weighing.	1. Filter disk not prewashed.
	2. Increasing weight may indicate (a) inadequate filter drying time and final or (b) dirty glassware contaminated filter.
	3. Decreasing weight may mean that the filters were not prewashed enough. Additional filter material was lost during analysis.
	4. Loss of filter material—possibly stuck to drying pan.
	5. Balance not calibrated at initial or final weighing or both.
Sample replicates do not agree within 5% of their average weight.	1. Chunky samples. Look for obvious visual differences between filter disks.

material remaining after the sample is dried is TDS. Total dissolved solids consist primarily of dissolved salts and colloidal (i.e., very small) particles.

If TSS is also being determined, a separate aliquot of sample should be analyzed using procedure for TSS procedure described in the previous section (Section 8.0).

9.2 Application

This method is appropriate for measuring TDS concentrations between 10 and 20 000 mg/L.

TABLE 3.14 Sample benchsheet for total suspended solids.

Method: _____ Time: _____

Date: _____

Analyst: _____ Location: _____

Oven temperature: _____

Sample date	Sample identification	Dish identification	Sample volume (mL)	A Tare weight (g)	B Dry weight (g)	B–A TSS residue (g)	TSS (mg/L)	C Ash weight (g)	B–C Ash residue (g)	TVSS (mg/L)	QC

9.3   Interferences

Highly mineralized waters that contain high concentrations of calcium, magnesium, chloride, and/or sulfate may be hygroscopic and will require prolonged drying, desiccation, and rapid weighing.

Samples with high concentrations of bicarbonate will require careful and possibly prolonged drying at 180 °C to ensure that all of the bicarbonate is converted to carbonate.

If there are too many dried solids in the evaporating dish, they may crust over and trap water mechanically. This water will not be driven off during drying. Total solids should be limited to about 200 mg.

9.4   Apparatus and Materials

The TDS method uses the same apparatus and materials as the TSS method (see Section 8.4 of this chapter).

9.5   Reagents

No reagents are required for this method other than Type III water.

9.6   Stock and Standards

For permit reporting, a certified standard will be needed. Certified standards may be purchased from a variety of sources.

9.7   Sample Collection, Preservation, and Holding Times

Samples should be collected in plastic or glass containers. If samples are not analyzed immediately, they should be cooled to ≤ 6 °C. There is no preservative other than cooling. Samples should be analyzed within 24 hours of collection and should not be held longer than 7 days under any circumstances.

9.8   Quality Assurance and Quality Control

When analyzing samples for process control, the operator–analyst may not have the time or resources to run all of the recommended quality assurance samples. Certified standards can be expensive. It is acceptable to decrease the analysis frequency for most or all QA/QC samples when analyzing process control samples; however, QA/QC samples should still be analyzed on a regular basis to ensure data quality. At a minimum, a blank and a duplicate should be analyzed with each batch. The operator–analyst must balance the benefit gained by running a particular QA/QC sample against available resources. Process control is only as good as the laboratory data on which it is based. For permit reporting, *Standard Methods* (APHA et al., 2017) recommends the following minimum quality assurance samples:

- Analyze all samples in replicate. Replicate results are averaged together to get a single reportable result for each sample.
- Blanks should always be lower than the MDL. When blanks are higher than this, it indicates that some or all of the samples in the batch may have been contaminated. A blank should be run once every 20 samples or once per batch, whichever is more frequent. For the TDS test, the blank consists of 100 mL of deionized water that is treated as if it were a regular sample. The initial dish weight and final dish weight should not differ by more than 0.5 mg or 4%, whichever is less.
- The certified standard should agree within 10% of its certified value or within the limits specified on the manufacturer's certificate. A standard should be run once per 20 samples or once per batch, whichever is more frequent.
- Analyze at least 10% of all samples in duplicate. Duplicates should agree within 5% of their average weight.
- It is good laboratory practice to analyze at least one standard and one duplicate for each set or batch of samples even if fewer than 20 samples are analyzed at a time.

9.9   Procedure

- **Step 1**—Clean the glassware:
  1. Wash all glassware, including evaporating dishes, with warm water and laboratory detergent.
  2. Rinse glassware with Type III water.
  3. Use immediately or air dry.
- **Step 2**—Prepare evaporating dishes:
  1. If only TDS will be determined, place the evaporating dish into a drying oven at 180 °C for 1 hour.
  2. If total volatile dissolved solids (TVDS) will be determined, place the evaporating dishes in the muffle furnace at 550 °C for 1 hour.
  3. Remove dishes from the oven or muffle furnace. Allow them to cool on a bench top until most of the heat is dissipated.
  4. Transfer dishes to a desiccator and store them until they are ready to use. Ensure that the dishes are completely cool before proceeding to Step 3.

- **Step 3**—Weigh evaporating dishes and record the weight on the benchsheet to the nearest 0.0001 g (0.1 mg).
- **Step 4**—Assemble the filtration apparatus. Set up the filter apparatus and water trap as described in Section 8.4.
- **Step 5**—Place a new, unused piece of filter disk on the filter apparatus.
- **Step 6**—Replace the filter funnel and secure it, if necessary.
- **Step 7**—While vacuum is applied, wash the filter disk with three successive 20-mL volumes of Type III water. Allow the water to completely drain between aliquots.
- **Step 8**—Remove the filter flask and discard the wash water.
- **Step 9**—Replace the filter flask.
- **Step 10**—Aliquot the sample:
  1. Select a sample volume that will leave between 2.5 and 200 mg of solids in the evaporating dish after drying. Total dissolved solids in raw wastewater range between 120 mg/L for low-strength domestic wastewater up to 860 mg/L for high-strength domestic wastewater (Metcalf & Eddy, Inc./AECOM, 2014). This translates to a sample volume between 1 L and 230 mL. Industrial wastewaters may have much higher or lower TDS concentrations.
  2. If more than 10 minutes are required to filter the sample, either the filter size should be increased or the sample volume should be decreased.
  3. Vigorously shake the sample.
  4. Measure the desired volume with a graduated cylinder.
  5. Add the measured sample to the filter funnel.
- **Step 11**—Rinse the graduated cylinder with Type III water. Add the rinsate to the filter funnel with the sample.
- **Step 12**—After the sample has completely passed through the filter, rinse the sides of the funnel and the filter with 10 mL of Type III water. Wait until the rinse water has passed completely through the filter.
- **Step 13**—Repeat Step 12 two additional times for a total of 30 mL of rinse water.
- **Step 14**—Remove the filter flask from the filtration assembly.
- **Step 15**—Transfer the total filtrate to a preweighed evaporating dish.
- **Step 16**—Rinse the filter flask with Type III water. Add the rinse water to the evaporating dish.
- **Step 17**—Repeat Steps 4 through 16 for all remaining samples.

- **Step 18**—Transfer all evaporating dishes to a drying oven or steam bath and evaporate to dryness. Adjust the temperature of the oven to a few degrees below the boiling point of the samples to prevent spattering.
- **Step 19**—When the dishes are completely dry, transfer them to a drying oven at 180 °C.
- **Step 20**—Dry for at least 1 hour.
- **Step 21**—Remove evaporating dishes from the drying oven and place them into a desiccator. Wait until they are completely cool before proceeding to Step 22.
- **Step 22**—Weigh each dish on an analytical balance and record the weight on the benchsheet to the nearest 0.0001 g (0.1 mg). *Standard Methods* (APHA et al., 2017) requires that Steps 21 and 22 be repeated until the weight no longer changes. This is defined as no more than 4% or 0.5 mg difference between weighings. In practice, the operator–analyst will quickly learn how long it takes for samples to dry to a consistent weight. For process control samples, consistency is more important than following the precise details of the method. If the same drying time is used for process control samples each time they are analyzed, results from one day can easily be compared to the next.
- **Step 23**—If volatile solids are needed, place the dish from the previous step in a muffle furnace at 550 °C until the sample is completely burned. The length of time for complete burning (ashing) depends on the size of the sample. For most samples, 15 minutes is typically sufficient to completely ash the sample. *Standard Methods* (APHA et al., 2017) recommends repeated ashing, cooling, and weighing cycles until a constant weight is achieved. This is not necessary for process control samples.
- **Step 24**—Remove the dish from the muffle furnace. Wear heavy, heat-resistant gloves and use tongs to transfer the dishes from the muffle furnace to a heat-resistant surface. A few ceramic tiles on the bench top will protect the laboratory counter from the heat.
- **Step 25**—Allow the dish to partially cool in air until most of the heat has dissipated.
- **Step 26**—Transfer the dish to a desiccator and allow it to cool to room temperature.
- **Step 27**—Weigh the dish and record the weight to the nearest 0.0001 g (0.1 mg).
- **Step 28**—Calculate the volatile solids in milligrams per liter.

9.10 Calculations

$$\text{Total dissolved solids,} \frac{\text{mg}}{\text{L}} = \frac{(A - B) \cdot 1000}{\text{Sample volume, mL}} \tag{3.14}$$

Where
   A = weight of dried solids plus the dish (mg) and
   B = weigh of dish (mg).

9.11 Reporting

Sample results are reported as milligrams per liter with three significant figures. The calculated result of 2432 mg/L is reported as 2430 mg/L and the calculated result of 18.34 mg/L is reported as 18.3 mg/L.

9.12 Troubleshooting and Tips

Table 3.15 contains troubleshooting tips for the TDS test.

TABLE 3.15 Troubleshooting for TDS test.

Problem	Probable cause
Standard result is lower than certified value.	1. Standard not shaken or mixed enough prior to aliquoting. 2. Standard poured too slowly or allowed to sit on bench after mixing. 3. Spattering of sample during dry step caused loss of sample volume.
Standard result is higher than certified value.	1. Dirty glassware. 2. Samples not dried long enough. 3. Oven temperature too low. 4. Standard may have been compromised. If previous uses were out of range low, the standard may have been concentrated in the bottle.

Blank changed by more than 0.5 mg between initial and final weighing.	1. Increasing weight may indicate
	  (a) inadequate drying time, (b) dirty glassware, (c) particulate matter from filter disk remaining in filter flask, or (d) initial weigh done on warm dish.
	2. Decreasing weight may indicate
	  (a) final weigh done on warm dish, (b) chip in sample dish (visual inspection), or (c) volatilization of residue present during initial weighing.

Sample replicates do not agree within 5% of their average weight.	Differences in sample volumes used.

9.13   Benchsheet

Table 3.16 shows a sample benchsheet for TDS.

10.0   VOLATILE AND FIXED SOLIDS IGNITED AT 550 °C

Method 2540 F-2015 of *Standard Methods* (APHA et al., 2017) includes volatile and fixed solids methods for the following:

- Total, fixed, and volatile solids in solid and semisolid samples (see Section 6.0 in this chapter);
- Total solids dried at 103 to 105 °C (see Section 7.0); and
- Suspended solids (total nonfilterable residue) (see Section 8.0).

11.0   SETTLEABILITY

The following sections comprise an expanded version of Method 2710 C-2011 of *Standard Methods*, 23rd ed. (APHA et al., 2017). This procedure is used for process control purposes only and is not listed as an approved reporting method in 40 CFR Part 136 (U.S. EPA, 2017). Method 2710 C-2011 recommends the use of a 1-L graduated cylinder, however; a 2-L Mallory-style settleometer is preferred for determining the settleability of MLSS.

11.1   General Description

The main purpose of the settleometer test is to give the operator–analyst an idea of how the sludge is settling in the secondary clarifier. The settleometer test simulates how activated sludge settles in the secondary clarifier. The settleometer test should be performed daily at activated sludge facilities. Information that should be recorded for long-term trend tracking includes the settled sludge volume (SSV) at 5 minutes, SSV at 30 minutes, and the time if and when the blanket "popped."

11.2   Apparatus and Materials

Three pieces of equipment are necessary to run settleability tests: a settleometer, a paddle, and a timer. The equipment is described as follows:

- Mallory settleometers—three, 2-L Mallory settleometers are suggested, one Mallory settleometer for the regular test, the second for dilution, and the third for any special test that may need to be run

TABLE 3.16 Sample benchsheet for total dissolved solids.

Method: _____

Date: _____ Time: _____

Analyst: _____ Location: _____

Oven temperature: _____

			A	B	B–A		C	B–C		
Sample date	Sample identification	Dish identification	Sample volume (mL)	Tare weight (g)	Dry weight (g)	TDS residue (g)	TDS (mg/L)	Ash weight (g)	Ash residue (g)	TDVS (mg/L)	QC

such as an extended-rise test or multiple dilutions. If Mallory set-tleometers are not available, 2-L beakers or similar devices may be substituted. A standard graduated cylinder should not be used because the narrow cylinder reduces settling velocity caused by excess friction with the walls. The base of the settleometer, where the wall meets the floor, should be square and not rounded.

- Paddle—a wide paddle is needed to gently stir, then quiet, the mixed liquor in the settleometer at the beginning of the test. The paddle should be made of smooth acrylic or a similar material, with a width slightly less than the diameter of the settleometer.

- Timer—the timer should be capable of sounding an alarm at the end of 5-minute intervals.

11.3   Sample Collection, Preservation, and Holding Times

It is important that mixed liquor samples collected for the settleability test are representative of mixed liquor flowing out of the aeration basin and into the secondary clarifier. Generally, this sample is collected either in a weir overflow box, or the secondary clarifier influent splitter structure, or at a similar point before its entry into secondary clarifiers. If the mixed liquor comes from two or more aeration basins, the operator–analyst should be sure to sample the mixed liquor after the sources have combined. However, it is also a good idea to periodically measure settleability in each individual aeration basin to check for differences in sludge condition between the basins. Finally, it is important to remember to collect the mixed liquor samples and not the scum and foam that may be present on the mixed liquor.

The necessity for collecting mixed liquor samples in wide-mouthed containers and starting the settleometer testing as soon as possible cannot be overemphasized. Indeed, testing should begin within 10 minutes of collecting. In addition, shaking and agitation of the samples should be minimized, especially during transportation to the testing site.

11.4   Procedure

Test conditions and sampling techniques can strongly influence test results. As such, the test should be set up in a location that is free from vibration and direct sunlight. If possible, the temperature should not vary a great deal from one test to another. The following steps should be followed for the settleability procedure:

- **Step 1**—Mix the mixed liquor sample carefully, then pour it into the settleometer carefully and rapidly with the least possible amount of additional aeration or turbulence. Stir the settleometer contents gently with a wide paddle to ensure a thorough mixing, then stop all movement of the mixed liquor with the paddle.

- **Step 2**—As the paddle is smoothly and carefully removed from the settleometer, start the timer that was previously set for 5-minute intervals. It is important to not leave the settleometer unattended, or the operator–analyst will miss out on some of the most important information to be gained from the settleometer test.

- **Step 3**—During the initial 5 minutes, the operator–analyst should observe the visual characteristics of the settling sludge. A conscientious operator–analyst will critically observe how the sludge particles agglomerate while forming the blanket. He or she will also observe whether the sludge compacts slowly and uniformly while squeezing clear liquid from the sludge mass, or whether tightly knotted sludge particles are simply falling down through the turbid effluent. The operator–analyst will observe how much and what type of straggler floc, if any, remains in the supernatant liquor above the main sludge mass.

- **Step 4**—The importance of conscientious, perceptive observation during the first 5 minutes cannot be overemphasized. During these important 5 minutes, the operator–analyst will acquire additional insight into sludge character and quality and will be in a much better position to evaluate what the settleometer test reveals.

- **Step 5**—At the end of the first 5 minutes, the operator–analyst should read and record the volume of the settleometer occupied by settled sludge. This is the most important reading taken during the test.

- **Step 6**—The operator–analyst should continue to read and record settled SSV at every 5-minute interval for the first 30 minutes and at 10-minute intervals for the second 30 minutes of the 1-hour test. The 60-minute observation is the most important number in the time series. It provides the check of final clarifier sludge blanket characteristic and is used to calculate operation control adjustments.

- **Step 7**—The operator–analyst should continue taking SSV readings until SSV changes become very small from reading to reading. If a very slow settling sludge (i.e., SSV greater than 900) is being tested, readings at 90, 120, 150, 180, and 240 minutes may be required.

- **Step 8**—After the last reading has been taken, the sample should be allowed to stand for several hours more. The operator–analyst should record the time at which previously settled sludge starts to swell and or rise to the surface of the settleometer. It is important to wash the settleometer and paddle with mild soap and to thoroughly rinse and dry it after the test.

11.5 Special Variations or Applications

There are essentially two variations of the test, both of which have been alluded to previously. The first is the rise test, which indicates that denitrification might be occurring in the secondary clarifier. The second is the dilution test for determining sludge age.

11.5.1 Rise Test

After conducting the standard 30 minute settleometer test, allow the beaker with the sample to set for several hours and observe the time that the sludge rises. This is an indication of denitrification. If the sludge rises too quickly (e.g., less than 90 minutes), observe the depth of sludge in the secondary clarifier and the sludge detention time. The lower the overflow rate, the higher the detention time. If this condition is occurring, pay close attention to the sludge blanket level in the core sampler. See if it is rising or expanded.

11.5.2 Dilution Test

The dilution test is conducted to determine whether there are too many solids (hindered settling) or too few solids in the aeration tank. It is important to rely on other tests, including visual observations and microbial tests, to make a judgment on necessary process control adjustments. The operator–analyst should run three settleometer tests at the same time. Undiluted MLSS should be placed in the first beaker, 50% MLSS and 50% secondary effluent in the second, and 25% MLSS and 75% secondary effluent in the third. The operator–analyst should conduct 60-minute settleabilities, prepare settling curves, and compare slopes. If the slopes are the same or parallel, this probably indicates settling problems caused by filamentous sludge. If an increase in settling rate occurs with the diluted samples (greater slope or rise over run), the MLSS is probably too high and wasting should be increased. The increase should not be drastic, but incremental each day. The operator–analyst should try gradually increasing the wasting over a 3-day period until the desired MCRT or food-to-microorganism ratio (F:M) and settling rates are achieved.

11.6 Calculations

- Observation of settling for the first 5 minutes
- Settling rate (volume of settled sludge vs. time)

- Time to rise
- Comparison of settling rates, which provides information to evaluate the age of the sludge, if the dilution test is being conducted.

11.7 Benchsheets

Tables 3.17, 3.18, and 3.19 provide sample benchsheets.

TABLE 3.17   Data collection sheet for settleometer test.

Location: _____

Method: _____

Date: _____
Sample ID: _____
Operator-Analyst: _____
Time of Test: _____   ATC (%) _____

$$SSC = \frac{(ATC) \times (1000)}{SSV}$$

Time	SSV, mL/L	SSC, %
0	1000	
5		
10		
15		
20		
25		
30		
35		
40		
45		
50		
55		
60		

Observations in the first 5 minutes:

Floc
- [ ] Granular
- [ ] Fluffy
- [ ] Compact
- [ ] Feathery

Particle Size
- [ ] Large
- [ ] Medium
- [ ] Small

Bio-solids/ Supernate Interface
- [ ] Well Defined
- [ ] Ragged

Supernatant
- [ ] Clear
- [ ] Cloudy

Straggler Floc
- [ ] Yes
- [ ] No

Observation after 30 minutes:
- [ ] Crisp/Sharp Edges
- [ ] Sponge-like
- [ ] Fluffy/Feathery
- [ ] Homogeneous

Date: _____
Sample ID: _____
Operator-Analyst: _____
Time of Test: _____   ATC (%) _____

$$SSC = \frac{(ATC) \times (1000)}{SSV}$$

Time	SSV, mL/L	SSC, %
0	1000	
5		
10		
15		
20		
25		
30		
35		
40		
45		
50		
55		
60		

Observations in the first 5 minutes:

Floc
- [ ] Granular
- [ ] Fluffy
- [ ] Compact
- [ ] Feathery

Particle Size
- [ ] Large
- [ ] Medium
- [ ] Small

Bio-solids/ Supernate Interface
- [ ] Well Defined
- [ ] Ragged

Supernatant
- [ ] Clear
- [ ] Cloudy

Straggler Floc
- [ ] Yes
- [ ] No

Observation after 30 minutes:
- [ ] Crisp/Sharp Edges
- [ ] Sponge-like
- [ ] Fluffy/Feathery
- [ ] Homogeneous

TABLE 3.18   Settling graph.

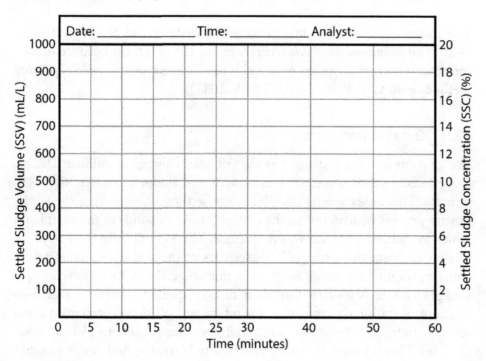

TABLE 3.19   Example settling graph.

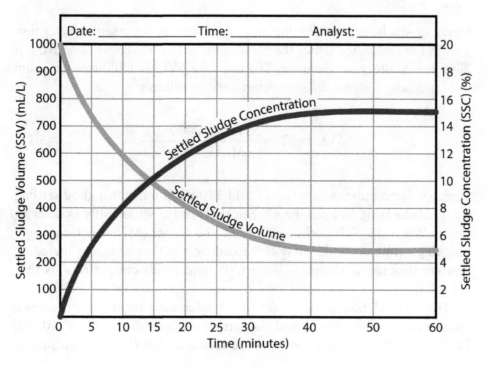

12.0 SLUDGE VOLUME INDEX

The following sections comprise an expanded version of Method 2710 D-2011 of *Standard Methods* (APHA et al., 2017). This procedure is used for process control purposes only and is not listed as an approved reporting method in 40 CFR Part 136 (U.S. EPA, 2017).

12.1 General Description

Sludge volume index is defined as the volume of sludge in milliliters (cubic inches) occupied by 1 g (0.04 oz.) of activated sludge after settling for 30 minutes. This index relates the 30-minute settling volume from the settleometer process control test to the concentration of solids in the sample on which the settleometer test was performed. The SVI will help the operator–analyst evaluate the settling characteristics of the activated sludge as the concentration of the solids in the system change. All WRRFs have unique sludge characteristics based on their design, specific influent characteristics, and operational parameters. Therefore, an SVI that represents a good settling sludge at one facility is generally not comparable to SVIs at other facilities. This is especially true if the facility is treating industrial or other unusual wastes.

12.2 Application

Operator–analysts often develop an SVI number from the settleometer test. The SVI is calculated using the 30-minute settled sludge volume from a 2000-mL Mallory settleometer. The units for SVI are milliliters per gram (mL/g). Sludge volume index is determined as follows:

$$SVI, \text{ mg/L} = \frac{SSV_{30} \times 1000 \text{ mg/g}}{MLSS \times \text{mg/L}} \qquad (3.15)$$

Sludge volume indexes between 75 and 150 mL/g indicate a good settling sludge. Increasing SVI numbers mean that sludge settleability is deteriorating. When the SVI reaches 200 mL/g, the sludge (MLSS) is technically "bulking." Bulking sludge may be caused by a variety of things including nutrient deficiency, filamentous bacteria, and solids concentrations that are too high.

Using the Mallory settleometer and the aforementioned test procedures, the settleometer test is run and the settled sludge volumes for 5, 10, 15, 20, 25, 30, 40, 50, and 60 minutes are recorded. For slow-settling sludge, additional readings may be required at the 90-, 120-, 150-, and 180-minute points or for as long as the sludge continues to settle. The corresponding settled sludge concentration (SSC) at each time is then calculated as follows:

$$SSC, \text{ mg/L} = \frac{MLSS \text{ mg/L} \times 1000, \text{ mL/L}}{SSV \text{ mL}} \qquad (3.16)$$

As SSV gets smaller with time, SSC gets larger as the sludge concentrates in the bottom of the settleometer. In an old, overoxidized sludge that settles quickly, maximum compaction occurs within about one-half hour. A medium-settleability sludge will typically compact to maximum within 1 hour. A young, under-oxidized or filamentous sludge may not compact maximally until 2 to 3 hours of settling time.

The settleometer test can be used to develop additional information about the system. If the settleometer is allowed to sit idly after the settling test has been completed, it can be determined whether or not significant nitrification, and the potential for denitrification, is occurring. As the settled sludge sits in the settleometer, the microorganisms run out of oxygen quickly; they begin to denitrify, converting nitrate to nitrogen gas. If the concentration of nitrate is high enough, sufficient nitrogen gas will be released to float the settled blanket or at least split the blanket and send some portion of it back to the top of the settleometer.

12.3 Interferences

It is important for the operator–analyst to note that the SVI calculation is only valid for MLSS concentrations with a normal range of 1000 to 4000 mg/L for activated sludge. Outside of this range, the SVI calculation can give meaningless results. For example, a MLSS with a concentration of 8000 mg/L and a 30-minute SSV of 950 mL will have an SVI of 119 mL/g. If the operator–analyst only considered the SVI number, he or she might think that the sludge was settling well, which is not the case. Conversely, this sludge is hardly settling at all. When creating daily log sheets for the settleometer and SVI tests, operator–analysts are encouraged to record the MLSS concentration, SSV, and SVI. These numbers are all meaningless separate from one another.

12.4 Required Data

- Thirty-minute settling volume in milliliters per liter from the settleometer test, and
- Mixed liquor suspended solids concentrations in milligrams per liter for the same sample on which the settleometer test was performed.

12.5 Calculations

The following equation should be used to calculate SVI:

$$SVI, \text{ mg/g} = \frac{SSV_{30} \times 1000 \text{ mg/g}}{MLSS, \text{ mg/L}} \tag{3.17}$$

Where

SSV_{30} = settled sludge volume after 30 minutes of settling and
MLSS = starting MLSS concentration.

Table 3.20 provides an SVI worksheet that can be used to calculate SVI.

12.6 Reporting

There are no reporting guidelines for SVI; this test is for process control only.

13.0 CENTRIFUGE METHOD FOR ESTIMATING SUSPENDED MATTER

13.1 General Description

In 1974, U.S. EPA developed the centrifuge method for estimating suspended solids in wastewater as a substitute for the gravimetric determination of TSS (West, 1974). This technique was intended to reduce the time and equipment required for process control of activated sludge systems and the skill level needed to operate a WRRF. The result of this analysis is given in percent solids by volume and is used to calculate solids inventory in "sludge units" by multiplying by tank volume. Sludge units then become a substitute for kilograms (pounds) of solids.

This method is empirical in nature, dependent on specific sludge characteristics, and subject to several limitations. Comparisons between facilities are made difficult by the highly specific nature of the test, which is dependent on sludge compactability. Therefore, the result of a centrifuge spin test depends on SVI. The same facility can give radically different centrifuge spin results for the same sludge concentration if sludge settleability has changed.

Sludge solids inventory is used to determine the critical core components of process control: solids retention time (SRT) or MCRT, wasting rates, return rates, and even aeration rates. It should follow that the laboratory techniques used to calculate solids inventory should be as accurate as possible. The operator–analyst may, therefore, want to consider daily TSS analyses in lieu of the estimation given by this procedure.

TABLE 3.20   Sludge volume index worksheet.

Location: _____

Method: _____

Date/Time	Sample ID	30-Minute Settling Volume, mL/L	MLSS, mg/L

This procedure is used for process control purposes only and is not listed as an approved reporting method in 40 CFR Part 136 (U.S. EPA, 2017).

13.2   Application

The laboratory centrifuge provides a quick and easy way to estimate activated sludge mixed liquor or return sludge suspended solids concentrations when used with the gravimetric method. The solids concentration is used in process control to determine SRT or MCRT and, ultimately, wasting rates.

Given the dynamic nature of biological systems, the centrifuge factor as determined in Step 1 in Section 13.7 will be reliable only as long as the biological system remains more or less stable. For practical purposes, the centrifuge factor should be reestablished at least once per week.

13.3 Interferences

Individual sludge characteristics, including the presence of filaments and inorganic matter, affect the compactability and percent volume of sludge.

13.4 Apparatus and Materials

- All equipment needed for gravimetric determination of suspended matter
- Laboratory centrifuge capable of 1200 to 2000 revolutions per minute (rpm)
- Tachometer for centrifuge
- Timer
- Centrifuge tubes graduated to 0.1 mL
- Pipet
- Pipet bulb

13.5 Sample Collection, Preservation, and Holding Times

Samples should be collected in plastic bottles and analyzed immediately. No preservation is possible. It is important for the operator–analyst to be aware that as a sludge sample ages in the bottle, biological activity continues to take place and the nature of the sludge will change.

13.6 Quality Assurance and Quality Control

One of several significant drawbacks of this method is that is that a certified standard does not exist. The operator–analyst may run duplicates by centrifuging two aliquots of the same sample.

13.7 Procedure

- **Step 1**—Determination of the centrifuge factor:
  1. Determine either mixed liquor or return sludge suspended solids gravimetrically.

2. Fill centrifuge tubes with a well-mixed portion of the same sample.

3. Load centrifuge according to the manufacturer's instructions.

4. Start the centrifuge and timer. Use a tachometer to adjust centrifuge revolutions per minute. Time and speed of centrifugation are not critical but must be the same for all determinations; 15 minutes at 1500 rpm is recommended.

5. After the centrifuge has stopped turning, remove centrifuge tubes. Read and record the volume (to the nearest 0.1 mL) of packed solids in the bottom of the tube. Calculate the centrifuge factor as follows:

$$\text{Centrifuge factor} = \frac{\text{Suspended solids, mg/L}}{\text{Volume of condensed solids, mL}} \qquad (3.18)$$

For example, given suspended solids = 1800 mg/L and volume of condensed solids = 1.5 mL,

$$\text{Centrifuge factor} = \frac{1800 \text{ mg/L}}{1.5 \text{ mL}}$$

$$\text{Centrifuge factor} = 1200 \text{ mg/L/mL}$$

• **Step 2**—Estimation of suspended solids. Fill centrifuge tubes with a well-mixed sample. Then, load and run the centrifuge as outlined in Step 1.

13.8   Calculations

$$\text{Suspended solids, mg/L}$$
$$= (\text{Centrifuge tube reading, mL}) \times (\text{Centrifuge factor, mg/L/mL}) \qquad (3.19)$$

For example, given the following:

Centrifuge tube reading	= 1.8 mL,
Centrifuge factor	= 1200 mg/L/mL,
Suspended solids (mg/L)	= (1.8 mL) (1200 mg/L/mL) = 2160 mg/L.

13.9   Reporting

Results should be reported to zero decimal places.

14.0 OXYGEN UPTAKE RATE AND SPECIFIC OXYGEN UPTAKE RATE

The following sections comprise an expanded version of Method 2710 B-2011 of *Standard Methods* (APHA et al., 2017).

14.1 General Description

The oxygen uptake rate (OUR) test and resulting respiration rate estimate how active the microorganisms are in the activated sludge process. The activity of the microorganisms is related to the amount of oxygen the organisms consume. The OUR test measures how much oxygen a sample of activated sludge consumes over a specific time period. The rate of oxygen uptake can vary due to either microorganism activity or the number of microorganisms. A large number of microorganisms will use more oxygen than a smaller number of microorganisms, even when they are respiring (breathing) at the same rate. The specific oxygen uptake rate (SOUR) test takes the OUR result and divides it by the concentration of MLSS in the activated sludge process. This normalizes the result to the amount of oxygen consumed by 1 gram of MLVSS per hour and allows operators to compare results from one day to the next, even though the MLVSS concentration may be changing. Using MLVSS in the calculation relates oxygen consumption to the mass of active microorganisms in the basin instead of the total MLSS, which includes inert, non-respiring material. The SOUR test is also referred to as the respiration rate.

14.2 Application

The OUR test can be used to determine the aerobic biological activity wherever the sample is taken; for example, the tail end of an aeration tank (most common) or along the aeration tank to track treatment progression, to test for toxic effects of influent wastewater and to demonstrate that digested biosolids are stabilized. The OUR test can also be used to determine the required time through the aeration tank for the microorganisms to stabilize the influent biochemical oxygen demand (BOD) and ammonia loads and become "activated"; that is, reach endogenous respiration rate. Endogenous respiration rate is the rate of bioactivity when the microorganism has stabilized all stored food and the rate of growth is less than the rate of cellular degradation; that is, the cells are starving.

14.3 Apparatus and Materials

- Calibrated and fully charged dissolved oxygen meter
- Two BOD bottles or two 1-L plastic bottles

- 7.6-cm (3-in.) section of 22-mm (7/8-in.) outer diameter plastic pipe that has been shaped on both ends to fit the mouth of a BOD bottle, to be used to transfer MLSS between BOD bottles (should fit snugly; length is approximate)
- A self-stirring BOD bottle probe or separate magnetic stirring device
- Stop watch or other timing device

Other data that will be required for the SOUR calculation includes VSS in grams per liter (ounces per gallon) of the activated sludge sample from which the OUR test was performed. For operational purposes, a TSS meter can be used for determining MLSS and last week's volatility percentage can be used for estimating the actual MLVSS value. Note that the SOUR calculation for aerobic digesters to meet the vector attraction reduction requirement given in the 503 Regulations (1.5 mg/L of dissolved oxygen per gram of biosolids per hour) uses total solids and not VSS and varies from the following procedure.

14.4 Procedure

- **Step 1**—Collect a fresh sample of activated sludge either from the tail end of the aeration tank or from other locations within the process to measure changes through the process.
  - NOTE: Perform the test within 15 minutes of collecting the sample for accurate results. Discard the sample and obtain a fresh sample if holding more than 15 minutes.
- **Step 2**—Fill a 300-mL BOD bottle with the activated sludge sample and insert a 7.6-cm (3-in.) section of shaped plastic pipe.
- **Step 3**—Insert the other end of the pipe into an empty BOD bottle. Pour the sample back and forth between the bottles until the dissolved oxygen is approximately 5 to 6 mg/L. Altitude may affect the maximum dissolved oxygen achievable. Do not shake the sample to aerate it because that can cause floc shear, which may expose new floc surfaces and make them available for oxygen uptake. Increasing floc surface area can cause SOUR results to differ from actual conditions in the activated sludge basin. If BOD bottles and/or the fitted plastic pipe are not available, two 1-L bottles may be used instead.
- **Step 4**—Remove the transfer pipe and fill one of the BOD bottles to overflowing. Some bubbles will gather at the top. Tilt the bottle and/or tap the sides of the bottle to dislodge air bubbles, then place the bottle into a Koozie or other material to help maintain constant temperature.

- **Step 5**—Use a spatula to work any remaining bubbles out of the sample. It is important to note that the activated sludge sample that remains after filling the BOD bottle can be used to perform the VSS test in Step 11 if a meter is not being used.

- **Step 6**—Insert the dissolved oxygen meter probe in the BOD bottle and begin stirring. Turn the dissolved oxygen meter to a 0 to 10 scale for manual read meters. Digital meters do not require scale adjustment.

- **Step 7**—Wait approximately 30 to 60 seconds for the dissolved oxygen meter reading to stabilize. It is important to note that the indicator needle or readout should be constantly dropping during this procedure.

- **Step 8**—Beginning at any given time, record the dissolved oxygen level of the sample in 30- or 60-second intervals. Do not record dissolved oxygen levels less than 1.0 mg/L. If the temperature varies by more than 0.1 °C from the start to the end of the test, the test should be done again with a fresh sample. Results are not valid when temperature is changing because OUR is affected greatly by temperature. and changes in OUR may be due primarily to temperature changes and not bioactivity.

- **Step 9**—Graph the results by plotting dissolved oxygen in milligrams per liter on the vertical axis and time in minutes on the horizontal axis.

- **Step 10**—Draw a straight line connecting the majority of the points. Extend the line so that it crosses the horizontal and vertical axes. It is important to note that the operator–analyst may discover that the line is not straight at the beginning and end of the test. The straight line is drawn to negate the effects of these curves. At the beginning, false values caused by undissolved bubbles can cause interference. The end of the curve flattens because of the limitation of the dissolved oxygen meter.

- **Step 11**—Determine the slope of the line, in milligrams per liter of oxygen per minute. The easiest way to do this is to use the points where the line crosses the axes. Divide the milligrams per liter of oxygen by the time in minutes. This is the OUR in mg/L/min.

- **Step 12**—Determine the VSS content of the activated sludge sample and express VSS concentration in grams per liter (g/L). This is done by dividing milligrams per liter of VSS by 1000.

- **Step 13**—Enter values in the equation and calculate the SOUR. The SOUR is expressed as milligrams of oxygen per hour per gram of VSS.

14.5   Calculations

There are two methods commonly used for calculating OUR and SOUR. Both methods should arrive at the same numerical results. In both instances, the respiration rate worksheet and graph will be used.

The first method requires drawing a straight line through the greatest number of points graphed. The line will cross both the horizontal (minutes) and vertical (mg/L) axes of the graph. The values at these two points are used in the OUR calculation. The milligrams per liter of oxygen value are divided by the minutes value and multiplied by 60 minutes per hour to derive milligrams per liter per hour (mg/L/h).

The second method is applicable when using the second or right-hand side of the graph to calculate OUR on an aerobic digester or extended aeration system. These typically flat lines may not conveniently cross the horizontal axis. The straight line through the majority of the points should still be drawn. The operator–analyst should pick a point on the line as time 0 and enter the milligrams per liter of oxygen value in the calculation box in the upper part of the graph. The operator–analyst should also pick a second point on the line as the end time and enter it. The points should be at least 5 minutes apart. Then, divide by the time span between the two points and multiply by 60 minutes per hour to derive the milligrams per liter per hour OUR, as follows:

$$OUR(mg/L/h) = \frac{(DO\ mg/L\ at\ time\ 2 - DO\ mg/L\ at\ time\ 1)}{Time\ at\ point\ 0 - time\ at\ point\ 2} \cdot 60\ min/hr \qquad (3.20)$$

$$SOUR\ (mg/h/g) = \frac{OUR(mg/L/h)}{MLVSS\ (g/L)} \qquad (3.21)$$

The respiration rate worksheet example can be used to perform respiration rate and OUR calculations.

14.6   Benchsheet

Table 3.21 shows a benchsheet for OUR.

15.0   DISSOLVED OXYGEN, WINKLER METHOD AZIDE MODIFICATION

The following sections comprise an expanded version of Method 4500-O C-2016 of *Standard Methods* Online (APHA et al., 2017).

TABLE 3.21 Respirometry graph.

Courtesy of Keith Radick

15.1 General Description

Dissolved oxygen levels in natural water and wastewater depend on the physical, chemical, and biochemical activities in the waterbody. The analysis for dissolved oxygen is a key test in water pollution and wastewater treatment process control. The Winkler method is no longer widely used. Dissolved oxygen probes, both the older-style Clark electrodes and the newer luminescent dissolved oxygen probes, are used instead. The probes are much faster than the Winkler method.

The Winkler method is based on the following principles:

- Oxygen in the water sample converts iodide ion (I^-) to iodine (I_2),
- The amount of iodine generated is determined by titration with sodium thiosulfate solution ($Na_2S_2O_3$),
- The endpoint is determined using starch as a visual indicator, and
- Oxygen concentration is calculated from the amount of sodium thiosulfate used in titration.

15.2   Application

The Winkler test method can be used to verify calibration of a dissolved oxygen meter, calibrate a dissolved oxygen meter, and test dissolved oxygen concentrations in water and wastewater. It can also be used as a measurement method in the BOD test.

15.3   Interferences

In the determination of dissolved oxygen, various materials cause interference. These include nitrite, iron salts, organic matter, excessive suspended matter, sulfide, sulfur dioxide, residual chlorine, chromium, and cyanide.

15.4   Apparatus and Materials

- Buret, graduated to 0.1 mL, with a 50-mL capacity
- Glass-stoppered BOD bottles with collars (300-mL size)
- Wide-mouth Erlenmeyer flask (250 mL)
- Measuring pipet (10 mL)
- Pipet bulb
- Graduated cylinder (250 mL)
- Optional dissolved oxygen and BOD volumetric flask graduated to deliver 201 mL

15.5   Reagents

- Manganese sulfate solution: Dissolve 480 g of manganese sulfate crystals ($MnSO_4 \cdot 4H_2O$) in 400 to 600 mL of deionized water and filter it through filter paper. After filtration, add enough deionized water to make 1 L of final volume. The final solution will be pink.

- Alkali-iodide-azide solution: Dissolve 10 g of sodium azide in 500 mL of deionized water. Add 480 g of sodium hydroxide and 750 g of sodium iodide and stir until dissolved. There will be white turbidity because of the sodium carbonate, but this is not harmful. It is important that the operator–analyst does not acidify this solution because toxic hydrazoic acid fumes may be produced. In addition, it is important to remember that the reaction is extremely exothermic and that the solution will be hot.

- Sulfuric acid, 36 N: Use concentrated reagent-grade sulfuric acid. It is important to handle this material carefully because it will burn hands and clothes. Affected areas should be rinsed with tap water to prevent injury.

- Sodium thiosulfate solution, 0.025 N: Dissolve exactly 6.205 g of sodium thiosulfate crystals ($Na_2S_2O_3 \cdot 5H_2O$) and 0.4 g of solid sodium hydroxide in freshly boiled and cooled deionized water. Bring the final volume up to 1 L with freshly boiled and cooled deionized water; prepare fresh weekly.

- Starch solution: Make a thin paste of 20 g of laboratory-grade soluble starch and 2 g of salicylic acid in a small quantity of deionized water. Pour this paste into 1 L of boiling deionized water. Allow this mixture to boil for a few minutes, then cool and settle overnight; remove the clear supernatant and save it, discarding the remainder.

- Phenylarsine oxide (PAO), 0.025 N: As an alternative to the 0.025-N sodium thiosulfate solution, U.S. EPA recommended PAO be used. This is available in standardized form from commercial sources.

15.6 Sample Collection, Preservation, and Holding Times

Samples should be collected carefully and should not be allowed to be agitated or remain in contact with air, as either condition may cause the gas content in the samples to change. Surface water samples should be collected in clean, 300-mL BOD bottles with tapered and pointed ground glass stoppers and flared mouths.

When sampling from a line under pressure, a glass or rubber tube should be attached to the tap and the other end should be placed in the bottom of the BOD bottle. The BOD bottle should be allowed to overflow 2 to 3 times its volume. The operator–analyst should ensure that there are no air bubbles in the bottle before and after placing the stopper. The water temperature should be recorded at the time of sampling.

Because the dissolved oxygen content of the sample is affected by temperature, barometric pressure, and dissolved oxygen concentrations, samples should be analyzed as quickly as possible after collection. This is particularly true for samples with a high oxygen demand such as activated sludge.

If it is not possible to analyze samples immediately, they may be preserved or "fixed" followed by storage at the temperature of collection and in the dark for up to 8 hours. Samples should be preserved by adding 0.7 mL of concentrated sulfuric acid and 1 mL of sodium azide solution; this will arrest biological activity. The analysis should be completed as soon as possible.

15.7 Procedure

- **Step 1**—Completely fill a 300-mL BOD bottle with the sample to be analyzed by siphoning the sample slowly into the bottle and allowing it to overflow for a period of time to displace the volume of the bottle two or three times (see Figure 3.18). The operator–analyst should ensure that no air is entrapped. In addition, dissolved oxygen samples taken from streams or lakes should be taken in a special dissolved oxygen sampler available in most laboratory supply houses.

- **Step 2**—Hold the tip of the pipet below the surface and add 1 mL of manganous sulfate solution and 1 mL of alkali-iodide solution. It is important to keep the tip of the pipet below the water surface to avoid introducing any dissolved oxygen by splashing. A brown precipitate, manganic hydroxide [$MnO(OH)_2$], will form in the bottle as shown in Figure 3.19.

- **Step 3**—Stopper the BOD bottle, being careful not to entrap any air bubbles. Mix the contents well by gentle inversion and wait for the precipitate to settle to half volume.

- **Step 4**—Remove the stopper.

- **Step 5**—Add 1 mL of concentrated sulfuric acid. It is important to note that as the water is displaced by the addition of acid, the acid may remain at the top when the stopper is replaced. The acid may then drip back onto the fingers of the operator–analyst when shaking the BOD bottle. As such, acid-resistant gloves should be worn.

- **Step 6**—Mix gently until the floc is no longer visible.

- **Step 7**—Allow the solution to stand for 5 minutes, but not in direct sunlight. The solution can stand safely for 2 hours in this condition.

- **Step 8**—Remove 99 mL from the BOD bottle by pouring it into a graduated cylinder. The operator–analyst should note that, in Steps 5 through 7, adding the acid dissolved the floc particles and liberated manganese III ions. The manganese ions, in turn, converted the iodide ion that was added previously to iodine. Steps 5 through 7 are shown in Figure 3.20.

- **Step 9**—Fill the buret with 0.025-N sodium thiosulfate solution.

- **Step 10**—Add drop-wise to the BOD bottle while mixing continuously until the yellow color almost disappears. The sample will be a very light straw color.

- **Step 11**—Stop the titration. Add 1 mL of starch solution to the sample and swirl to mix. Starch turns blue–black in the presence of iodine. Adding the starch solution makes it easier to see the final endpoint.

- **Step 12**—Continue titrating until the blue color just disappears.

- **Step 13**—Record the start and end volumes from the buret.

- **Step 14**—Calculate the dissolved oxygen concentration.

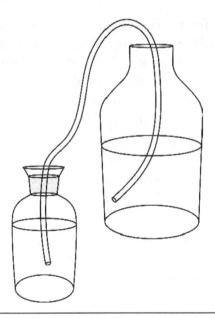

FIGURE 3.18   Siphon sample into BOD bottle taking care to avoid air bubbles.

15.8   Calculations

The dissolved oxygen present is expressed in milligrams per liter and is equal to the total number of milliliters of 0.025-N sodium thiosulfate or PAO solution used in the titration. One milliliter of sodium thiosulfate or phenylarsine solution is equal to 1 mg/L of dissolved oxygen. For example,

- Start volume = 100.0 mL
- End volume = 92.6 mL
- Volume sodium thiosulfate used = 7.4 mL
- Dissolved oxygen concentration = 7.4 mg/L.

FIGURE 3.19   Manganous hydroxide precipitation. *Courtesy of the Carbondale Wastewater Treatment Plant, Carbondale, Colorado*

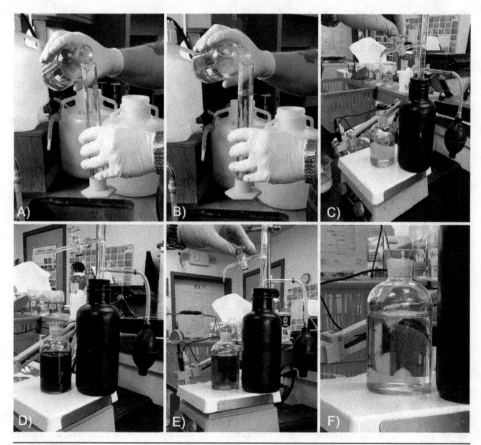

FIGURE 3.20   Multiple steps of the Winkler method. *Courtesy of the Carbondale Wastewater Treatment Plant, Carbondale, Colorado*

15.9 Reporting

According to *Standard Methods* (APHA et al., 2017), an experienced operator–analyst can maintain a precision of ±50 μg/L with visual endpoint detection. This translates to 7.00 mg/L ± 0.05 mg/L. Results should be reported in milligrams per liter of dissolved oxygen to either the most precise measurement of the buret or to three significant figures.

16.0 BIOCHEMICAL OXYGEN DEMAND (5-DAY BIOCHEMICAL OXYGEN DEMAND)

The following sections comprise expanded versions of Method 4500-O G-2016 for the dissolved oxygen–membrane electrode method and Method 5210 B-2016 for 5-day BOD (BOD$_5$) of *Standard Methods* Online (APHA et al., 2017).

16.1   General Description

This test determines the amount of organic material present in wastewater by measuring the amount of oxygen consumed by microorganisms as they degrade or "eat" the organic constituents of the waste. The test consists of measuring dissolved oxygen before and after a 5-day incubation period of the sample at 20 °C to determine the amount of oxygen used for a known amount of sample. With wastewater and facility effluent, dilution of the sample is made with standard dilution water or seeded dilution water and the dissolved oxygen is determined before and after the 5-day incubation period; BOD is then calculated. If the sample being tested has been chlorinated, UV-disinfected, ozonated, adulterated, and/or acidified, the reader should refer to Section 16.3 and Section 16.9, Steps 3, 4, and 5, on pretreatment and oxygen demand determination with seeding.

16.2   Application

The BOD test is used primarily for National Pollutant Discharge Elimination System (NPDES) reporting, but it is also used for determining process loading, F:M, and treatment efficiency. Table 3.22 summarizes BOD values for various wastewater processes.

16.3   Interferences

Caustic alkalinity, mineral acid, free chlorine, heavy metals, and toxic organics are among the factors that can influence test accuracy.

The extent of oxidation of nitrogenous compounds (nitrification) during the 5-day incubation period depends on the presence of microorganisms that can carry out this oxidation. Such organisms are typically not found in raw wastewater or primary effluent in sufficient numbers to oxidize ammonia in the 5-day test. Currently, many biological treatment effluents contain significant numbers of nitrifying organisms. Because nitrification can occur during the BOD test with these samples, inhibition of nitrification is recommended for the following sample types: secondary effluent, samples seeded with secondary effluent, and samples of polluted water. Many states require the reporting of BOD results as carbonaceous BOD (CBOD), which would require inhibition of nitrification for all sample types required to be reported as such. Before reporting, the operator–analyst should verify whether the discharge permit calls for BOD or CBOD.

Other potential interferences include the following:

- Probes with membranes or optical sensors respond to oxygen partial pressure, which is a function of both altitude (barometric pressure) and dissolved inorganic salts.

TABLE 3.22 Biochemical oxygen demand ranges for various wastewater processes.

Treatment process	Expected BOD$_5$ range
Raw influent domestic wastewater	110 to 350 mg/L
Primary clarifier effluent	Primary clarifiers can remove between 50% and 70% of influent TSS and between 25% and 40% of influent BOD.
Organic loading rates for lagoons	Aerobic, low rate: 6.7 to 13.5 g BOD/m²·d (60 to 120 lb BOD/d/ac) Aerobic, high rate: 9.0 to 17.9 g BOD/m²·d (80 to 160 lb BOD/d/ac) Anaerobic: 22.4 to 56.0 g BOD/m²·d (200 to 500 lb BOD/d/ac)
Roughing filter organic loading rate	Low-rate rock: 0.07 to 0.22 kg BOD/m³·d (4.4 to 13.1 lb BOD/1000 cu ft) Intermediate-rate rock: 0.24 to 0.48 kg BOD/m³·d (15 to 30 lb BOD/1000 cu ft) High-rate rock: 0.4 to 2.4 kg BOD/m³·d (25 to 150 lb BOD/1000 cu ft) High-rate plastic: 0.6 to 3.2 kg BOD/m³·d (38 to 200 lb BOD/1000 cu ft) Roughing: >1.5 kg BOD/m³·d (94 lb BOD/1000 cu ft)
Activated sludge F:M	Conventional: 0.2 to 0.4 kg BOD/kg MLVSS (0.2 to 0.4 lb BOD/lb MLVSS) Extended aeration: 0.05 to 0.15 kg BOD/kg MLVSS (0.05 to 0.15 lb BOD/lb MLVSS)
Activated sludge organic loading rate	Conventional: 0.3 to 0.6 kg BOD/m³·d (20 to 40 lb BOD/1000 cu ft/d) Extended aeration: 0.08 to 0.24 kg BOD/m³·d (5 to 15 lb BOD/1000 cu ft/d)

Source: Metcalf & Eddy, Inc./AECOM (2014); WEF (2017).

- ○ Many new meters will autocorrect for barometric pressures. If the meter does not autocorrect, the operator–analyst may need to enter the barometric pressure or altitude manually.

- ○ Conversion factors may be calculated from dissolved oxygen concentration versus salinity. This must be developed experimentally.

- ○ Gases that can pass through or react with the oxygen probe may interfere.

- Hydrogen sulfide can interfere by reacting with and coating the membrane anode.

- High concentrations of oil and grease may plug the membrane or optical probe.

16.4 Apparatus and Materials

- Optical or membrane dissolved oxygen probe with meter
- Biochemical oxygen demand incubation bottles, 300-mL ± 3-mL capacity, with ground glass stoppers
- Overcaps for BOD bottles
- Incubator capable of maintaining a temperature of 20 ± 1 °C
- Erlenmeyer flasks (250-mL, wide-mouth)
- Wide-tip volumetric pipet for measuring samples with suspended solids
- Suitable volumetric glassware for sample transfer and measurement (standards are prepared with Class A pipets and volumetric flasks; high solids wastewater will use Class B wide-tip pipets or serological pipets)
- Pipet bulb

16.5 Reagents

- Water: The source of water is not restricted by the method and may be distilled, tap, or receiving stream water so long as it is free of biodegradable organics and substances that may inhibit the test such as chlorine or heavy metals.

- Prepare stock buffer solutions for making BOD dilution water. Storing buffer solutions at ≤ 6 °C will prevent bacteriological growth and prolong their useful life but is not required by *Standard Methods*. Alternatively, buffer solution pillows and stock solutions can be purchased premade from several different vendors:

  - Phosphate buffer solution: Dissolve 8.5 g of monobasic potassium phosphate (KH_2PO_4), 21.75 g dibasic potassium phosphate (K_2HPO_4), 33.4 g dibasic sodium phosphate heptahydrate crystals ($Na_2HPO_4 \cdot 7H_2O$), and 1.7 g ammonium chloride (NH_4Cl) in about 500 mL of deionized water and dilute to 1 L. The pH should be 7.2 without further adjustment (American Chemical Society [ACS] grade or better).

- ○ Magnesium sulfate solution: Dissolve 22.5 g of magnesium sulfate crystals ($MgSO_4 \pm \cdot 7H_2O$) in deionized water and dilute to 1 L (ACS grade or better).

- ○ Calcium chloride solution: Dissolve 27.5 g of anhydrous calcium chloride ($CaCl_2$) in deionized water and dilute to 1 L (ACS grade or better).

- ○ Ferric chloride solution: Dissolve 0.25 g of ferric chloride ($FeCl_3 \cdot 6H_2O$) in deionized water and dilute to 1 L (ACS grade or better).

- Dilution water: Add 1 mL each of phosphate buffer, magnesium sulfate, calcium chloride, and ferric chloride solutions to each liter of deionized water. If 5 L of dilution water is desired, add 5 mL of each solution. Store the dilution water at a temperature as close to 20 °C as possible. This water should not show a reduction in dissolved oxygen of more than 0.2 mg/L from the start to the end of the BOD_5 test. Dilution water should be shaken or aerated with a small aquarium pump to saturate it with dissolved oxygen.

  - ○ It is important to note that some laboratories have had success with putting a clean carboy of dilution water *without nutrients* into the BOD incubator overnight. The water will be at the right temperature. Adding the nutrients in advance can sometimes cause bacteria and algae to grow in the carboy. If this happens, the carboy should be cleaned with a dilute bleach solution and rinsed thoroughly. In addition, it is important to note that the salts added to the dilution water do two important things. First, the salts make the dilution water about the same salt concentration as the inside of the bacteria. This is important because if there was not any salt in the dilution water, the bacteria would absorb more and more water in an attempt to make their internal and external salt concentrations equal. Eventually, the increased water absorption would cause the bacterial cell walls to rupture, killing the bacteria. A large number of dead bacteria will result in less oxygen depletion and a result that is biased low. The second thing the dilution water does is provide important nutrients such as nitrogen, phosphorus, and iron that the bacteria need to live, thrive, and survive. These nutrients might not be present in the samples in large enough quantities to give the bacteria the best growing conditions, so additional nutrients are added to ensure optimum conditions for bacterial survival.

- Seed solution: Secure an unchlorinated sample of raw wastewater or primary clarifier effluent 24 hours before setting up the BOD test.

Collect 1 L of unchlorinated sample and allow it to stand at room temperature overnight. Pour off the clear portion of the sample and use it for the seed. Commercial seed is also available. If a commercial dry seed is used, however, it must be rehydrated in the laboratory with BOD dilution water.

- Sodium sulfite solution, approximately 0.025 N: Dissolve 1.575 g of sodium sulfite (Na_2SO_3) in 1 L of deionized water. This solution is not stable; prepare daily when dechlorination of wastewater samples is required.

- Nitrification inhibitor: Use reagent-grade 2-chloro-6-(trichloromethyl) pyridine. Commercial products can also be used.

  ○ Sodium hydroxide solution, 1 N: Required only when pH adjustment of wastewater samples is required.

  ○ Sulfuric acid solution, 1 N: Required only when pH adjustment of wastewater samples is required.

16.6 Stock and Standards

Dry reagent-grade glucose and reagent-grade glutamic acid (glucose and glutamic acid solution [G/GA]) at 103 °C for 1 hour. Store these in a desiccator until needed. Add 0.1500 g (± 0.0001 g) of glucose and 0.1500 g (± 0.0001 g) of glutamic acid to 1 L of volumetric flask and dilute to volume with water. This solution is good for 7 days if kept sterile and refrigerated at <6 °C ± 2 °C. The solution should be replaced if it becomes cloudy or shows signs of bacterial or chemical contamination (ACS grade or better).

A commercial solution can be purchased, but concentrations may vary. The final concentration must be the same as the aforementioned solution. The solution should be kept sterile and refrigerated at <6 °C ± 2 °C. The solution should be replaced if it becomes cloudy or shows signs of bacterial or chemical contamination.

16.7 Quality Assurance and Quality Control

The BOD procedure requires different types of QA/QC samples to ensure sample results that are both accurate and precise (the operator–analyst should read this section thoroughly before starting the BOD test). Required QA/QC samples include reagent blanks, seed controls at different dilutions, two G/GA standards, multiple dilutions per sample, and a laboratory duplicate.

Blanks should be prepared to represent each type of test being performed within the batch. Seed blanks are also called *seed controls*. The following blanks should be prepared:

- Unseeded BOD blank—a 300-mL bottle of dilution water

- Seeded BOD blanks—typically, a 6-, 9-, and 12-mL sample of seed added to three separate BOD bottles and diluted with BOD and CBOD dilution water

- Unseeded CBOD blank—a 300-mL bottle of dilution water with nitrification inhibitor

- Seeded CBOD blank—typically, a 10-, 20-, and 30-mL sample of seed added to three separate BOD bottles and BOD and CBOD diluted with dilution water

Table 3.23 shows expected dissolved oxygen depletion for blanks. The dilution water blank should exhibit <0.2 mg/L of dissolved oxygen depletion over 5 days. If the dilution water blank changes more than this amount, the cleanliness of the BOD bottles needs to be checked. High dilution water blanks may also result from trace organics and metals present in the dilution water. The container used for mixing dilution water may also be contaminated; therefore, all equipment and bottles should be cleaned thoroughly.

The dissolved oxygen uptake in each sample bottle that is a result of adding seed should be between 0.6 and 1.0 mg/L of dissolved oxygen depletion over 5 days. If the seed blank changes more than this amount and the dilution water blank is acceptable, the volume of seed added to each BOD bottle for the next analysis should be decreased. If the seed blank changes by more than this amount and the dilution water blank is also high, all of the BOD bottles should be recleaned and checked to see that the dissolved oxygen meter is properly calibrated with the right barometric pressure. The amount of seed added to sample bottles should be adjusted up or down so that the G/GA standard is within acceptable limits even if it causes the seed blank contribution to be higher than 1.0 mg/L.

TABLE 3.23   Expected dissolved oxygen depletion for blanks.

Blank	Depletion expected	Notes
Unseeded BOD	<0.2 mg/L	If outside this range, review procedures.
Seeded BOD	<1 mg/L and >0.6 mg/L	
Unseeded CBOD	<0.2 mg/L	
Seeded CBOD	>0.6 mg/L and <1 mg/Land	

Note. The range given for seeded BOD and CBOD indicates the desired level of depletion in samples that can be attributed to seed. Depending on the version of *Standard Methods* consulted, the recommended ranges vary significantly.

Source: APHA et al. (2017).

16.7.1   Prepare a Standard

Transfer 6 mL of the glutamic acid-glucose solution to a BOD bottle. Add the desired volume of seed solution and dilute with BOD and CBOD dilution water. The results obtained for these checks should read 198 ($\pm$ 30.5) mg/L BOD. The procedures should be checked if any variation is obtained.

16.7.2   Replicates

Standard Methods (APHA et al., 2017) requires a minimum of two valid dilutions for each sample. The method recommends that five dilutions be prepared for each sample, but allows fewer bottles to be prepared when the operator–analyst is familiar with the sample. The RPD between valid replicates of a given sample should be within $\pm 20\%$. Relative percent difference is calculated by dividing the difference between two sample results by their average.

16.7.3   Duplicates

Laboratory duplicates should be run on each batch type. Therefore, a laboratory duplicate will be set for each BOD and CBOD set of seeded and unseeded samples. Laboratory duplicates are separate subsamples of a larger sample and should be set up with multiple dilutions, just like the original sample. Results from multiple dilutions may be averaged together to obtain a single result. Results for a sample and its laboratory duplicate should not be averaged together. Only the sample result is reported. The duplicate is for QA/QC purposes only. The RPD of the duplicates should be <20%.

For sample results to be considered valid, the following criteria must be met:

- The sample must deplete at least 2 mg/L dissolved oxygen over 5 days, and
- There must be at least 1 mg/L of dissolved oxygen remaining in the sample bottle.

Sample results that do not meet these criteria should not be used for reporting purposes. If it is not possible to collect another sample and all other QA/QC samples are within limits, samples that do not meet depletion requirements may have to be reported. In this instance, where none of the sample aliquots deplete at least 2 mg/L of dissolved oxygen, the operator–analyst should calculate the BOD_5 result for the largest aliquot and report the BOD_5 result as less than that number. For example, a 240-mL sample aliquot only depletes 1.5 mg/L of dissolved oxygen to give a reported BOD_5 of <2 mg/L. In the future, the operator–analyst should use a larger aliquot, if possible. In a situation where all of the sample aliquots deplete to the point where there is <1 mg/L dissolved oxygen in each sample bottle, the BOD_5 result should be calculated for the smallest aliquot and the BOD_5 result should be reported as greater than that number. In the future, the operator–analyst should use smaller aliquots.

16.8   Sample Collection, Preservation, and Holding Times

Samples for BOD analysis should be collected in glass or plastic containers. The samples should be cooled to ≤6 °C when they are not analyzed immediately. Samples should be analyzed within 6 hours of collection but may be held for up to 48 hours. In the case of composite samples, the hold time begins when the last aliquot of the composite sample is collected.

16.9   Procedure

- **Step 1**—Glassware cleaning. Biochemical oxygen demand bottles must be extremely clean. Most of the time, washing with hot water and a laboratory detergent is enough to thoroughly clean the BOD bottles (it is important to remember to never stick a bottle brush inside the bottle as it may scratch the glass, which will make it more difficult to thoroughly clean the bottles in the future). If dilution water blanks are higher than 0.2 mg/L of oxygen depletion, a more rigorous cleaning procedure may be needed. A satisfactory cleaning solution is an acid-dichromate mixture made by dissolving 60 to 65 g of sodium or potassium dichromate in 30 to 35 mL of hot water and, after allowing this mixture to cool slowly, adding concentrated sulfuric acid to make 1 L of solution. Bottles can be cleaned by adding some of the solution to an empty bottle, shaking gently, and allowing it to stand overnight. The solution may be reused until it begins to change color to green.
  - It is important to note that acid dichromate must be handled with caution because it will damage clothes and most metals, and will cause skin burns if not washed off promptly. It is recommended that safety glasses, rubber gloves, and aprons be worn and that caution be taken whenever acid solutions are used. A 10% hydrochloric acid solution can be a safer solution to the acid-dichromate mix. The cleaning solutions must be rinsed thoroughly from glassware six times with tap water and then two or three times with deionized water.
  - It is also important to note that the acid-dichromate solution contains hexavalent chromium, which is extremely toxic. If it is not

completely rinsed from BOD bottles, it may interfere with the BOD test.

- **Step 2**—Prepare seed solution and buffered dilution water. The operator–analyst should ensure that the dilution water is at 20 °C. Seed solution and dilution water should be prepared 24 hours ahead of time. Nutrients should not be added to the dilution water until immediately before use. It is important to note that, if a commercially available seed is being used, the manufacturer's preparation directions should be followed. In many instances, commercial seeds must mature for at least 1 hour before use and should be discarded within 6 hours of rehydrating.

- **Step 3**—Check the pH of each sample to be analyzed. If the sample pH is not between 6.5 and 7.5 s.u., adjust the sample pH with 1N sodium hydroxide or 1N sulfuric acid accordingly. The amount of acid or base added cannot dilute the sample by more than 0.5%. The pH of the dilution water should not be affected by the lowest sample dilution.

- **Step 4**—Check previously chlorinated samples for the presence of residual chlorine:

  1. Carefully measure 100 mL of well-mixed sample into a 250-mL Erlenmeyer flask.

  2. Add a few drops of potassium iodide (KI) to the sample and dissolve the crystals.

  3. Add 1 mL of concentrated sulfuric acid and mix well.

  4. Add five drops of starch. If a blue color is not produced, chlorine is absent and the BOD of the sample may be determined without dechlorination.

  5. If a blue color is produced, dechlorinate any samples with chlorine residual by adding one or more drops of a 5% sodium thiosulfate solution. Check for chlorine residual after each drop is added. Adding excess sodium thiosulfate can bias BOD results.

  6. Samples that are dechlorinated must be seeded.

- **Step 5**—Remove excess dissolved oxygen from supersaturated samples. Samples that contain more than 9 mg/L of dissolved oxygen at 20 °C may be encountered in cold water or in water where photosynthesis occurs. To prevent oxygen loss from supersaturation during the 5-day incubation period, reduce the amount of dissolved oxygen in the sample by bringing the sample to approximately 20 °C in a partially filled bottle while agitating it by vigorous shaking or aerating with compressed air.

- **Step 6**—Adjust temperature of all samples and dilution water to 20 °C.
- **Step 7**—Determine the dilutions required.

1. For samples that are routinely analyzed, this step is not necessary. Prior to setting up the BOD test, run a COD analysis on the sample if possible.

2. Estimate the BOD_5 result by multiplying the COD result by 0.5.

3. Calculate the range of aliquots to be used according to the following formulas:

   - $BOD_5(est) = (2 \text{ mg/L dissolved oxygen used})/(\text{aliquot size}/300 \text{ mL})$ and

   - $BOD_5(est) = (5 \text{ mg/L dissolved oxygen used})/(\text{aliquot size}/300 \text{ mL})$.

   - If the COD for the unknown sample is 300 mg/L, then the estimated BOD_5 is 150 mg/L and the aliquot sizes are 4 and 10 mL. Just to be sure, aliquots for the BOD_5 test should be between 2 and 20 mL. If it is not possible to run a COD analysis on the sample, the operator–analyst should set up a wide range of dilutions that encompass the entire range of possible results.

4. Set up the BOD analysis as described in Steps 8 through 27 of this section. After 24 hours, read back the dissolved oxygen on the unknown samples to check their depletion. With the BOD test, almost 75% of oxygen depletion occurs in the first 24 hours. If the samples have not used more than 1.5 mg/L of dissolved oxygen or if they have used more than 4 mg/L, a new batch of BOD should be set up using different aliquots. Checking depletion after 24 hours is not a substitute for setting up a wide range of sample dilutions. A minimum of three dilutions is needed and five is recommended for an unknown sample. In some instances, even more dilutions may be warranted.

   - It is important to note that when performing intermediate checks, it is critically important to ensure that the dissolved oxygen probe is clean and rinsed thoroughly between sample bottles. Failure to rinse may result in cross-contamination. After reading the dissolved oxygen, the stopper should be replaced and the operator–analyst should check to make sure the water seal is intact. If needed, a few drops of deionized water can be added over the stopper.

5. According to *Standard Methods* Online method 5210 B-2016 for 5-day BOD, the largest aliquot size that should be used for BOD analysis is 240 mL into a 300-mL BOD bottle. Otherwise, there

will not be enough buffered dilution water or seed in the bottle. There are commercially available packets of concentrated nutrients made specifically for adding nutrients to BOD bottles with sample volumes larger than 240 mL. If these are available, a larger sample volume may be used; however, nutrients must be added directly to the BOD bottle. It is important to note that the hold time for BOD samples is 48 hours. If the samples are checked after 24 hours, there is still time to reset.

- **Step 8**—The BOD bottles should be labeled or the bottle numbers on the BOD benchsheet should be recorded. The BOD bottles should be randomized. The dilution water blank and seed blanks must randomly cycle through all of the BOD bottles in the laboratory. It is important for the operator–analyst to remember that the purpose of the dilution water blank is to check for contamination. It cannot fulfill this function if it is always set up in the same clean BOD bottle.

- **Step 9**—Begin adding samples to the BOD bottles. Record sample volumes and any dilutions made on the BOD benchsheet. For sample volumes of 25 mL and smaller, a wide-mouth pipet should be used. For sample volumes of 25 mL or greater, an appropriately sized graduated cylinder should be used. A dilution water blank, at least two seed blanks, and two standards should be included with each analytical batch. For every sample, at least two dilutions should be made, and preferably three. For unknown samples, *Standard Methods* (APHA et al., 2017) recommends using five dilutions. Typical aliquot sizes for routine samples are illustrated in Table 3.24. Only pipet the aliquot volume required for a single BOD bottle; multiple aliquot volumes should not be transferred in a pipet draw. This will lead to solids settling out and sample bias.

TABLE 3.24   Sample aliquot sizes for routine wastewater samples.

Sample type	Recommended aliquot
300 mg/L standard	6.0 mL
Facility influent	3.0 and 6.0 mL
Primary clarifier influent and effluent	7.5 and 9.0 mL
Trickling filter effluent	15 and 30 mL
Secondary clarifier effluent	90 and 120 mL
Tertiary unit processes	210 and 240 mL
Facility final effluent	150, 200, and 240 mL
Septic samples	Dilute 1 : 100, then 1 to 15 mL

- ○ If the expected sample concentration is greater than 560 mg/L, the sample should be diluted (1 part sample to 9 parts dilution water) before adding 3 to 6 mL of the diluted sample to a BOD bottle for analysis. The minimum aliquot volume transferred to a BOD bottle will be 3 mL, as set by *Standard Methods* (APHA et al., 2017). Table 3.25 provides estimated sample sizes for BOD determination.

- **Step 10**—For samples that need to be seeded, a wide-mouth pipet should be used to add 2 mL seed solution to each sample bottle. Seed also needs to be added to the three seed blank bottles and to each standard bottle. Seed solution should not be added to the dilution water blank.

- **Step 11**—If analyzing for CBOD, a nitrification inhibitor should be added to each sample bottle. Nitrification inhibitor should not be added to the standard. If both BOD and CBOD samples are being analyzed, two sets of dilution water and seed blanks should be prepared. Nitrification inhibitor should be added to one set of blanks. Figure 3.21 shows an example of initial bottle setup with required quality control samples.

- **Step 12**—Calibrate the dissolved oxygen meter according to the manufacturer's instructions. Be sure to check that the barometric pressure

TABLE 3.25 Estimated sample sizes for BOD determination.

Sample added to 300-mL bottle (mL)	Minimum BOD, mg/L	Maximum BOD, mg/L
3	210	560
6	105	280
9	70	187
12	53	140
15	42	112
18	35	94
21	30	80
24	26	70
27	24	62
30	21	56
45	14	37
60	11	28
75	8	22
150	4	12

FIGURE 3.21   Biochemical oxygen demand sample bottle setup. *Courtesy of Indigo Water Group*

has been corrected for altitude. For example, in Denver, this should be set for 630 mm mercury, not 760 mm mercury.

- **Step 13**—Fill the first BOD bottle to the top of the neck with aerated dilution water. Add the dilution water slowly, taking care not to entrain air bubbles. Shake bottles gently to release any air bubbles that do form.

- **Step 14**—Place the dissolved oxygen probe holder into the first BOD bottle and insert a stir bar and the dissolved oxygen probe. Transfer the whole assembly to the stir plate. Read and record the initial dissolved oxygen concentration in milligrams per liter on the benchsheet. The dilution water should be saturated with dissolved oxygen. At 20 °C and sea level, the dilution water should have a dissolved oxygen concentration of approximately 9.5 mg/L. Initial dissolved oxygen measurements for samples will depend on the altitude of the laboratory and the percentage of sample in the BOD bottle. As a rule of thumb, initial dissolved oxygen measurements should be above 7 mg/L. It is important for the operator–analyst to look carefully at the end of the dissolved oxygen probe and check for air bubbles. Sometimes, air bubbles become stuck on the end of the probe. A quick flick of the probe can knock them loose. In addition, many dissolved oxygen probes are equipped with their own stirring mechanisms. In this instance, a stir plate and stir bar are not necessary. Many probes are also equipped with integrated holders.

- **Step 15**—After reading, remove the probe and probe holder from the bottle and rinse them well with deionized water.

- **Step 16**—Stopper and cap the BOD bottle. Ensure that there are no air bubbles in the bottle. If an air bubble is seen, uncap the bottle and add more dilution water.

- **Step 17**—Repeat Steps 12 to 15 for each BOD bottle. To prevent cross-contamination, the operator–analyst should read the dilution water blanks first, then proceed from clean to dirty based on the expected BOD concentration of the sample.
- **Step 18**—Transfer all BOD bottles in the batch to the BOD incubator. Record the date and time the bottle were put into the incubator on the benchsheet.
- **Step 19**—Incubate at 20 °C for 5 days plus or minus 4 hours.
- **Step 20**—Remove the BOD bottles from the BOD incubator. Record the date and time on the benchsheet.
- **Step 21**—Check each bottle for entrained air bubbles. Bottles with visible air bubbles may not be used for calculating a BOD result; any such bottles should be noted on the benchsheet.
- **Step 22**—Calibrate the dissolved oxygen meter according to the manufacturer's instructions. The operator–analyst should be sure to check that the barometric pressure has been adjusted for altitude as required.
- **Step 23**—Uncap and unstopper one bottle at a time. As soon as the bottles are opened, they will begin to absorb oxygen from the air. These should be tightly stoppered until right before the operator–analyst is ready to measure the dissolved oxygen in that bottle.
- **Step 24**—Place the dissolved oxygen probe holder into the first BOD bottle and insert the dissolved oxygen probe. Transfer the whole assembly to the stir plate. Read and record the final dissolved oxygen concentration in milligrams per liter on the benchsheet.
- **Step 25**—After reading, remove the probe and probe holder from the bottle and rinse well with deionized water.
- **Step 26**—Repeat Steps 21 through 23 for each BOD bottle.
- **Step 27**—Calculate results.

16.10 Calculations

For each test bottle that has at least 2.0 mg/L dissolved oxygen depletion and at least 1.0 mg/L dissolved oxygen remaining, BOD should be calculated as follows:

$$BOD_{5,\,mg/L} = \frac{[(DO_i - DO_f) - (S_i - S_f)F]}{\dfrac{\text{Sample volume, mL}}{300 \text{ mL}}} \tag{3.22}$$

Where

DOi = dissolved oxygen of diluted sample immediately after preparation (mg/L),

DOf = dissolved oxygen of diluted sample after 5-day incubation at 20 °C (mg/L),

Si = seed blank dissolved oxygen immediately after preparation (mg/L),

Sf = seed blank dissolved oxygen after 5-day incubation at 20 °C (mg/L), and

F = fraction of seed added to the sample compared to the amount of seed added to the seed blank.

16.11 Reporting

The MDL for the BOD test is 2 mg/L. Sample results that are calculated lower than 2 mg/L should be reported as <2 mg/L. For sample results to be valid and reportable, all of the following criteria must be met:

- Dilution water blank is depleted to less than 0.2 mg/L. It is important to note that if all other QA/QC parameters are within limits and the dilution water blank is depleted more than 0.2 mg/L, results may still be reported; however, the results of the dilution water blank should also be reported along with an explanation.
- Standard must be 198 mg/L ± 30 mg/L.
- Individual sample bottles must have used at least 2 mg/L of dissolved oxygen and must have at least 1 mg/L of dissolved oxygen remaining.
- Replicates should agree within ±20% RPD. Relative percent difference is calculated by dividing the difference in sample results by their average and multiplying by 100.
- Replicate results should be averaged together to get one final result for each sample.
- Duplicates should agree within ±20% RPD.
- Duplicate results are never averaged together with the original sample. There should be one result for the sample and another result for the duplicate.

16.12 Benchsheet

Table 3.26 shows a benchsheet for BOD.

TABLE 3.26 Benchsheet for 5-day biochemical oxygen demand.

Standard Methods 5210 B

				Location:
Setup date		Operator-Analyst		
Read date		Operator-Analyst		
Bottle no.				
Sample identification	Blank	Seed blank	GA concentration = _____ mg/L	
Sample date			Lot no.	
Sample pH				
Seed added (mL)				
Aliquot (mL)				
Initial dissolved oxygen				
Final dissolved oxygen				
Difference				
Difference – seed				
BOD5 – mg/L				
Average				
Bottle no.				
Sample identification				
Sample date				
Sample pH				
Seed added (mL)				
Aliquot (mL)				
Initial dissolved oxygen				
Final dissolved oxygen				
Difference				
Difference – seed				
BOD5 – mg/L				
Average				

Quality control parameters:

1. Sample must deplete or use at least 2 mg/L of dissolved oxygen to be used for calculation purposes.
2. A residual of 1 mg/L of dissolved oxygen must remain at the end of the test for the result to be valid.
3. Standard recovery should be with +/- 15% of the true value.
4. Replicate aliquots of the same sample that meet the above criteria must also agree within +/- 20% RSD.

203

17.0 CHLORINE RESIDUAL—AMPEROMETRIC TITRATION

The following sections comprise an expanded version of Method 4500-Cl D-2011 of *Standard Methods* Online (APHA et al., 2017).

17.1 General Description

Chlorine in water solutions is not stable. As a result, its concentration in samples will decrease rapidly. Exposure to sunlight or other strong light or to agitation will further reduce the quantity of chlorine in solutions. Therefore, samples to be analyzed for chlorine cannot be stored, and tests must be started immediately after grab sampling. Excessive light and agitation must be avoided.

17.2 Application

Chlorine residual is monitored primarily on WRRF effluents because of the effect of chlorine on the environment and final use. The analysis can be used to verify instrumentation, which controls chemical equipment including dechlorination if zero residual for discharge is required. The analysis can also be used in conjunction with bacteria monitoring to identify whether sufficient solids removal or chlorine contact time is achieved.

17.3 Interferences

Color, turbidity, iron, manganese, and nitrates are all factors that may interfere with the accuracy of this test.

17.4 Apparatus and Materials

- Volumetric pipet (10 mL)
- Pipet bulb
- Flask (500 mL)
- Graduated cylinder (1 and 20 mL)
- Glass stirring rod
- Amperometric titration apparatus (electrode, agitator, and buret)
- Beaker (250 mL)

17.5 Reagents

- Sulfuric acid (1.8 M): Carefully add 10 mL of concentrated H_2SO_4 to 90 mL of distilled water. This reaction is extremely exothermic and the solution will be hot.

- Acetate buffer: Dissolve 243 g of sodium acetate ($NaC_2H_3O_2 \cdot H_2O$) or 146 g anhydrous sodium acetate in 400 mL of distilled water. Carefully add 462 mL of concentrated acetic acid and dilute to 1000 mL. The reaction is extremely exothermic and the solution will be hot.

- Starch indicator: Dissolve 5 g of soluble starch in 1.0 L of boiling distilled water. Preserve with 1.25 g of salicylic acid.

17.6   Stock and Standards

- Standard 0.1-N potassium biiodate, $KH(IO_3)_2$: Dry primary standard-grade $KH(IO_3)_2$ at 103 + 2 °C for 2 hours. Weigh out exactly 3.249 g. Transfer to a 1000-mL, Class A volumetric flask. Fill the flask about one-third full with distilled water. Swirl to dissolve $KH(IO_3)_2$. Dilute to 1000 mL.

- Potassium biiodate, 0.005 N: Dilute 50.0 mL of the primary 0.1-N standard to 1000 mL with distilled water.

- Stock phenylarsine oxide, (PAO) 0.025 N (available commercially).

- Stock iodine, 0.1 N (available commercially).

17.7   Sample Collection, Preservation, and Holding Times

A 50-mL grab sample can be collected from the wastewater stream and taken directly to the laboratory for immediate analysis. There can be no holding time or preservation of the sample. Stock PAO needs to be standardized fresh for each analysis. Stock iodine solution also needs to be standardized daily.

17.8   Procedure

- **Step 1**—Standardize working PAO solution for the test:
  1. Dissolve approximately 2 g potassium iodide (KI) in 100 to 150 mL of deionized water.
  2. Carefully add 10 mL H_2SO_4 solution followed by 20 mL of 0.005 N $KH(IO_3)_2$ solution.
  3. Place the mixture in a dark place for 5 minutes.
  4. Dilute to 300 mL.
  5. Titrate 0.00564-N PAO solution to a pale straw color.
  6. Add a small amount of indicator and wait for a blue color to develop.

7. Continue the titration with the PAO solution until the blue color disappears.

8. Run a duplicate analysis (duplicates should agree within ±0.05 mL).

9. Record the actual volume of PAO used to eliminate the blue color for calculating the normality of PAO.

- **Step 2**—Calculate the normality of PAO, as follows:

$$\text{PAO Normality} = \frac{(\text{Ideal volume of PAO, mL})(\text{Normality of } KH(IO_3)_2, \text{equiv/L})}{\text{Actual volume of PAO, mL}} \quad (3.23)$$

For example, given the following:

Ideal volume of phenylarsine oxide   = 20 mL,
Normality of $KH(IO_3)_2$ = 0.005 N (equiv/L), and
Actual volume of phenylarsine oxide = 19.8 mL, then
Normality of phenylarsine oxide   = 20 (0.005)/19.8 = 0.051 N (equiv/L).

- **Step 2**—Titration of chlorine (Cl_2) residual:

1. Place 200 mL of sample in a 250-mL beaker. If the residual is expected to be greater than 2 mg/L, use 100 mL of sample.

2. Add 1 mL of KI solution and 1 mL acetate buffer to the beaker.

3. Place the beaker on the titrator with the electrode submerged.

4. Turn on the agitator electrode assembly.

5. Titrate 0.00564-N PAO until the meter's needle comes to a complete rest.

6. Subtract the last drop from the number added for a more accurate result of volume of PAO.

17.9 Calculations

When a 200-mL sample is used, no calculation is necessary. The buret reading is in milligrams per liter. For 100-mL samples, the following formula should be used:

$$\text{Chlorine, mg/L} = \frac{(\text{Volume of PAO, mL})(\text{Volume adjusting constant})}{\text{Volume of sample, mL}} \quad (3.24)$$

For example, given the following:

Volume of phenylarsine oxide = 4.5 mL,
Volume of adjusting constant = 200 mg/L, and
Volume of sample = 100 mL, then
Chlorine residual, mg/L = (4.5 mL)(200 mg/L)/100 mL = 9 mg/L.

17.10 Reporting

For many facilities, chlorine residual is a discharge permit limit and is reported on the monthly discharge monitoring report (DMR). Chlorine residual may also be recorded on daily operations log sheets for process control. Chlorine is used in WRRFs for disinfection, odor control, and destruction of filamentous organisms. Chlorine residual is commonly written by tenths (one decimal) in milligrams per liter, such as 0.5 mg/L.

17.11 Troubleshooting and Tips

An important safety issue to keep in mind is to always add acid to water and to never add water to acid. See preparation of diluted sulfuric acid (H_2SO_4) and sodium hydroxide (NaOH) and of the acetate buffer in Section 17.5.

18.0 AMMONIA-NITROGEN—ION-SELECTIVE ELECTRODE METHOD

The following sections comprise an expanded version of Method 4500-NH_3 D or E-2011 of *Standard Methods* Online (APHA et al., 2017).

18.1 General Description

Ammonia-nitrogen (NH_3-N) is one of the four forms of nitrogen (N) of interest commonly present in wastewater and measured in the laboratory. Raw wastewater contains anywhere from 12 to 50 mg/L of ammonia-nitrogen. Ammonia exists in two forms in wastewater: an ionized form (ammonium ion), shown as NH_4^+ and as free ammonia, shown as NH_3. Free ammonia exists as a dissolved gas. The percentages of NH_4^+ and NH_3 present depend on the pH of the sample with NH_3 dominating above pH 9.3. Ammonia-nitrogen concentrations will be greatly reduced in the effluent or discharge of facilities that nitrify.

Ammonia-nitrogen is highly produced in wastewater by the following two ways:

- Deamination, or the taking away of a nitrogen and hydrogen group typically through degradation of organic nitrogen-containing compounds, and

- Hydrolysis of urea, or the breakdown of urea during its reaction with water.

The ammonia ion-selective electrode (ISE) method measures ammonia concentrations in aqueous matrices. As stated in *Standard Methods* Online (APHA et al., 2017), Method 4500 NH_3 D-2011, the ISE method uses a hydrophobic gas-permeable membrane to separate the sample solution from an electrode with an internal fill solution of ammonium chloride. Dissolved ammonia [$NH_{3(g)}$ and $NH_4^+{}_{(aq)}$] is converted into NH_3 (gas) by raising pH to above 11 with a strong base (NaOH). As the ammonia gas diffuses through the membrane, the fixed level of chloride in the internal fill solution is sensed by a reference electrode. Potentiometric measurements are made with a pH meter having an expanded millivolt scale or with a specific ion meter.

In general, direct determination of ammonia-nitrogen is limited to good-quality nitrified wastewater. Because of the varying quality of the sources of wastewater for which this method was developed, a preliminary distillation step, as described in Method 4500 NH_3 B from *Standard Methods* online (APHA et al., 2017) or Method 4500 NH_3 D-97 may be required by the state or regional regulating body.

18.2   Application

Ammonia-nitrogen reflects the progress of the breaking down of nitrogenous compounds in wastewater. Throughout treatment, the goal is to decrease the amount of ammonia-nitrogen. If the ammonia-nitrogen concentration does not decrease throughout the process, inadequate treatment is evident. Ammonia-nitrogen measurement is helpful in routinely determining process changes and efficiency during various stages of treatment. In addition, the daily determination of ammonia-nitrogen levels in the effluent is required on the NPDES permit for most, if not all, states.

18.3   Interferences

Readings are temperature-dependent. It is critically important that all samples and standards be allowed to come to room temperature before beginning the analysis.

Amines serve as a positive interference in ammonia-nitrogen testing and may be enhanced with the addition of acid. Mercury and silver interferences are present but can be eliminated with the use of sodium hydroxide and ethylene-diaminetetraacetic acid (NaOH/EDTA) as the buffer, rather than just NaOH. If a sample is suspected of containing residual chlorine, it should be treated with a dechlorination agent. In addition, almost all anion, cation, and dissolved species present will not interfere when using the ISE technology.

18.4 Apparatus and Materials

- High-performance ammonia ISE, Thermo Orion model 9512HPB-NWP (*Standard Methods* [APHA et al., 2017] lists model 95-10, but that model is currently not offered) or equivalent models such as EIL 8002-2 and Beckman 39565 (see Figure 3.22)
- Appropriate electrode membranes (from the manufacturer)
- Electrometer compatible with electrode of choice; pH meter with expanded millivolt (mV) scale capable of 0.1-mV resolution between -700 mV and $+700$ mV
- Magnetic stirrer, thermally insulated with a Teflon-coated stirring bar
- Stir plate
- Class A graduated cylinders, 100 mL for sample measurement
- Beakers (150 mL)
- Double-walled beaker (2000 mL)
- Class A volumetric flasks (various sizes)
- Class A volumetric pipets (various sizes)

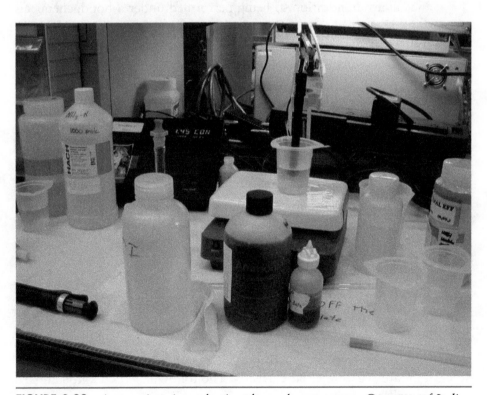

FIGURE 3.22   Ammonium ion-selective electrode apparatus. *Courtesy of Indigo Water Group*

- Class A graduated pipets (various sizes)
- Pipet bulb
- Micropipets and tips, adjustable 10 μL to 1000 mL
- pH strips
- Dropper
- Residual chlorine test strips, hand-held meter, or residual chlorine setup, and so on (to test for chlorine interferences)

18.5   Reagents

- High-grade, ammonia-free reagent water.
- NaOH/EDTA solution, 10 N: Dissolve 400 g of NaOH in 800 mL of reagent water in a 2000-mL double-walled beaker. Add 45.2 g of ethylene-diaminetetraacetic acid, tetrasodium salt, and tetrahydrate ($Na_4EDTA \cdot 4H_2O$) and stir until dissolved under a hood, as this reagent will produce heat and give off vapors. It is important for the operator–analyst to note that the reaction taking place while making this reagent is extremely exothermic and will produce a lot of heat and fumes. In addition to being performed under a hood, chemicals should be added slowly. Once cool, transfer the reagent to a 1000-mL volumetric flask and bring it up to volume. This reagent can alternatively be commercially purchased.
- Sodium thiosulfate (dechlorinating agent): Dissolve 3.5 g of sodium thiosulfate ($Na_2S_2O_3 \cdot H_2O$) in reagent water and dilute it to 1000 mL in a volumetric flask. Prepare fresh weekly.
- Ammonia electrode filling solution: This solution is purchased and available from the same manufacturer as the ISE probe. It is important to ensure that the solution is compatible with the probe in use.
- Ammonia electrode storage solution: The operator–analyst should refer to the electrode user guide of the probe.
- Concentrated sulfuric acid.

18.6   Stock and Standards

Two stocks should be prepared from separate sources (i.e., different vendor or lot number). The first will be used for the calibration curve and the second for the check standard. A second source is used to help make sure that the stock standard was prepared correctly. The following stock and standard solutions will be needed for the ammonia by ISE analysis:

- Stock ammonium chloride solution (1000 mg/L): Dissolve 3.819 g of anhydrous NH_4Cl (dried at 100 °C for about 1 hour and completely cooled in desiccator for about 30 minutes) in water and dilute to 1000 mL with reagent water. The stock is typically stable for up to 1 year.

- Standard ammonium chloride (NH_4Cl) solutions: Prepare by diluting the stock solution. Various concentrations can be made for the curve and quality control check standards. These concentrations are made by diluting the 1000-mg/L stock (see Section 9.3 of Chapter 2 for instructions to make standard dilutions); concentrations should be made fresh every day. Note that the concentrations chosen for the calibration curve should differ from one another by an order of magnitude. For example, select a group of standards with concentrations of 0.1, 10, 100, and 1000 mg/L or a group with concentrations of 0.2, 20, and 200 mg/L. The range of the calibration curve (i.e., lowest standard to highest standard) should bracket the expected ammonia-nitrogen concentrations in the samples. Samples with concentrations higher than the highest standard will need to be diluted. Samples with concentrations lower than the lowest standard should be reported as less-than values. Prepare the calibration curve from the first source 1000-mg/L ammonium chloride stock (see Table 3.27 for the recommended curve preparation). The 0.1-mg/L standard should be prepared from a 10-mg/L intermediate standard.

- Calibration verification: Prepare the calibration verification standards from the second source ammonium chloride 1000-mg/L stock using the method described in Section 9.3 of Chapter 2.

- Laboratory-fortified blank (LFB): Prepare the LFB from the second source ammonium chloride 1000-mg/L stock using the method described in Section 9.3 of Chapter 2.

TABLE 3.27 Standard preparation for ammonia ion-selective electrode method.

Final standard concentration (C_2)	Concentration of standard being used (C_1)	Volume of standard needed to make final concentration (V_1)	Final volume of solution (V_2)
0.10 mg/L	10 mg/L	1.0 mL	100 mL
1.00 mg/L	1000 mg/L	0.1 mL	100 mL
10.0 mg/L 100 mg/L	1000 mg/L	1.0 mL 10 mL	100 mL
	1000 mg/L		100 mL

Record in a logbook how the standards were prepared. If the facility is audited by regulators, they may want to verify the preparation of the standards.

18.7   Sample Collection, Preservation, and Holding Times

For sample collection,

- Samples are collected as a 24-hour composite, as per the NPDES permit
- Process samples can be collected as a grab or composite
- Samples can be collected in either plastic or glass containers
- Minimum sample size is 500 mL

In terms of preservation, for composite samples, a container that will hold the composited sample should be preserved with concentrated H_2SO_4, predetermined to completely preserve the sample to pH <2. The sample preservation must be initiated upon collection of the first aliquot. Composited samples must be cooled to ≤6 °C throughout the sampling process.

Grab samples (process control) should be preserved with concentrated H_2SO_4 and stored at ≤6 °C if not analyzed within 24 hours. If the sample is to be analyzed within 24 hours, it should be stored at ≤6 °C until analysis is performed. Unpreserved samples have a 24-hour hold time and preserved samples have a 28-day hold time.

18.8   Quality Assurance and Quality Control

As discussed in Section 11.1 of Chapter 2, quality control is used to measure overall method performance. Quality control is mandatory for compliance reporting. At a minimum, the instrument must be calibrated before analyzing process control samples, and check samples should be analyzed periodically. The following quality control samples are recommended by *Standard Methods* Online (APHA et al., 2017) for the ammonia-nitrogen ISE method:

- Calibration curve: The calibration curve needs to cover the expected concentration range of samples and meet acceptable slope requirements. Samples must not be reported below the lowest calibration point or above the highest point of the calibration curve. Because the lowest point of the curve is typically the lowest concentration of analyte that can accurately be measured, it is wise to set the reporting limit at the lowest point of the curve. If a sample is over the range of the calibration curve, it can be diluted and reanalyzed. *Standard*

Methods (APHA et al., 2017) suggests a calibration range of 0.1, 1.00, 10.0, 100, and 1000 mg/L for the method. The 1000-mg/L calibration is not needed for wastewater matrices, but the 0.1-mg/L calibration is needed as the effluent concentrations are typically low level. In addition, a blank is not used in this calibration curve for this method.

The generation of a calibration curve for the ISE method is required for every day of use or if electrode maintenance has been performed, especially if the operator–analyst is reporting for regulatory purposes. Additionally, a new calibration curve must be performed if any of the check standards or blanks fail.

- Calibration verification (check standard): A check standard is made from the second source 1000-mg/L ammonium chloride stock. The first check standard (initial calibration verification [ICV]) is the first piece of information after a successful calibration is completed to reflect the performance of the curve. Every check standard after that (continuing calibration verification [CCV]) is analyzed to confirm ongoing success of the calibration curve. Check standards can be made at a number of concentrations, although it is good laboratory practice to choose a low-level concentration for ICV and concentrations that are in the middle of the curve or as close to the concentrations of the samples themselves for the CCV. Check standards should be performed at a frequency of 1 per group of 10 samples and at the end of a batch. Although acceptance criteria are not stated in *Standards Methods* Online (APHA et al., 2017), they are generally ±10% (90% to 110%);

- Reagent blank: For this method, the reagent blank, or the *method blank*, should be less than the reporting limit. If the blank is higher than the MDL but less than the reporting limit, with samples higher than the reporting limit, a comment must be made to identify that the blank was found to have detectible limits of ammonia-nitrogen. If the blank is higher than the reporting limit, the source of contamination must be identified and eliminated before continuing. The reagent blank should be analyzed immediately after the (initial) check standard and at a frequency of 1 per batch or a group of 20 samples thereafter.

- Laboratory-fortified blank (laboratory control standard): An LFB is made from the second source 1000-mg/L ammonium chloride stock. A check standard should be analyzed at a concentration that is in the middle of the curve at a frequency of 1 per batch of 20 samples. Control limits for percent recovery acceptance are determined by creating a control chart plotting several results. The control chart can be updated periodically.

- Duplicates: Duplicates should be analyzed at a frequency of at least 1 per batch of 20 samples. The absolute RPD between duplicates must be calculated. A control chart should be generated to establish acceptance criteria.
- Laboratory-fortified matrix (LFM) (matrix spike): An LFM should be analyzed at a frequency of at least 1 per batch of 20 samples. A percent recovery must be calculated. Generally, it is acceptable to take the acceptance criteria generated for the LFB because sometimes the matrix of the sample can interfere with the addition of the spike.

For quality control calculations, the operator–analyst should refer to Section 12 of Chapter 2.

18.9   Procedure

- **Step 1**—Sample preparation:
  1. Bring 500 mL of sample to ambient temperature.
  2. Dechlorination is performed on samples suspected of having chlorine (effluent, contact chlorine basin, etc.). Check for chlorine by using residual chlorine strips, titration, or a meter (e.g., from Hach Company [Loveland, Colorado]). Follow individual instructions for chlorine meters.
  3. After residual chlorine has been quantified, add the appropriate amount of dechlorinating agent to the samples. Use 1 mL of reagent to remove 1 mg/L of residual chlorine in a 500-mL sample.
  4. Pour 100 mL of total volume using a graduated cylinder into a 150-mL beaker.
- **Step 2**—Sample measurement. Samples should be analyzed in the following order: a minimum of three calibration standards, ICV, reagent blank, sample 1, sample 2, and so on, up to 20 samples, followed by a CCV. The CCV should be analyzed once every 20 samples and again at the end of the analytical run. The operator–analyst may choose to run CCVs more frequently, especially when instrument drift is a concern. All samples must be bracketed by either an ICV and CCV or two CCVs. Additionally, one laboratory duplicate and one LFM (spike) should be analyzed for each batch or every 20 samples, whichever is more frequent.
  1. Before sample analysis can begin, the ammonia meter must be calibrated. Steps 5 through 10 should be followed for each standard. Be sure to follow the manufacturer's instructions for storing the calibration curve results in the meter. An acceptable slope, as per the manufacturer's instructions, must be achieved before

generating data. If the slope is out of range, the meter must be recalibrated.

2. Follow Steps 5 through 10 for the ICV standard. The ICV measured value should be within ±10% of the true value. If it is not, stop and determine the cause of the problem and begin again starting with the calibration curve.

3. Follow Steps 5 through 10 for the reagent blank. The measured result should be less than the lowest standard used.

4. Analyze the first sample after all opening quality control has been analyzed and is acceptable (curve, initial check standard, reagent blanks, and LFB).

5. Immerse the ISE electrode in sample.

6. Add a stir bar and stir the sample without creating a vortex.

7. Check the membrane for any air bubbles. Eliminate any air bubbles by gently tapping the electrode against the beaker or flicking the beaker with a finger. Bubbles on the membrane can affect response.

8. Add NaOH/EDTA buffer to the sample. Check the pH with a pH strip or pH meter to ensure that the sample is >11 s.u. If the sample is not at the proper pH, add more buffer in 1-mL increments until it is achieved. Keep track of how much buffer is added. This will be recorded on the benchsheet.

9. Record the meter result once it has stabilized. If the sample is over the calibration range, it should be diluted and reanalyzed.

10. Remove the electrode from the sample and rinse. Then blot it dry, making sure not to scratch or puncture the membrane.

11. Immerse in the next sample. Repeat these steps until all samples and required quality controls have been analyzed.

12. Once every 20 samples, analyze a CCV sample followed by a reagent blank.

- **Step 3**—Power down the instrument and store the probe as per the manufacturer's instructions.

- **Step 4**—Reduce raw data on the benchsheet by applying any dilution factors or additional buffer used (see Section 18.10).

- **Step 5**—Hand wash or use a dishwasher to clean sampling and analysis equipment as specified in Section 4.2 of Chapter 2.

18.10 Calculations

The ammonia ion-selective electrode generates an electrical signal, measured in millivolts, when exposed to ammonia. The electrical signal generated correlates to ammonia concentration. During calibration, the operator–analyst exposes the ion-selective electrode to increasing concentrations of ammonia as each standard solution is analyzed. Calibration procedures vary by manufacturer, but the end result is a calibration curve stored within the instrument that correlates millivolts generated by a standard or sample to its ammonia concentration. Most meters display ammonia concentration as a direct readout but may also be configured to display millivolts as desired by the operator–analyst.

Assuming the meter displays the ammonia-nitrogen concentration in milligrams per liter, the displayed result is recorded on the benchsheet as raw data. The analyst must also account for sample dilutions, if made, and for the amount of buffer solution added to the sample. The following formula is used to calculate the final result from the raw data generated by the meter:

$$\text{NH}_3\text{-N mg/L} = A \times B \times [(100 + D)/(100 + C)] \qquad (3.25)$$

Where

A = dilution factor,
B = result from the meter reading (mg/L),
C = volume of buffer added to each calibration standard (mL), and
D = volume of buffer added to the sample (mL).

The method calls for 1 mL of NaOH/EDTA buffer. When 1 mL is used, the formula is simply $\text{NH}_3\text{-N mg/L} = A \times B$. The operator–analyst will only need to apply the full formula if more than 1 mL of buffer is added to the samples.

18.11 Reporting

Results should be reported on the benchsheet in milligrams per liter and to three significant figures (i.e., 0.123, 1.23, 12.3, 123). Ammonia-nitrogen results should be reported to one decimal place on the DMR; if the result is less than 0.1 mg/L, however, report the result to two decimal places. In any instance, results that are less than the established reporting limit should be reported as less than the reporting limit value.

18.12 Troubleshooting and Tips

Table 3.28 provides a troubleshooting guide. Other tips for troubleshooting are as follows:

- Perform meter and electrode maintenance routinely (meter checks, changing electrode filling solution, membrane replacement). This will

TABLE 3.28   Troubleshooting guide for ammonia ion-selective electrode method.

	Issue	Action(s)
Slope	Slope outside Acceptable window of −57 mV ± 3 mV	• Verify or remake calibration working standards and/or stock solution. • Change electrode membrane and inner filling solution. • Perform a meter check. • Check appearance of inner probe body, coiling, and tip, and replace if needed (if inner probe is replaced, a new MDL study must be performed).
Standards	Recovery not within limits	• Verify slope. • Verify or remake working standards/stock solution. • Change electrode membrane and inner filling solution. • Perform a meter check.
Blank	Blank over acceptable limit	• Ensure that glassware or source of blank is not contaminated.
CCV (continuing calibration verification)	CCV recovery not within limits	• Verify slope. • Verify or remake working standards and stock solution. • Change electrode membrane and inner filling solution. • Perform a meter check.
Sample analysis	Drift, unstable readings, or slow electrode response	• Ensure that all connections are properly connected. • Change electrode membrane/inner filling solution. • Perform a meter check. • Check appearance of inner probe body, coiling, and tip, and replace if needed (if inner probe is replaced, a new MDL study must be performed). • Replace membrane on units having a loose membrane. • NOTE: If significant rehabilitation is done on the probe, a new standard curve needs to be prepared.

(continued)

TABLE 3.28   Troubleshooting guide for ammonia ion-selective electrode method.

Issue	Action(s)
Low ammonia concentration	• Make sure the stirrer is not too fast. • Make sure that sample readings are prompt after the addition of buffer. • Indication that pH is <11 standard units (add more buffer).
Unusual results	Reanalyze to verify result.

prevent the operator–analyst from having to troubleshoot most problems in the middle of an analysis.

- Do not add buffer prior to immersing the electrode in the sample or create a vortex while stirring the sample. This will result in a premature loss of ammonia.

- It is good laboratory practice to analyze samples from "clean" to "dirty"; for example, the effluent should be analyzed before the influent. This will highly reduce the possibility of carryover.

- The meter may have a mechanism that tells you when it is stable. If the result is visually unstable, it is important to wait a little bit longer. If not, a higher or lower result than intended may be reported.

- If the reading keeps declining or bottoming out at zero, the pH is probably not greater than 11 s.u. Check the level with a pH strip and add buffer in 1-mL increments until this is achieved.

- Do not use expired standards or reagents.

18.13 Benchsheet

The following items should be included on the benchsheet (see also Table 3.29 for a sample benchsheet):

- Operator–analyst's initials
- Date and time of analysis
- Sample date(s)
- Sample name or other type of identification
- Preparation date or other type of identification (if generated) of 1000-mg/L ammonium chloride stocks

TABLE 3.29 Sample benchsheet.

XYZ Wastewater Treatment Facility
Ammonia-Nitrogen (NH3-N)
Standard Methods, 4500 NH3 D, 21st Edition

Instrument Model and Serial #: _Thermo Orion 920 A / 56Y093856_   Analysis Date / Time: _11/9/2010_   Analyst: _JD_

Electrode Model and Serial #: _Thermo Orion 95-12 / 918475736_   Batch #: _1234567_

Intermediate Standard		Calibration Curve	
1st source 10 mg/L Intermediate Standard Prep Date:	_11/9/2010_	0.100 mg/L Curve Standard Prep Date:	_11/9/2010_
		1.00 mg/L Curve Standard Prep Date:	_11/9/2010_
2nd source 10 mg/L Intermediate Standard Prep Date:	_11/9/2010_	10.0 mg/L Curve Standard Prep Date:	_11/9/2010_

NaOH/EDTA Buffer Prep Date: 5/2/2010

Slope = —   57.8   mV/dec   Acceptable Range is -54 to - 60 mV/dec (this may vary per electrode manufacturer)

QC Prep Date	QC Check	Meter Reading	Final Conc. (mg/L)		
11/9/2010	Initial Calibration Verification (ICV) 0.100 mg/L	0.1047	0.105	ICV Rec= 105 % (90-110%)	
NA	Method Blank	0.0002	0.000		
11/9/2010	Laboratory Fortified Blank (LFB) 1.00 mg/L	1.124	1.12	LFB Rec= 112 % (85-115%) *(according to lab control chart of LFB)	

Collection Date	SAMPLE	Dilution Factor	mLs Buffer	Meter Reading	Final Conc. (mg/L)	Comments/QC Calc
11/8/2010	Effluent	1x	1	0.2334	0.233	
11/8/2010	Effluent Duplicate	1x	1	0.2221	0.222	Avg=0.230 mg/L, %RPD=9%
11/9/2010	Final Clarifier A1	1x	1	1.253	1.25	
11/9/2010	Final Clarifier A2	1x	2	0.9923	0.990	added additional buffer
11/9/2010	Final Clarifier A3	1x	1	0.6527	0.653	
11/9/2010	Final Clarifier B1	1x	1	0.8740	0.874	
11/9/2010	Final Clarifier B2	1x	1	0.5523	0.552	
11/9/2010	Final Clarifier B3	1x	1	0.6799	0.680	
11/9/2010	Influent A	20x	1	2.348	23.5	
11/9/2010	Influent A + 2.00 mg/L Spike	20x	1	4.451	44.5	% recovery=106%
11/9/2010	Influent B	20x	1	3.286	32.9	
11/9/2010	Influent C	20x	1	4.559	45.6	

QC Prep Date	QC Check	Meter Reading	Final Conc. (mg/L)		
11/9/2010	Continuing Calibration Verification (CCV) 5.00 mg/L	5.257	5.26	CCV Rec= 105 % (90-110%)	

Collection Date	SAMPLE	Dilution Factor	mLs Buffer	Meter Reading	Final Conc. (mg/L)	Comments/QC Calc
11/9/2010	Final Clarifier C1	1x	1	1.369	1.37	
11/9/2010	Final Clarifier C1 Duplicate	1x	1	1.333	1.33	Avg=1.35 mg/L, %RPD=3%
11/9/2010	Final Clarifier C2	1x	2	1.452	1.44	added additional buffer
11/9/2010	Final Clarifier C3	1x	1	1.5026	1.5	
11/9/2010	Final Clarifier C3 + 1.00 mg/L Spike	1x	1	2.489	2.49	% recovery=105%

QC Prep Date	QC Check	Meter Reading	Final Conc. (mg/L)		
11/9/2010	Continuing Calibration Verification (CCV) 3.00 mg/L	2.899	2.90	CCV Rec= 97 % (90-110%)	

- Preparation date or other type of identification (if generated) of standards
- Preparation date of NaOH/EDTA buffer
- Method reference
- Instrument and electrode model and serial number
- Location of analysis being performed
- Batch number for reporting results in the data system

19.0 NITRATE-NITROGEN—ION-SELECTIVE ELECTRODE METHOD

The following sections comprise an expanded version of Method 4500-NO$_3$ D-2016 of *Standard Methods* Online (APHA et al., 2017).

19.1 General Description

Like ammonia-nitrogen (NH$_3$-N), nitrate-nitrogen (NO$_3$-N) is another species of nitrogen that is of great interest in wastewater treatment. Unlike NH$_3$-N, NO$_3$-N is found only in small amounts in the influent, but increases in concentration as treatment progresses, resulting in concentrations of up to 30 mg/L in the effluent. Because bacteria feed on the organic nitrogenous material in wastewater, nitrate is the final stop of the nitrogen oxidation process.

As stated in *Standard Methods* (APHA et al., 2017), the principle behind the NO$_3$-N ion electrode is a selective sensor that develops a potential across a thin, porous, inert membrane that holds in place a water-immiscible liquid ion exchanger. The electrode response to NO$_3$-N ion activity is between 0.14 to 1400 mg/L NO$_3$-N/L. The lower limit of detection is determined by the small but finite solubility of the liquid ion exchanger.

19.2 Application

Because nitrogen as total organic nitrogen is a measure of the proteins and their intermediate decomposition production, it shows the strength of the wastewater. At certain stages of aeration, nitrate-nitrogen is produced through the rapid oxidization of nitrite-nitrogen (NO$_2$-N). Because nitrates are the final and most stable forms of oxidized nitrogen, the degree of treatment can be determined by performing NO$_3$-N analysis. Increasing nitrate concentrations in the effluent are a clear indication that there is a high degree of treatment.

19.3   Interferences

Potential interference can be minimized by the use of the buffer solution listed in Section 19.5. The silver sulfate (Ag_2SO_4) buffer solution removes Cl^-, Br^-, I^-, S_2^-, and CN^-; sulfamic acid removes NO_2-N; the buffer's pH 3 eliminates HCO_3 and maintains a constant pH and ionic strength; and $Al_2(SO_4)_3$ complexes organic acids.

19.4   Apparatus and Materials

- Nitrogen probe apparatus—double-junction reference electrode; Thermo Orion model 90-02 or equivalent, as per *Standard Methods* (APHA et al., 2017)
- Nitrate-sensing ion electrode, such as Thermo Orion 9307BNWP
- Appropriate electrode membranes (from the manufacturer)
- Electrometer compatible with electrode of choice; pH meter with expanded millivolt (mV) scale capable of 0.1-mV resolution
- Magnetic stirrer, thermally insulated with TFE-coated stirring bar
- Stir plate
- Beakers (50 mL)
- Double-walled beakers (2000 mL)
- Class A volumetric flasks (various sizes)
- Class A volumetric pipets (various sizes)
- Class A graduated pipets (various sizes)
- Pipet bulb
- Micropipets and tips adjustable 10 μL to 1000 mL
- Dropper

19.5   Reagents

- High-grade reagent water.
- Sodium hydroxide solution, 0.10 N: Dissolve 4.0 g of NaOH in 800 mL of reagent water in a 2000-mL double-walled beaker. Stir to dissolve under a hood, as this reagent will produce heat and give off vapors. Once cool, transfer to a 1000-mL volumetric flask and bring up to volume.
- Buffer solution: In a 2-L double-walled beaker, dissolve 17.32 g of $Al_2(SO_4)_3 \cdot 18H_2O$, 3.43 g of Ag_2SO_4, 1.28 g of H_3BO_3, and 2.52 g of

sulfamic acid (H_2NSO_3H) in about 800 mL of reagent water. Adjust to pH 3.0 with a pH meter by slowly adding 0.10 NaOH with a dropper. Transfer to a 1-L volumetric flask to bring up to volume. Store buffer in an amber glass jar. Because of the silver content of this solution, this reagent and the samples to which it is added may qualify as hazardous waste. Check local regulations prior to disposal.

- Reference electrode filling solution—dissolve 0.53 g of $(NH_4)_2SO_4$ in reagent water and dilute to 100 mL. Filling solution can also be commercially purchased for certain electrode manufacturers.

- Electrode storage solution: The operator–analyst should refer to the electrode user guide of the probe.

19.6   Stock and Standards

Two stocks should be prepared from separate sources (i.e., different vendor or lot number). The first will be used for the calibration curve and the second for the check standard. The following stocks and standard solutions will be required for the nitrate-nitrogen ion selective electrode method:

- Stock potassium nitrate solution (1000 mg/L): Dry 8 to 9 g of potassium nitrate (KNO_3) in an oven at 105 °C for 24 hours. Cool in desiccator for approximately 30 minutes. Dissolve 7.218 g in reagent water and dilute to 1 L. Preserve with 2 mL of $CHCl_3$ (chloroform). This stock is stable for at least 6 months.

- Standard potassium nitrate solutions: Various concentrations can be made for the curve and quality control check standards. These concentrations are made by diluting the 1000-mg/L stock (see Section 9.3 of Chapter 2 for instructions on how to make standard dilutions). These solutions are stable for several weeks.
  - Calibration curve: Prepare the calibration curve from the first source 1000-mg/L potassium nitrate stock every day of use (see Table 3.30 for the recommended curve preparation).
  - Calibration verification: Prepare check standards from the first source potassium nitrate 1000-mg/L stock daily, using the dilution formula described in Section 9.3 of Chapter 2.
  - Laboratory fortified blanks: Prepare LFBs from the second source potassium nitrate 1000-mg/L stock daily, using the dilution formula described in Section 9.3 of Chapter 2.

TABLE 3.30   Standard preparation for nitrate ion-selective electrode method.

Final standard concentration (C_2)	Concentration of standard being used (C_1)	Volume of standard needed to make final concentration (V_1)	Final volume of solution (V_2)
1.00 mg/L	1000 mg/L	0.1 mL	100 mL
10.0 mg/L	1000 mg/L	1.0 mL	100 mL
50.0 mg/L	1000 mg/L	5.0 mL	100 mL

19.7   Sample Collection, Preservation, and Holding Times

For sample collection,

- Process samples can be collected as a composite or grab sample
- Samples can be collected in either a glass or plastic container
- Minimum sample size is 100 mL

For preservations,

- Refrigerate to $\leq 6\ °C$
- Hold time of 48 hours

19.8   Quality Assurance and Quality Control

As discussed in Section 11.1 of Chapter 2, quality control is used to measure overall method performance. The following are quality controls used for the nitrate-nitrogen ion electrode method:

- Calibration curve: The calibration curve needs to be developed in such a way as to capture the concentration range of samples and to meet acceptable slope requirements. Samples must not be reported below the lowest calibration point or above the highest point of the calibrations curve. Because the lowest point of the curve is typically the lowest amount that an analyte can accurately measure, it is wise to set the reporting limit at the lowest point of the curve. If a sample is over the range of the calibration curve, it can be diluted and reanalyzed. *Standard Methods* (APHA et al., 2017) suggests a calibration range of 1.00, 10.0, and 50.0 mg/L. In addition, a blank is not used in this calibration curve for this method.

- ○ The generation of a calibration curve for the ISE method is required for every day of use. Additionally, a new calibration curve must be performed if any of the check standards or blanks fails.
- Calibration verification (check standard): A check standard is made from the first source 1000-mg/L potassium nitrate. The first check standard (ICV) is the first piece of information after a successful calibration is completed to reflect the performance of the curve. Every check standard after that (CCV) is analyzed to confirm ongoing success of the calibration curve. Check standards can be made at a number of concentrations, although it is good laboratory practice to choose concentrations that are in the middle of the curve or as close to the concentrations of the samples themselves. Check standards should be performed at a frequency of one per group of 10 samples and at the end of a batch. Although acceptance criteria are not stated in *Standards Methods*, they are generally ±10% (90% to 110%).
- Reagent blank: For this method, the reagent blank, or the *method blank*, should be less than the reporting limit. If the blank is higher than the MDL but less than the reporting limit, with samples higher than the reporting limit, a comment must be made to identify that the blank was found to have detectible limits of nitrate-nitrogen. If the blank is higher than the reporting limit, the source of contamination must be identified and eliminated before continuing. The reagent blank should be analyzed immediately after the (initial) check standard and at a frequency of one per batch or a group of 20 samples thereafter.
- Laboratory-fortified blank (laboratory control standard): An LFB is made from the second source 1000-mg/L potassium nitrate stock. A check standard should be analyzed at a concentration that is in the middle of the curve at a frequency of 1 per batch of 20 samples. Control limits for percent recovery acceptance are determined by creating a control chart plotting several results. Twenty data points are typically sufficient, and the control chart can be updated periodically.
- Duplicates: Duplicates should be analyzed at a frequency of at least 1 per batch of 20 samples. The absolute RPD between duplicates must be calculated. A control chart should be generated to establish acceptance criteria.
- Laboratory-fortified matrix (matrix spike): An LFM should be analyzed at a frequency of at least 1 per batch of 20 samples. A percent recovery must be calculated. Generally, it is acceptable to take the acceptance criteria generated for the LFB because sometimes the

matrix of the sample can interfere with the addition of the spike. The true value for the matrix spike is as follows:

$$\frac{[\text{Spike volume (mL)} \times \text{Concentration of standard (mg/L)}]}{[\text{Sample volume (mL)} + \text{Spike volume (mL)} + \text{Buffer volume (mL)}]} \qquad (3.26)$$

- Quality control calculations: The operator–analyst should refer to Section 11.3 of Chapter 2 for percent recovery and RPD calculations.

19.9 Procedure

Samples should be analyzed in the following order: a minimum of three calibration standards; ICV; reagent blank; sample 1, sample 2, and so on, up to 20 samples; followed by a CCV. The CCV should be analyzed once every 20 samples and again at the end of the analytical run. The operator–analyst may choose to run CCVs more frequently, especially when instrument drift is a concern. All samples must be bracketed by either an ICV and CCV, or two CCVs. Additionally, one laboratory duplicate and one laboratory-fortified matrix (spike) should be analyzed for each batch or every 20 samples, whichever is more frequent. The procedure is as follows:

- **Step 1**—Bring all standards and samples to room temperature.
- **Step 2**—Before sample analysis can begin, the nitrate meter must be calibrated. Follow the directions under Step 4 for each standard. Be sure to follow the manufacturer's instructions for storing the calibration curve results in the meter. The calibration slope must be between −54 and −60 mV. If the slope is out of range, the meter must be recalibrated.
- **Step 3**—Follow the directions under Step 4 for the ICV standard. The ICV measured value should be within ±10% of the true value. If it is not, the operator–analyst should stop, determine the cause of the problem, and begin again starting with the calibration curve.
- **Step 4**—Follow the directions under Step 4 for the reagent blank. The measured result should be less than the lowest standard used.
- **Step 5**—Analyze the first sample after all opening quality controls have been analyzed and are acceptable (curve, initial check standard, reagent blanks, and LFB).
  1. Measure 10 mL of sample (total volume) with a graduated pipet and place it in a 50-mL beaker.
  2. Add 10 mL of nitrate buffer with a graduated pipet to the sample.

3. Immerse electrode tip into the sample.

4. Add a stir bar and stir sample at a moderate rate.

5. Record the meter result once it has stabilized. If the sample is over the calibration range, it should be diluted and reanalyzed.

6. Remove electrodes from the sample and rinse and shake dry. Do not wipe or rub the electrode.

7. Immerse in the next sample. Repeat steps until all samples and required quality controls have been analyzed.

- **Step 6**—Power down the instrument and store the probe as per the manufacturer's instructions.

- **Step 7**—Reduce raw data on the benchsheet by applying any dilution factors or additional buffer used (see Section 18.10).

- **Step 8**—Step 8—hand wash or use a dishwasher as specified in Section 4.2 of Chapter 2.

19.10 Calculations

The nitrate ion-selective electrode generates an electrical signal, measured in millivolts, when exposed to nitrate. The electrical signal generated correlates to nitrate concentration. During calibration, the operator–analyst exposes the ion-selective electrode to increasing concentrations of nitrate as each standard solution is analyzed. Calibration procedures vary by manufacturer, but the end result is a calibration curve stored within the instrument that correlates millivolts generated by a standard or sample to its nitrate concentration. Most meters display nitrate concentration as a direct readout but may also be configured to display millivolts as desired by the operator–analyst.

Assuming the meter displays the nitrate-nitrogen concentration in milligrams per liter, the displayed result is recorded on the benchsheet as raw data. The analyst must also account for sample dilutions, if made.

19.11 Reporting

Results should be reported on the benchsheet in milligrams per liter and to three significant figures (0.123, 1.23, 12.3, etc.).

19.12 Troubleshooting and Tips

Table 3.31 provides a troubleshooting guide for the nitrate ion-selective electrode method. Other troubleshooting tips are as follows:

- Meter and electrode maintenance (e.g., meter checks and changing the electrode filling solution) should be performed routinely. This will

TABLE 3.31 Troubleshooting guide for nitrate ion-selective electrode method.

	Issue	Action(s)
Slope	Unacceptable slope	• Verify or remake calibration working standards and/or stock solution. • Change electrode membrane/inner filling solution. • Perform a meter check. • Check performance of the electrode (refer to instruction manual) and replace if needed.
Standards	Recovery not within limits	• Verify slope. • Verify or remake working standards/stock solution. • Change electrode tip and inner filling solution. • Perform a meter check.
Blank	Blank over acceptable limit	• Ensure that glassware or source of blank is not contaminated.
CCV (Continuing Calibration Verification)	CCV recovery not within limits	• Verify slope. • Verify or remake working standards/stock solution. • Change electrode membrane/inner filling solution. • Perform a meter check.
Sample analysis	Drift, unstable readings, or slow electrode response	• Ensure that all connections are properly connected. • Change electrode tip/inner filling solution. • Perform a meter check. • Check performance of the electrode (refer to instruction manual) and replace if needed.
	Unusual results	• Reanalyze sample. • Perform a standard check at the concentration to confirm result.

prevent the operator–analyst from having to troubleshoot most problems in the middle of an analysis.

- It is good laboratory practice to analyze samples from "clean" to "dirty" (low concentrations to higher concentrations). For most analytes, the effluent should be analyzed before the influent. This will greatly reduce the possibility of carryover and cross-contamination. For nitrate, the highest concentrations may be in the final effluent or immediately after an anoxic zone.

- Even though the meter has a mechanism that indicates when it is stable, it is important to keep an eye on the result. If the result is visually unstable, it is important to wait a little bit longer, or a higher or lower result than originally intended may be reported.

- Standard and samples should be allowed to reach the same temperature for precise measurement.

- Check the electrode for air bubbles. If bubbles are present, reimmerse the electrode in the solution and gently tap it.

19.13 Benchsheet

Table 3.32 provides a sample benchsheet for the nitrate ion-selective electrode method. In addition, the following are required items for a benchsheet:

- Operator–analyst's initials
- Date and time of analysis
- Sample date(s)
- Sample name or other type of identification
- Preparation date or other type of identification (if generated) of 1000 mg/L of potassium nitrate stocks
- Preparation date or other type of identification (if generated) of standards
- Preparation date of nitrate buffer
- Method reference
- Instrument and electrode model and serial number
- Location of the analysis being performed
- Batch number for reporting the result in the data system

20.0 PHOSPHORUS

The following sections comprise an expanded version of Method 4500-P-2011 of *Standard Methods* Online (APHA et al., 2017).

TABLE 3.32   Sample benchsheet for nitrate ion-selective electrode method.

XYZ Wastewater Treatment Facility
Nitrate-Nitrogen (NO3-N)
Standard Methods, 4500 NO3 D, 21st Edition

Instrument Model and Serial #:	Thermo Orion 920 A / 772Z89384	Analysis Date / Time: 11/9/2010	Analyst: JD
Reference Electrode Model and Serial #:	Thermo Orion 90-02 / 64U87936	Batch #: 1234567	
Sensing Electrode Model and Serial #:	Thermo Orion 9307BNMP / 587934800		

Calibration Curve

1.00 mg/L Curve Standard Prep Date:	6/26/2010
5.00 mg/L Curve Standard Prep Date:	6/26/2010
50.0 mg/L Curve Standard Prep Date:	6/26/2010

Nitrate Buffer Prep Date: 5/2/2010

Slope = — 57.8   mV/dec   Acceptable Range is -54 to - 60 mV/dec *(this may vary per electrode manufacturer)*

QC Prep Date	QC Check	Meter Reading	Final Conc. (mg/L)	
11/9/2010	Initial Calibration Verification (ICV) 1.00 mg/L	0.9574	0.957	ICV Rec= 96 % (90-110%)
NA	Method Blank	0.2786	0.279	
11/9/2010	Laboratory Fortified Blank (LFB) 10.0 mg/L	10.578	10.6	ICV Rec= 106 % (90-110%) *(according to lab control chart of LFB)

Collection Date	SAMPLE	Dilution Factor	Meter Reading	Final Conc. (mg/L)	Comments/QC Calc
11/8/2010	Effluent	1x	27.57	27.6	
11/8/2010	Effluent Duplicate	1x	27.82	27.8	Avg=27.8mg/L, %RPD=1%
11/9/2010	Final Clarifier A1	1x	25.50	25.0	
11/9/2010	Final Clarifier A2	1x	23.11	23.1	
11/9/2010	Final Clarifier A3	1x	22.59	22.6	
11/9/2010	Final Clarifier B1	1x	24.01	24.0	
11/9/2010	Final Clarifier B2	1x	22.78	22.8	
11/9/2010	Final Clarifier B3	1x	23.58	23.6	
11/9/2010	Final Clarifier C1	1x	19.22	19.2	
11/9/2010	Final Clarifier C2	1x	22.04	22.0	
11/9/2010	Final Clarifier C3	1x	20.99	21.0	
11/9/2010	Final Clarifier C3 + 9.95 mg/L Spike	1x	30.22	19.2	% recovery=93%

QC Prep Date	QC Check	Meter Reading	Final Conc. (mg/L)	
11/9/2010	Continuing Calibration Verification (CCV) 30.0 mg/L	29.57	29.6	CCV Rec= 99 % (90-110%)

Collection Date	SAMPLE	Dilution Factor	Meter Reading	Final Conc. (mg/L)	Comments/QC Calc

QC Prep Date	QC Check	Meter Reading	Final Conc. (mg/L)

20.1   General Description

Phosphorus is an essential nutrient for cellular activity and growth. It is found in three forms, orthophosphate (PO_4^{3-}), polyphosphate (orthophosphate chains), or organically bound phosphorus. Ortho forms predominate in most natural and wastewaters. Sources in wastewater include body wastes, food residues, and detergents. Phosphorus is removed from the waste stream biologically by wasting microorganisms containing organically bound phosphorus and chemically by precipitation and adsorption with metal salts, most commonly aluminum sulfate (alum) or ferric chloride (ferric). In enhanced biological phosphorus removal (EBPR) systems, the addition of an anaerobic basin favors the cultivation of microorganisms that can store an excess amount of phosphorus. Phosphorus discharged to the receiving water can cause algal blooms and subsequent oxygen depletion.

In this phosphorus removal procedure (APHA et al., 2017), the phosphorus concentration is determined colorimetrically. The absorbance is measured and compared to a calibration curve of known standards. Ammonium molybdate and antimony potassium tartrate react in an acid medium with dilute solutions of phosphorus to form an antimony-phospho-molybdate complex. This complex is reduced to an intensely blue-colored complex by ascorbic acid. The color is proportional to the phosphorus concentration. Only orthophosphate forms a blue color in this test. Samples for total phosphorus must be digested before analysis. Polyphosphates and organic phosphorus compounds must be converted to the orthophosphate form by a persulfate digestion. Samples can be filtered through a 0.45-μm filter for determination of dissolved orthophosphate or total dissolved phosphorus.

20.2   Application

Phosphorus is analyzed for permit reporting and process control. Most permits require final effluent and receiving water phosphorus concentration limits. Normal domestic wastewater ranges from 2 to 20 mg/L total phosphorus, of which approximately 25% will be organically bound phosphorus and 75% will be orthophosphate. In a conventional activated sludge process, a good nutrient balance is 100:5:1 (BOD:N:P). Phosphorus deficiencies may lead to the flourishing of filamentous bacteria such as *Thiothrix* spp. and Types 021N, 0041, and 0675.

Biological phosphorus removal systems should produce effluent at concentrations of 1 mg/L or less. Enhanced biological phosphorus removal systems are often capable of effluents with 0.1 mg/L. These BPR and EBPR systems require a BOD:P of 20:1 to 25:1. Nitrate concentrations greater than 1 mg/L in anaerobic selector zones will prevent the release of phosphorus and inhibit the BPR process.

It is important for the operator–analyst to be aware that facility sidestream processes, such as digester supernatant or sludge dewatering centrates and filtrates, can have extremely high phosphorus concentrations averaging 100 mg/L total phosphorus. If ignored, these sidestreams can be detrimental to BPR systems. Therefore, consistent sidestream monitoring and flow management is necessary for optimum process performance.

Generally, chemical phosphorus treatment is needed to achieve concentrations below 0.1 mg/L. In chemical phosphorus removal processes, influent and effluent total phosphorus concentrations are determined to calculate removal efficiencies and optimize chemical use. Calculating the ratio of moles of aluminum or iron dosed to moles of influent phosphorus or moles of phosphorus removed, and plotting this ratio against effluent phosphorus concentration, provides a useful operational tool for controlling chemical doses (Figure 3.23). Finally, although a stoichiometric dose can be calculated, actual chemical doses are highly specific to the individual wastewater.

20.3 Interferences

Arsenates react with the molybdate reagent to give a blue color similar to that formed with phosphate; concentrations as low as 0.1 mg/L will interfere with the phosphorus determination. Hexavalent chromium and nitrite interfere to give results about 3% low at concentrations of 1 mg/L and 10% to 15% low at concentrations of 10 mg/L.

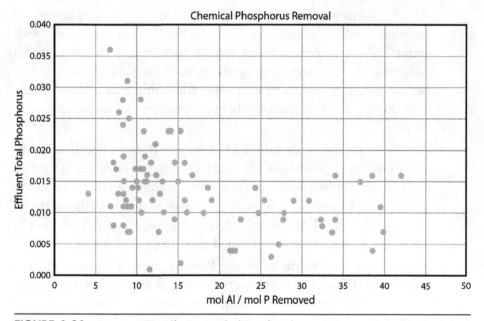

FIGURE 3.23 Process sample control chart for determining molar chemical feed ratios. *Courtesy of Chris Maher*

20.4   Apparatus and Materials

- Erlenmeyer flasks (125 mL)
- Graduated cylinder (50 mL, Class A, and marked TD)
- Measuring ceramic spoon for ammonium persulfate (0.4 g)
- Volumetric pipets (1, 5, 10, 25, and 50 mL)
- Graduated pipet (10 mL)
- Pipet bulb
- Eyedroppers with bottles (two) for phenolphthalein and 5-N H_2SO_4
- Pump-style auto-pipetor with 8-mL capacity
- Buret (50 mL) with a stand for titration with 6-N NaOH
- Hot plate in the fume hood
- Spectrophotometer with 1- and 5-cm cells and adapters. A path length of 5 cm allows for more accurate results at low levels. Not all spectrophotometers have this option and, when they do, it is an added expense. Not all facilities will have low-level samples. The use of the 5-cm path length for low-level samples should be considered optional.

20.5   Reagents

- Sulfuric acid (5 N): Prepare 5 N sulfuric acid solution by diluting 140 mL of concentrated H_2SO_4 (96% to 98%; 36 N) to a final volume of 1000 mL with distilled water; fill a 1000-mL volumetric flask with about 700-mL deionized water. Slowly add 140 mL of concentrated sulfuric acid. The reaction is exothermic, so the solution will be hot. Let the solution cool to room temperature, dilute to a final volume of 1000 mL, and mix thoroughly by stoppering and inverting several times.

- Potassium antimony tartrate (PAT) solution: Prepare potassium antimony tartrate [$K(SbO)C_4H_4O_6 \cdot 1/2\ H_2O$] solution from 1.3715 g PAT and enough distilled water to reach a final volume of 500 mL; dissolve 1.3715 g PAT in about 400 mL of deionized water in a 500-mL volumetric flask. After PAT is completely dissolved, dilute the solution to 500 mL, mix, and store in a labeled glass-stoppered bottle.

- Ammonium molybdate solution: Prepare ammonium molybdate, $(NH_4)_6Mo_7O_{24} \cdot 4H_2O$, solution from 20 g of ammonium molybdate and enough distilled water to reach a final volume of 500 mL; dissolve 20 g of ammonium molybdate in about 400 mL of deionized water in a 500-mL volumetric flask. Dilute to 500 mL, mix, and store in a labeled glass-stoppered bottle.

- Ascorbic acid solution: Prepare ascorbic acid solution from 1.76 g of ascorbic acid and enough distilled water to achieve a final volume of 100 mL; dissolve 1.76 g of ascorbic acid in about 80 mL of deionized water in a 100-mL volumetric flask. Dilute to 100 mL, mix, and refrigerate at <6 °C. The solution will be stable for 5 days.

- Combined reagent: Make the combined reagent fresh for each analytical run. Begin by removing the PAT and ammonium molybdate solutions from the refrigerator and let them warm up to room temperature. Prepare ascorbic acid solution, as needed, for the number of samples. Combine reagents in the auto-pipet in the order given in Table 3.33. Mix after adding each reagent. The combined reagent is only good for 4 hours.

- Strong acid solution: Prepare strong acid solution by diluting 300 mL of concentrated H_2SO_4 (96% to 98%; 36 N) with enough distilled water to achieve a final volume of 1000 mL; fill a 1000-mL volumetric flask with about 600 mL of deionized water. Slowly add 300 mL of concentrated sulfuric acid. The reaction is exothermic, so the solution will be hot. Let the solution cool to room temperature, dilute it to 1000 mL, and mix.

- Phenolphthalein solution: Phenolphthalein disodium salt equals 0.5 g and distilled water equals 100 mL; dissolve 0.5 g of phenolphthalein disodium salt in about 80 mL of deionized water in a 100-mL volumetric flask. Dilute to 100 mL and mix.

- Sodium hydroxide, 6 N: NaOH equals 120 g and distilled water equals 500 mL; fill a 500-mL volumetric flask with about 400 mL of

TABLE 3.33   Combined reagent table for phosphorus analysis.

Add in this order		1	2	3	4	
No. of samples	Batches	5N H_2SO_4	PAT	AM	Ascorbic acid (g)	DI H_2O for ascorbic
11	1	50	5	15	0.528	30
23	2	100	10	30	1.056	60
30	2.5	125	12.5	37.5	1.32	75
36	3	150	15	45	1.584	90
42	3.5	175	17.5	52.5	1.848	105
48	4	200	20	60	2.112	120

deionized water. Put a stir bar in the flask and stir on the magnetic stir plate. Slowly add 120 g of NaOH pellets through a funnel. The reaction is exothermic, so the solution will be hot. Put the volumetric flask in an ice bath if the solution is needed quickly and let it cool to room temperature. Retrieve the stir bar while rinsing the retriever and bar; dilute to 500 mL and mix.

- Ammonium persulfate: Reagent-grade ammonium persulfate, crystal (CAS NO: 7727-54-0).

20.6 Stock and Standards

- Stock solution (100 mg/L): Potassium dihydrogen phosphate (KH_2PO_4) equals 0.2195 g and distilled water equals 500 mL; dissolve 0.2195 g of pre-dried KH_2PO_4 (approximately 0.5 g) at 105 °C for 1 hour, allow to cool in a desiccator in about 400 mL of deionized water, and dilute to a final volume of 500 mL.
- Standard phosphate solution (1 mg/L): Stock solution equals 10 mL and distilled water equals 1000 mL; add 10 mL of 100-mg/L stock phosphorus solution to approximately 800 mL of deionized water in a 1000-mL volumetric flask. Dilute to a final volume of 1000 mL.

Table 3.34 provides a chart of working standard dilutions.

20.7 Sample Collection, Preservation, and Holding Times

Phosphorus samples should be collected in plastic or glass bottles that have been acid washed with 1:1 hydrochloric or nitric acid. Do not touch the insides of the containers or lids, especially with bare fingers.

If testing for orthophosphate, the sample should be filtered through a 0.45-μm filter and stored unpreserved at 4 °C in an acid-washed glass

TABLE 3.34 Working standard dilutions.

Standard concentration C_1 (ppm) 1	Volume of standard needed V_1 (mL) 0	For standard dilution concentration of C_2 (ppm) 0	Dilute to final volume of V_2 (mL) 50
1	1	0.02	50
1	5	0.1	50
1	10	0.2	50
1	25	0.5	50
1	50	1.0	50

bottle. If the operator–analyst is testing only for total phosphorus, 1 mL of concentrated H_2SO_4 per liter of sample should be added (pH <2) and stored in an acid-washed plastic or glass bottle at ≤ 6 °C. Alternatively, the sample should be allowed to freeze at or below 10 °C. The operator–analyst should be aware that phosphorus may be adsorbed onto the surface of plastic bottles.

The hold times are 48 hours for orthophosphate samples and 28 days for total phosphorus samples. Hold time begins at the end of composite collection period.

20.8   Quality Assurance and Quality Control

- Method blank: For each analysis, a method blank of deionized water carried through the entire procedure should be run. The method blank is used to zero the spectrophotometer.

- Continuing calibration verification: The CCV should be prepared identically to the 0.2-mg/L standard dilution. The CCV should be determined to be 80% to 120% of the true value or between 0.16 and 0.24 mg/L.

- Duplicate: A sample should be randomly selected and a duplicate volume should be analyzed. The RPD should be calculated between the samples.

- Laboratory-fortified blank: A second stock and standard solution should be prepared from a second source of potassium dihydrogen phosphate (KH_2PO_4) (i.e., different lot number or manufacturer). This set of stock and standard should be used to prepare the LFB identically to the 0.5-mg/L standard dilution.

- Laboratory-fortified matrix: The second set of stock should be used to prepare the LFM. The LFM will vary depending on the concentration of the sample being fortified. The operator–analyst should refer to Section 12.3 of Chapter 2 for guidance. For a 50-mL sample, no more than 1 mL of standard should be added as the spike.

20.9   Procedure

- Step 1—Calibration curve:
  1. Remove PAT, ammonium molybdate, and the standard from the refrigerator and let them warm up to room temperature.
  2. Turn on the spectrophotometer and set it to 880 nm. If the calibration curve is for total phosphorus, preheat the hot plate to a medium-high level.

3. Set up six numbered 125-mL Erlenmeyer flasks on the laboratory bench. Make up the standard dilutions given in Table 3.34 as follows:

   (a) Pour approximately 100 mL of 1 mg/L phosphorus standard into a clean, dry beaker.

   (b) Using volumetric pipets, pipet the given amount of standard into the flask; swirl to mix.

   (c) Using the 50-mL Class A graduated cylinder marked TD, measure the given amount of deionized water and pour consistently into the flask.

4. For ortho-phosphorus curves, proceed to Step 3. For total phosphorus curves, proceed to Step 4. Standard dilutions should be analyzed per these procedures, then a calibration curve should be plotted as described in Section 20.10. A new calibration curve should be created whenever a new standard solution is prepared.

- **Step 2**—Sample preparation:

1. Line up the samples (e.g., final effluent, receiving water, tertiary effluent, secondary effluent, primary effluent, and influent) on a laboratory bench in order from cleanest (i.e., lowest phosphorus concentration) to dirtiest (i.e., highest phosphorus concentration). In front of the sample bottles, set up numbered 125-mL Erlenmeyer flasks. Include flasks for the necessary quality control samples.

2. Prepare 50-mL samples diluted as necessary to give <1 mg/L P. Secondary clarifier effluent is typically diluted by 5 (use a 10-mL sample) and influent is typically diluted by 10 (use a 5-mL sample). Tertiary and final effluent and receiving water are typically full strength (use a 50-mL sample).

3. Using a 50-mL Class A graduated cylinder marked TD, add 50 mL of deionized water to the blank, then add dilution water as needed to the corresponding flasks including the quality control samples.

4. Use a 10-mL volumetric pipet to add 10 mL of standard to the standard flask.

5. Use a 10-mL graduated pipet to add clarifier and influent samples, rinsing the pipet between additions.

6. Working from cleanest to dirtiest, measure the full-strength samples as follows:

   (a) Gently invert the sample five times.

   (b) Rinse the graduated cylinder with the sample.

    (c) Measure 50 mL and transfer it to the flask.

    (d) Clean the graduated cylinder by rinsing it with deionized water three times, discarding rinses into sink.

    (e) Repeat for each sample.

    7. For orthophosphate, continue to Step 3. For total phosphorus, go to Step 4.

- **Step 3**—Analyze samples for orthophosphate:

  1. Add 1 drop of phenolphthalein. If the sample turns pink, add 5-N H_2SO_4 drop-wise until the sample is clear.

  2. Use the auto-pipettor to add 8 mL of combined reagent and swirl to mix.

  3. Let color develop for at least 10 minutes, but no more than 30 minutes.

  4. Fill 1-cm cells with blank, CCV, and high-range samples (clarifier and influent). Insert the blank and press "0 ABS." Read out and record absorbance of CCV and all samples.

  5. Change the cell adapter in the spectrophotometer.

  6. Fill 5-cm cells with blank, CCV, and low-range samples. Insert the blank and press "0 ABS." Read out and record absorbance of the CCV and all samples.

  7. Calculate orthophosphate using the calibration curve.

- **Step 4**—Analyze samples for total phosphorus:

  1. Add 1 mL of strong acid to each flask using a 10-mL graduated pipet.

  2. Add 0.4 g of ammonium persulfate, which is a level ceramic spoonful.

  3. Swirl each flask to mix.

  4. Place flasks on the hot plate, turn on the fume hood, and gently boil down to about 10 mL, which is near the widest point on the 125-mL Erlenmeyer flask. This should take approximately 45 minutes. Do not allow the flasks to boil to dryness.

  5. While samples are boiling down, make the combined reagent.

  6. As samples reach the 10-mL mark, remove them from heat and let them cool. The last sample removed should cool for 15 minutes. Turn off the hot plate.

  7. Place flasks back in order on the bench.

8. Rinse the sides of each flask with two rotations of a deionized wash bottle. This should give a total volume of about 30 mL.

9. Add one drop of phenolphthalein.

10. Slowly titrate to a consistent light pink with 6-N NaOH.

11. Add 5-N H_2SO_4 one drop at a time, swirling in between, to turn samples clear again.

12. Return samples to a final volume of 50 mL, working from cleanest to dirtiest:

    (a) Pour the sample from the flask into a 50-mL graduated cylinder.

    (b) Dilute to the 50-mL mark using a deionized wash bottle.

    (c) Pour the sample back into the flask in a consistent manner.

    (d) Clean the graduated cylinder by rinsing it with deionized water three times. Discard the rinse water into the sink.

    (e) Repeat for each sample.

13. Use the auto-pipetor to add 8 mL of combined reagent and swirl to mix.

14. Let color develop for 20 minutes.

15. Fill 1-cm cells with blank, CCV, and high-range samples (clarifier and influent). Insert the blank and press "0 ABS." Read out and record absorbance of the standard and all samples.

16. Change the cell adapter in the spectrophotometer.

17. Fill 5-cm cells with blank, CCV, and low-range samples. Insert the blank and press "0 ABS." Read out and record absorbance of CCV and all the samples.

18. Calculate total phosphorus using the calibration curve.

- **Step 5**—Glassware cleanup: Glassware should be cleaned immediately after performing analysis. Samples and leftover combined reagent can typically be disposed of down the drain; however; the antimony (tin) contained in PAT may qualify any waste containing combined reagent as hazardous. Therefore, the operator–analyst should check local regulations. If another phosphorus analysis is going to be performed immediately, a triple rinse of deionized water will suffice. Otherwise, all glassware should be acid washed with 1 : 1 HCl (see Section 4.2.1 of Chapter 2 for instructions on acid washing glassware).

    1. If the glassware is reserved only for phosphorus and stored upside down or full of deionized water, acid washing may be needed only occasionally. The operator–analyst should be aware that many detergents contain phosphate.

20.10 Calculations

Data from the calibration curve is used to calculate the phosphorus concentration of the samples. A calibration curve must be developed for both the 1- and 5-cm path lengths. These curves can be produced manually on graph paper, or a computer spreadsheet program can be used. Spreadsheets are preferable because they can easily perform a linear regression of the data and give an equation used to calculate the sample concentrations. Data should be entered into the spreadsheet as in Table 3.35.

Both columns of data should be highlighted and charted in an x–y format to produce a graph like that in Figure 3.24. Under the chart menu, the operator–analyst should select the option to add a trend line and then select a linear or linear regression trend. The operator–analyst should also opt to display the equation and R value (see Figure 3.25).

Solving the equation of the line for "x" (mg/L) yields the equation to find sample concentrations from the absorbance "y" in the equation for the line A as follows:

$$\frac{(A - 0.0007)}{0.6142} = \text{Sample concentration mg/L} \qquad (3.27)$$

This equation can easily be entered into a spreadsheet so that the operator–analyst only has to enter absorbances and the program returns concentrations. For diluted samples, it is important to multiply by the dilution factor. Of course, the actual numbers used here are only an example; each operator–analyst will have his or her own calibration curve and equation.

20.11 Reporting

- Results are reported in milligrams per liter as phosphorus
- Significant figures = three figures
- The MDL = 0.01 mg/L for the 5-cm path length

TABLE 3.35   Standard concentration versus absorbance in phosphorus method.

Standard dilution concentration, ppm	Absorbance
0.000	0.000
0.020	0.013
0.100	0.067
0.200	0.128
0.500	0.321
1.000	0.642

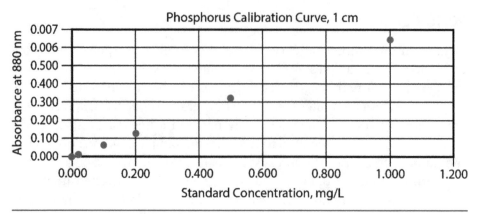

FIGURE 3.24   Phosphorus calibration curve.

20.12 Troubleshooting and Tips

- Contaminated glassware is the bane of the phosphorus test. Phosphorus is found in many detergents and cleaners, in tap water, in dust that can settle in uncovered glassware, and in dozens of other sources one may not even consider. The lower the concentration of phosphorus in the sample, the more critical it becomes to have absolutely clean glassware.

- Reserve a set of glassware to be used only for phosphorus analysis. Clean the glassware immediately after use and store it in an enclosed cabinet.

- Strive for consistency in technique within the analysis and between the calibration curve and the analysis. Errors made in the calibration

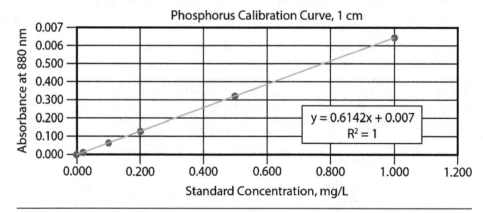

FIGURE 3.25   Completed phosphorus calibration curve with linear regression.

curve will be eliminated if they are also made in the analysis. Work slowly and methodically when titrating samples to get a consistent light-pink tint. This will ensure that all samples are at a similar pH. In addition, be consistent when transferring samples from flasks to graduated cylinders and back.

- Trying to see the phenolphthalein color change against a black laboratory bench can be difficult. It can help to lay out samples and flasks on a strip of white paper towel.

20.13 Benchsheet

Table 3.36 provides a sample benchsheet for phosphorus.

21.0 TEMPERATURE

The following sections comprise an expanded version of Method 2550 B-2010 of *Standard Methods* Online (APHA et al., 2017).

21.1 General Description

The temperature of wastewater indicates the amount of thermal energy it contains. It is measured in degrees Celsius or degrees Fahrenheit.

Accurate temperature readings are necessary to calculate dissolved oxygen saturation values, correlate biological activity, and for many other purposes. The reported temperature should represent the process or waterbody being monitored at the time of sampling. Therefore, the temperature must be taken at the sampling point.

21.2 Application

In general, the rate of biological activity depends on temperature. As temperature increases, microorganisms accelerate consumption of organics and use of oxygen in the wastewater. The reaction rates approximately double with every 10 °C (18 °F) increase in temperature until higher temperatures begin to inhibit biological activity.

In areas of the country subject to wide temperature swings between summer and winter, WRRF operator–analysts must be aware of the effect that changes in temperature will have on the activity of the microorganisms; the warmer the temperature, the higher the activity. Table 3.37 shows expected temperature ranges for wastewater processes.

TABLE 3.36 Sample benchsheet for phosphorus test.

Method: _____ _____ Date/time digestion: _____

Date: _____ _____ Date/time analysis: _____

Analyst: _____ _____

Sample date	Sample identification	A Sample volume (mL)	B Total volume (mL)	B/A = C Dilution factor	D Instrument reading (ABS/% T)	E Blank value	D–E Corrected instrument reading	F Concentration (mg/L)	F × C Final concentration (mg/L)	Quality control
	Reagent blank	✕	✕	✕		✕	✕	✕		

TABLE 3.37 Expected temperature ranges for wastewater processes.

Process	Typical temperature ranges	
	°F	°C
Influent	65 to 85	18 to 29
Effluent	60 to 95	16 to 35
Receiving waters	60 to ambient temperature	16 to ambient temperature
Digester (recirculated sludge prior to heat exchanger and supernatant)	60 to 100	16 to 38
Mesophilic digestion	90 to 100 °F	32 to 38 °C
Thermophilic digestion	131 °F or higher	55 °C or higher

Source: WEF (2017); WEF et al. (2018).

21.2.1 Ponds

Optimum temperature for operating ponds is 20 °C, except for anaerobic ponds for which optimum temperature is 30 °C. Oxygen solubility is greater at lower temperatures than at higher temperatures.

21.2.2   Facility Influent

A significant increase in temperature for a short period of time typically indicates the presence of an industrial discharge. A significant drop in temperature often indicates intrusion of stormwater. In areas of the country subject to wide temperature swings between summer and winter, WRRF operator–analysts must be aware of the effect that changes in temperature have on microorganism activity; the warmer the temperature, the higher the activity.

The maximum amount of oxygen in water or wastewater is temperature-dependent. Colder water is capable of containing more dissolved oxygen than warmer water. However, colder water may actually contain less dissolved oxygen than warmer water, depending on conditions in the water.

21.2.3   Clarification (Primary and Secondary)

Wastewater temperature influences the rate of settling. Warm weather increases the rate of biological activity, thus diminishing the freshness of wastewater in the collection system and promoting gasification in the settling basins and slow settling. On the other hand, the lower viscosity of warm water compared to that of cold water allows particles to settle faster. At a water temperature of 27 °C (80 °F), the settling rate exceeds that at 10 °C (50 °F) by nearly 50%.

Primary tank efficiency decreases during the winter months because of high wastewater viscosity at cold temperatures. Another reason for poorer winter performance is the increased density of cold wastewater. As water density increases, settling rates decrease.

Temperature differences, and associated density differences, between tank contents and incoming flow may create density currents within the basin. These currents can interfere with effective particle settling character-istics, thereby impairing settling performance.

Removal of grease and scum (floatables) is affected by temperature. At summer temperatures, grease and scum may tend to stay in suspension, and some may enter the settled sludge instead of floating to the surface. Lower wastewater temperatures require floatable solids to be skimmed more frequently.

21.2.4   Secondary Treatment

Nitrification occurs over a wide range of temperatures; a reduction in temperature decreases the reaction rate. In warmer climates, nitrification has been observed at MCRT values of 3 days or fewer, whereas in colder climates MCRT values greater than 20 days may be required to achieve effective nitrification. Denitrification processes are temperature-sensitive, decreasing with decreased temperature.

21.2.5   Filtration

Filter media expansion is dependent on the backwash rate, which, in turn, depends on the water temperature and the filter media's size and specific gravity. A rise in water temperature from 10 to 20 °C (50 to 68 °F) requires the wash rate to increase about 30% to maintain a given expansion of silica sand. The amount of filter aid needed increases as water temperature decreases.

21.2.6   Anaerobic Digestion

Temperature-related stress is caused by a change in digester temperature of more than 1 or 2 °C (2 or 3 °F) in fewer than 10 days, which reduces the biological activity of methane-forming microorganisms. If the methane formers are not quickly revived, the acid formers, which are unaffected by temperature change, continue to produce volatile acids that will eventually consume available alkalinity and cause the pH to decline. Digester temperatures should not be changed more than 1 °F (0.6 °C) per day to allow the microorganisms to acclimate to the new environment. The most typical causes of temperature stress are overloading solids and exceeding the instantaneous capacity of the heating system. Most heating systems can eventually heat the digester contents to operating temperature, but not without a harmful temperature variation.

Another cause of temperature stress is operating the digester outside its optimum temperature range. For example, a mesophilic digester has an optimum temperature range of 32 to 38 °C (90 to 100 °F). At temperatures lower than 32 °C (90 °F), the biological process slows. At temperatures above 38 °C (100 °F), the digester efficiency is not improved and the system is wasting energy. Thermophilic digesters operate at temperatures between 55 and 60 °C (131 and 140 °F).

The liquid temperature in an aerobic digester significantly affects the rate of volatile solids reduction that increases as temperature increases. As with all biological processes, the higher the temperature, the greater the efficiency. At temperatures lower than 10 °C (50 °F), the process is less effective. In most aerobic digesters, temperature is a function of ambient weather conditions and is not controlled.

21.2.7   Biosolids Dewatering

21.2.7.1   Polymer Makeup

Typically, dilute solutions of emulsion or dry cationic polymers are used in the dewatering process. The polymers are temperature-sensitive and generally degrade at temperatures higher than 50 °C (120 °F). When making up a polymer solution, the water temperature should be between 60 and 90 °F (16 to 32 °C).

21.2.7.2   Centrifuge

Centrifuge performance is linked to sludge temperature, that is, "warmer" is better up to a limit of 60 °C (140 °F). Centrifuges operate on the density difference between solids and liquids. As the temperature rises, the density of water decreases. The density of the solids is not significantly affected by temperature. Numerous tests have shown that every increase of 5 °C (9 °F) raises the cake solids by 2%. The upper limit is approximately 60 to 65 °C (140 to 150 °F) when the polymer begins to break down. Despite this being known for years, few, if any, facilities take advantage of it because it takes a lot of energy to heat the sludge and because of concern that odors may increase.

21.3 Apparatus and Materials

- A National Institute of Standards and Technology (NIST) thermometer for calibration of other thermometers. However, it is important to note that mercury thermometers are being phased out and it is no longer possible to purchase NIST-calibrated mercury thermometers. Although mercury thermometers may still be used in the laboratory, electronic thermometers are gradually replacing them
- Mercury-filled Fahrenheit thermometer
- Mercury-filled Celsius thermometer
- Thermoresistor probe thermometer

Typical laboratory equipment should include a mercury-filled Celsius thermometer with a range of about 0 to 100 °C (32 to 212 °F). This will suffice for most general purposes. The scale should be subdivided into 0.5 or 1 °C (0.5 or 1 °F) for ease in reading.

Three types of thermometers are commonly used: (1) total immersion, (2) partial immersion, and (3) thermoresistor. Each of these thermometers is shown in Figure 3.26. Total immersion thermometers must be immersed completely in the liquid to the depth of the etched circle around the stem. The thermometer must be immersed below the scale level to yield the correct temperature. Removing the thermometer from the liquid causes rapid temperature changes. Partial immersion thermometers, on the other hand, must be immersed in the water to the depth of the etched circle that appears around the stem below the scale level.

Electronic thermometers use a thermoresistor to measure resistance as the temperature changes. Once the probe is placed in the sample, it heats up to the temperature of the sample. The thermoresistor inside the thermometer then expands, increasing its resistance. Then, a microcontroller measures the resistance and converts it into a temperature.

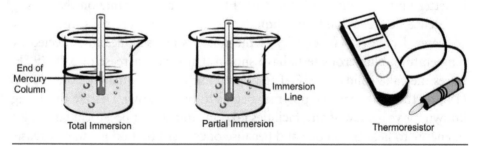

Total Immersion   Partial Immersion   Thermoresistor

End of Mercury Column

Immersion Line

FIGURE 3.26   Thermometers.

21.4   Sample Collection, Preservation, and Holding Times

At least one liter of sample should be taken and analyzed immediately.

21.5   Procedure

- **Step 1**—Either test in situ or grab a 1-L sample.
- **Step 2**—Place the thermometer or thermometer probe to the proper depth in the sample.
- **Step 3**—Once the reading has stabilized, record the value.

The operator–analyst should note that there is no special cleaning required for glassware cleanup.

21.6   Calculations

$$°C = \frac{(°F - 32 °F)}{1.8} \tag{3.28}$$

$$°F = (°C)(1.8) + 32$$

21.7   Reporting

Temperature should be recorded to the nearest fraction of a degree that can be estimated.

21.8   Benchsheet

Temperature measurements are typically performed as part of a series of tests to evaluate process performance and the data are recorded on a comprehensive worksheet. As such, a single benchsheet is not typically used.

22.0   TURBIDITY

The following sections comprise an expanded version of Method 2130-B-2011 (U.S. EPA Method No. 180.1, Revision 2.0) of *Standard Methods Online* (APHA et al., 2017).

22.1   General Description

Turbidity is the optical property that causes light to be scattered and absorbed rather than transmitted with no change in direction or flux level through the sample. Turbidity in water is caused by suspended and colloidal matter such as clay, silt, finely divided organic and inorganic matter, and plankton and other microscopic organisms. Turbidity testing provides a measure of the matter suspended in water or wastewater that scatters or otherwise interferes with the passage of light through the water.

22.2   Application

Turbidity does not correlate directly with suspended solids concentrations because (1) the size and quantity of particles can vary, and (2) color can interfere with the turbidity measurement. It is possible for water to have low TSS and have high turbidity. In this instance, the particles would be small and fine without a lot of total mass. Together, they cause turbidity but would not result in a high TSS measurement. It is also possible for water to have high TSS and low turbidity. In this instance, the solids particles would be large, but there would be a lot of clear water between particles. The solids would have a lot of mass but, because a lot of light would pass through the clear voids, turbidity would be low. Table 3.38 provides expected turbidity ranges for wastewater processes.

A relationship between TSS and turbidity can be made for a given system under a particular set of operating conditions. If conditions change significantly, the relationship must be reestablished. To establish the relationship, multiple grab samples should be taken and analyzed for suspended solids

TABLE 3.38   Expected turbidity ranges for wastewater processes.

Process	Turbidity without chemical addition, NTU	Turbidity with chemical addition, NTU
Aerated/facultative lagoon effluent	4–20	N/A
Contact stabilization	2.5–8	0.1–2
Conventional activated sludge effluent	1–5	0.1–2
Extended aeration effluent	0.5–3	0.1–2
Filtration effluent		
Conventional	2	N/A
Membrane	2–10	N/A
MBR effluent	<0.2	N/A
Trickling filter effluent		
High-rate	4–10	0.1–2
Two-stage	2.5–8	0.1–2

over a suitable range of turbidity values. The two values should be plotted against each other on an "x–y" graph. Then, linear regression can be used to determine the relationship and an equation can be developed ($y = 0.9586$ $x + 2.5038$) that estimates solids given the turbidity. This relationship is shown in Figure 3.27. The general equation for a straight line is $y = mx + b$, where m is the slope of the line and b is the intercept. The variables y and x indicate any single point on the line.

Using the equation that was developed, the secondary effluent TSS can be predicted based on the effluent turbidity. This is plotted in Figure 3.28.

An examination of the two sets of data shows that for this facility, the "estimated" solids are approximately 2 mg/L higher than the actual solids. In this example, a turbidity of 5 nephelometric turbidity units (NTU) translates to a TSS concentration of 9.5 mg/L. If the desired effluent TSS is less than 10 mg/L, the operator–analyst can be confident that when the turbidity measures 9.5 NTU, the effluent quality is satisfactory.

22.3 Interferences

- Dirty or scratched glassware
- Air bubbles in the sample
- Debris or rapidly settling coarse sediment
- Condensation on the sample cell
- Particles that are made up of light-absorbing materials such as activated carbon

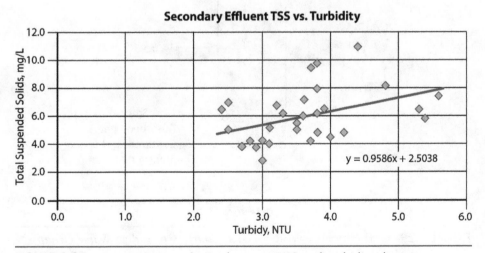

FIGURE 3.27 Creating a correlation between TSS and turbidity data.

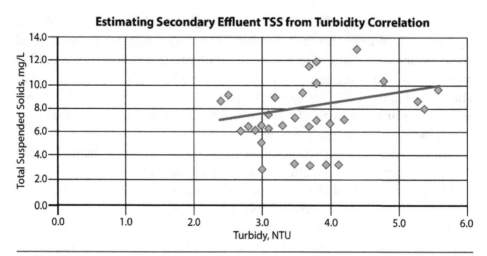

FIGURE 3.28 Estimate of clarifier effluent TSS using turbidity.

22.4 Apparatus and Materials

- Turbidity meter—This consists of a nephelometer with a light source for illuminating the sample and one or more photoelectric detectors with a readout device to indicate intensity of light scattered at 90 deg to the path of incident light. A schematic of a typical turbidity meter is shown in Figure 3.29. The instrument should be designed to minimize stray light reaching the detector in the absence of turbidity and to be free from significant drift after a short warm-up period.

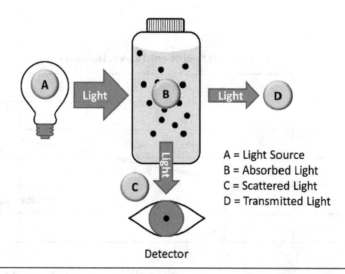

FIGURE 3.29 Turbidity meter components. *Courtesy of Indigo Water Group*

- Sample cells—Select sample cells or tubes of clear, colorless glass or plastic. Keep cells meticulously clean, both inside and out, and discard them if they are scratched or etched. The cells should never be handled where the instrument's light beam will strike them. Use tubes with sufficient extra length, or with a protective case, so that they may be handled properly.

  ○ Fill cells with samples that have been agitated thoroughly and allow sufficient time for bubbles to escape. Minor differences between sample cells significantly affect measurements. Best results are obtained by using matched pairs of cells or the same cell for both standardization and sample measurement.

  ○ Minor imperfections and scratches that may contribute to stray light can be masked by coating the outside of the cell with a thin layer of silicone oil. Use silicone oil with the same refractive index as glass. Avoid excess oil because it may attract dirt and contaminate the sample compartment of the instrument. Using a soft, lint-free cloth, spread the oil uniformly and wipe off excess oil. The cell should appear to be nearly dry with little or no visible oil.

- Lint-free tissue.

22.5 Reagents

Secondary standards are standards that the manufacturer (or an independent testing organization) has certified will give instrument calibration results equivalent (within certain limits) to the results obtained when the instrument is calibrated with the primary standard. Secondary standards made with suspensions of microspheres of styrene-divinylbenzene copolymer typically are as stable as concentrated formazin and are much more stable than diluted formazin. These suspensions can be instrument-specific; therefore, it is important to use only suspensions formulated for the type of nephelometer being used. Secondary standards provided by the instrument manufacturer (sometimes called "permanent" standards) may be necessary to standardize some instruments before each reading and other instruments only as a calibration check to determine when calibration with the primary standard is necessary.

All secondary standards, even the aforementioned permanent standards, change with time. Therefore, the operator–analyst should replace them when their age exceeds the shelf life. It is important to note that not all secondary standards have to be discarded when comparison with a primary standard shows that their turbidity value has changed. In some instances, the secondary standard should simply be relabeled with the new turbidity value. Always follow the manufacturer's directions.

22.6   Sample Collection, Preservation, and Holding Times

Turbidity sampling containers and their lids should be scrupulously cleaned with a laboratory-grade detergent and rinsed with distilled water before use. Samples should be measured immediately to prevent temperature changes, particle flocculation, and sedimentation from changing sample characteristics. Samples should be gently agitated before being poured. If storage is required, it is permissible to refrigerate or cool the sample to ≤6 °C to minimize microbiological decomposition of solids. Samples for turbidity analysis may be held for up to 48 hours before analysis. However, for best results, the turbidity should be measured immediately without altering the original sample conditions.

22.7   Quality Assurance and Quality Control

A series of standards should be used to calibrate the turbidity meter in each range of interest. Sample cells should be free of contamination, etches, and scratches.

22.8   Procedure

To obtain the most accurate results when measuring turbidity, the manufacturer's instructions should be followed. The operator–analyst should check at least one standard in each instrument range to be used. It is important to make sure the nephelometer gives stable readings in all sensitivity ranges used.

Directions for determining turbidity using a nephelometric turbidity meter are as follows:

- **Step 1**—Turn on the turbidimeter. An example of a turbidimeter is shown in Figure 3.30.
- **Step 2**—Select the formazin standard vial for the expected turbidity. For typical secondary effluent, choose the 20 NTU standard.
- **Step 3**—Hold the vial by the cap and wipe to remove fingerprints and water spots (see Figure 3.31). Using a lint-free cloth, clean the glass surface of the standard without leaving any fingerprints.
- **Step 4**—Apply a thin bead of silicone oil from the top to the bottom of the cell, just enough to coat the cell with a thin layer of oil (see Figure 3.32). Use the oiling cloth to spread the oil uniformly. Then,

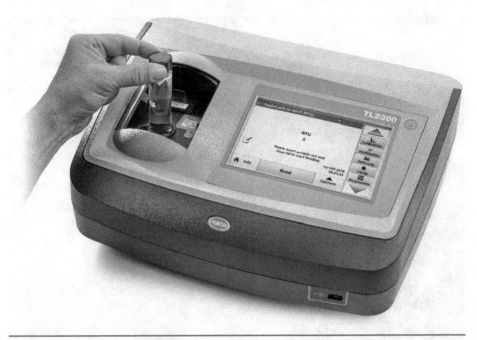

FIGURE 3.30   Turn on turbidimeter. *Courtesy of Hach Company*

FIGURE 3.31   Use lint-free cloth and clean glass surface. *Courtesy of Indigo Water Group*

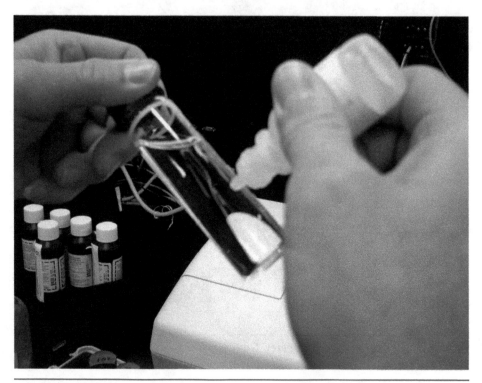

FIGURE 3.32   Apply a thin bead of silicone oil to the cell surface. *Courtesy of Indigo Water Group*

wipe off the excess. The cell should appear nearly dry with little or no visible oil.

- **Step 5**—Place the standard in the turbidimeter compartment (see Figure 3.33) and close the cover to prevent the entry of light.
- **Step 6**—Press the enter key. The display counts down from 60 to 0 and displays the measurement. The measurement should be the same as the standard (i.e., about 20 NTUs). If the measurement is not the same, the turbidimeter calibration needs to be checked.
- **Step 7**—Collect a representative sample in a clean container. Fill the sample cell to the line (about 30 mL) and place a cap on the vial. Fill the sample cell carefully to avoid creating bubbles and to ensure that the sample is homogeneous (see Figure 3.34).
- **Step 8**—After filling the vial, dry it with a soft, lint-free absorbent paper to remove external condensation caused by variation in temperature.
- **Step 9**—Repeat Steps 3 and 4.

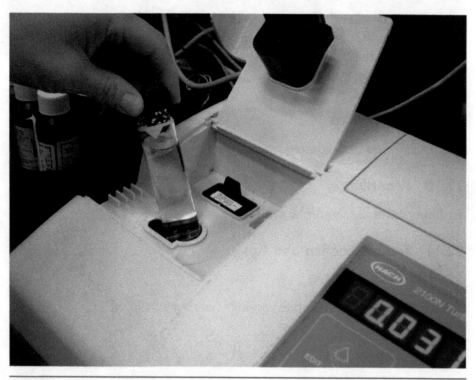

FIGURE 3.33 Place vial in instrument. *Courtesy of Indigo Water Group*

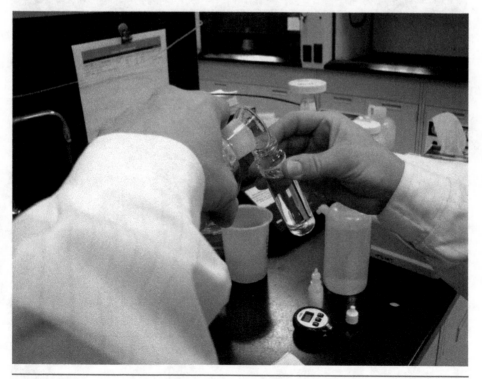

FIGURE 3.34 Fill sample vial. *Courtesy of Indigo Water Group*

- **Step 10**—Gently invert the sample cell and then insert it into the turbidimeter cell compartment. Close the lid to prevent light from entering the compartment.
- **Step 11**—Read the turbidity value (see Figure 3.35).

22.9   Calculations

Calculations are not required unless samples require dilution.

22.10   Reporting

Turbidity readings should be reported as per Table 3.39.

22.11   Troubleshooting and Tips

22.11.1   Troubleshooting

Table 3.40 provides a troubleshooting chart for turbidity.

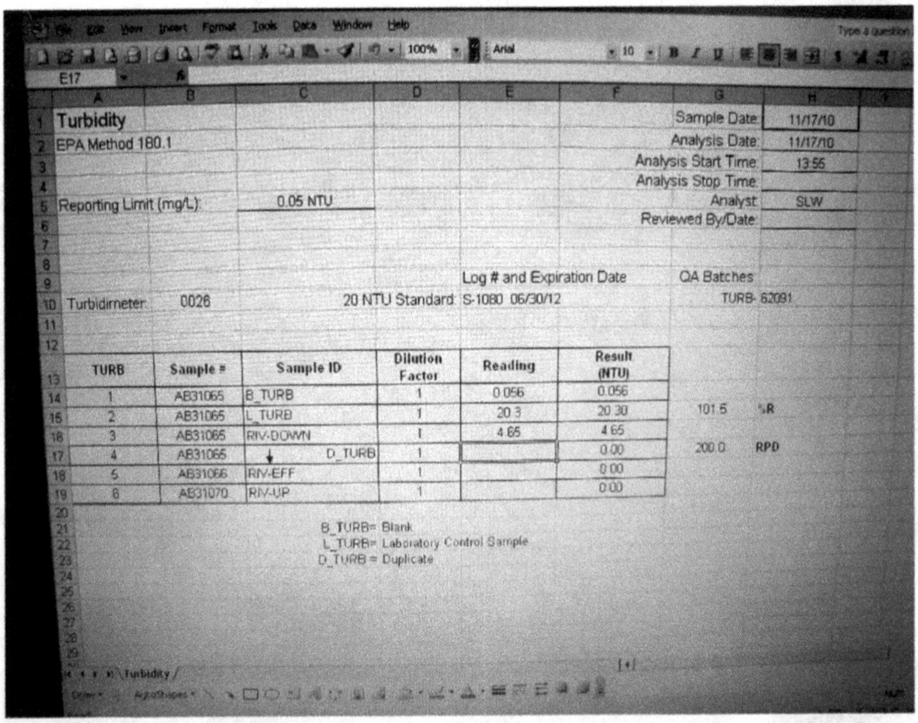

FIGURE 3.35   Record value. *Courtesy of Indigo Water Group*

TABLE 3.39   Report turbidity readings.

Turbidity range, NTU	Report to the nearest NTU
0–1	0.5
1–10	0.1
10–40	1
40–100	5

Source: Standard Methods, Method 2130 B-2011 (APHA et al., 2017).

22.11.2 Tips

- Dirty glassware and the presence of air bubbles give false results.

- "True color," that is, water color attributed to dissolved substances that absorb light, causes measured turbidities to be low. This effect typically is not significant in treated water.

- Letting a sample stand for a period of time to remove air bubbles may settle turbidity-causing particulates and, as a result, the sample

TABLE 3.40   Troubleshooting chart for turbidity.

Observation	Response
Reading is erratic.	Bubbles are present in the sample cell. Tap the cell to dislodge the bubbles.
	Condensation may be present on the sample cell holder.
	Rapidly settling solids.
The turbidity appears high or low for the sample.	Bubbles are present in the sample cell. Tap the cell to dislodge the bubbles.
	Sample cell surface may be fouled. Clean cell surface with lint-free cloth.
	Rapidly settling solids can cause turbidity measurements to drop from high to low.
Blanks do not read within range.	Check calibration solutions for expiration.
	Check accuracy against that of another turbidity meter.
Reading first appears stable and then drifts up unexpectedly.	Wipe cell dry with soft, lint-free cloth.

temperature may change. These events may change sample turbidity, resulting in a nonrepresentative measurement.

- Condensation may occur on the outside surface of a sample cell when a cold sample is being measured in a warm, humid environment; this interferes with turbidity measurement. Remove all moisture from the outside of the sample cell before placing the cell in the instrument. If fogging recurs, warm the sample slightly by letting it stand at room temperature or by partially immersing it in a warm water bath for a short period of time. It is important to make sure that samples are again well mixed.

22.12 Benchsheet

Turbidity measurements are typically performed as part of a series of tests to evaluate process performance and the data are recorded on a comprehensive worksheet. As such, a single benchsheet is not used. However, when analyzing multiple samples, a form such as that shown in Table 3.41 may be used.

TABLE 3.41   Sample benchsheet for turbidity.

Method: _____

Date: _____

Time: _____

Operator-Analyst: _____

Location: _____

mple identificat	Sample date	Measurement range	Turbidity range	Turbidity reading (NTU)	Comments

23.0 OZONE: INDIGO COLORIMETRIC METHOD

The following sections comprise an expanded version of Method $4500\text{-}O_3$ B-2011 of *Standard Methods* Online (APHA et al., 2017). This procedure is used for process control purposes only and is not listed as an approved reporting method in 40 CFR Part 136 (U.S. EPA, 2017).

23.1   General Description

Ozone is a powerful oxidant that is used in water treatment and WRRFs to enhance disinfection and removal of color and taste. When used to disinfect wastewaters, a residual amount of ozone can react with bromide, if present, to form bromate, which is a regulated compound.

23.2   Application

The scope of this application cover samples analyzed for water and wastewater. In this method, when dissolved ozone is present, it decolorizes the solution of blue potassium indigo trisulfonate. The discolored solution is then read on a spectrometer at 600 nm. The spectrophotometer method will measure down to 2 µg/L of ozone.

23.3   Interferences

Results obtained for ozone analyzed by the spectrometric method vary as a consequence of the complexity of the reactions involved and the instability of ozone. The following substances can interfere with the accuracy of the indigo colorimetric method for ozone:

- Hydrogen peroxide
- Chlorine
- Bromine
- Manganese

23.4   Apparatus and Materials

- Spectrophotometer capable of reading 600 nm
- Data acquisition source (computer and spreadsheet for computation)
- Analytical balance
- Erlenmeyer flasks
- Matched cuvettes
- Graduated cylinders

- Appropriate personal protective equipment (gloves, glasses, etc.)
- Kimwipes® to dry the cuvette
- Water bottles
- Volumetric flasks (varying sizes)
- Racks for test tubes and/or flasks

23.5   Reagents

- Concentrated phosphoric acid
- Potassium indigo trisulfonate
- Malonic acid
- Potassium dihydrogen phosphate
- Glycine
- Indigo solution, available from the Hach Company in ready-to-use form or from other certified vendors in vacuum-sealed packages. If the ready-to-use solution is not available, stock indigo solution can be prepared as follows:
  - Add 400 to 500 mL of deionized water to a 1-L volumetric flask.
  - Add 1 mL of concentrated phosphoric acid.
  - Add 0.77 g of potassium indigo sulfate trisulfonate.
  - Stir until completely dissolved.
  - Bring solution to mark.
  - Place the solution in an amber bottle and store it in the dark.

23.6   Stock and Standards

Prepared ozone standards in different concentrations (0 to 0.75 mg/L) are available from Hach Company and other certified vendors. Ready-to-use check secondary standards kits include four 10-mL (1-in.) sealed glass cells with a blank and three different standards within the typical testing range, instructions, a plastic case, and a certificate of analysis.

23.7   Quality Assurance and Quality Control

- Prepare a reagent blank.
- Prepare a calibration curve using the desired concentration that will bracket expected sample concentrations.
- Analyze duplicate and method blanks with the samples.

23.8   Sample Collection, Preservation, and Holding Times

Grab samples are required for this analysis. Sample collection must be done carefully because ozone is a transient entity and care must be taken to prevent ozone from escaping the sample. Samples should be collected without turbulence or stirring and should be analyzed immediately for best results. A maximum delay of 4 hours is permissible for drinking water samples. Hold times are variable for other sample matrix types.

Sample transfer to a different container must be limited to the minimum to prevent loss during this process. For each sample to be collected, three clean 100-mL flasks with 10 mL of indigo solution added to each glass should be used and labeled appropriately. The sample number should be recorded on the data collection form.

Next, pour the collected sample carefully without agitation in the flask until the color of the indigo solution fades slightly. Gently swirl the solution to ensure good mixing of the sample with the raw water sample. If the blue tint is completely bleached out of the sample, the process needs to be restarted and the sample is no longer viable. Finally, wipe excess water off each sample flask with a dry Kimwipe® or paper towel.

23.9   Procedure

- **Step 1**—Turn on the spectrophotometer and allow 15 minutes for warm up and stabilization.
- **Step 2**—Set the wavelength to 600 nm.
- **Step 3**—Set the instrument to read absorbance or concentrations depending on the choice.
- **Step 4**—While the instrument is warming up, prepare the sample for analysis. Select a clean test tube cuvette and fill it with sample solution from each flask. For sample collection steps, the operator–analyst should refer to Section 23.8. Each sample should have been collected in three separate 100-mL flasks containing 10 mL of indigo solution.
- **Step 5**—Record the volume used on the data entry form.
- **Step 6**—Prepare a reagent water blank that is free of ozone contaminant.
- **Step 7**—Prepare at least three standards to generate the calibration curve. The concentrations used will depend on the expected concentrations of the samples. A calibration curve might consist of standards at 0.25, 0.50, and 1.0 mg/L. The calibration curve should bracket the range of expected sample results.

- **Step 8**—Add 10 mL of indigo solution to each of the flasks containing a standard and the flask containing the blank.
- **Step 9**—Carefully fill each of the flasks from Step 8 (already containing 10 mL of indigo blue solution) to the 100-mL mark. Make sure there are no air bubbles in the flask.
- **Step 10**—For the blank and each standard, fill a 10-cm sample cuvette and dry with a Kimwipe®. It is important to remember to not handle the tube below the sample line.
- **Step 11**—Open the test chamber of the spectrophotometer and place the sample cuvette in the chamber with the vertical white line facing the groove on the right side of the chamber.
- **Step 12**—Close the lid of the chamber.
- **Step 13**—Read and record the absorbance value of the blank on the data collection form.
- **Step 14**—Repeat Step 13 for each standard.
- **Step 15**—When the calibration curve has been successfully analyzed, repeat the procedure for each of the samples.

The operator–analyst should note the following:

- **Step 16**—Waste indigo solution from the blank or samples may be poured into either the liquid waste container or down the sink.
- **Step 17**—Depending on the cuvette size and type, when it is removed, the spectrophotometer will typically read about -0.045 if it has been zeroed properly. *Standard Methods* (APHA et al., 2017) recommends using a 10-cm sample cuvette.
- **Step 18**—Alternatively, the indigo method can also be used with the AccuVac Ampul (Hach Company) ozone measurement process. With AccuVac, all the required chemicals are prepackaged under vacuum into a cuvette-ampule device. The sealed glass tip of this device is submerged in the ozone-containing water sample and the tip is broken open to allow the water to enter the device and dissolve the chemicals therein. The cuvette-ampule is used directly for the spectrophotometric determination of dissolved ozone.

23.10 Calculations

The dissolved ozone concentration is calculated from the difference in light absorption of an untreated blank and of a treated sample, as follows:

$$\text{Dissolved } O_3, \text{ mg/L} = \frac{\Delta \text{ absorption} \times 100}{f \times b \times v} \qquad (3.29)$$

Where

> b = path length of the cuvette in cm,
> v = the volume of the ozone-containing sample added in mL (normally 90 mL) to raise the total volume of the sample plus the indigo solution to 100 mL, and
> f = 0.42.

An alternative calculation is

$$Ozone\ Residual,\ \frac{mg}{L} = \frac{(ABS\ blank \times blank) - (ABS\ Sample - TV\ sample)}{0.42 \times SV \times b}$$

(3.30)

Where

> ABS = absorbance of blank or sample;
> TV = total volume of sample, or blank, plus indigo (mL);
> SV = volume of the sample (mL);
> b = path length of cell = 1 cm; and
> 0.42 = constant.

24.0 ACTIVATED SLUDGE SYSTEM MICROSCOPIC EXAMINATION

24.1 General Description

The objective in the operation of the activated sludge process is to provide a suitable environment for the growth of a stable microbiological population that removes organics and other pollutants and develops good sludge settling characteristics. The biological population is a mixture of heterotrophic, autotrophic, and facultative bacteria, along with different types of protozoans and metazoans. Bacteria are the workhorses of the activated sludge system and are primarily responsible for purifying wastewater. Activated sludge is formed by aerating the liquid wastes in the presence of bacteria until the bacteria have stabilized the organic matter. Activated sludge is not formed by special bacteria, but is a normal process for all bacteria in the presence of available food. The food in question is the food in the aeration basin(s), as opposed to the influent load. Because the volume of aeration tanks is generally large relative to influent flow, it is difficult to increase the food concentration when the bacteria are constantly taking it out unless an extreme flow or slug load enters the facility.

This procedure is used for process control purposes only and is not listed as an approved reporting method in 40 CFR Part 136 (U.S. EPA, 2017).

24.2 Application

As bacteria grow in the system, different types of protozoans and metazoans grow along with them (see Figure 3.36). When the concentration of food is high, the bacteria grow rapidly, and certain types of protozoans and metazoans dominate. As the bacteria continuously remove the food, they purify the wastewater and their growth slows down. With less food available and slower-growing bacteria, different types of protozoans and metazoans dominate. The activated sludge process is controlled by stabilizing growth and controlling the relationship between the food and organisms in the system. The operator–analyst controls the growth rate by controlling the amount of sludge in the system, or the age of the system. This is required because the operator–analyst must treat what comes in and, generally, cannot control the food. By controlling the age of the system, the operator–analyst stabilizes the biology and controls the types and relative dominance of the organisms in the system. Figure 3.36 shows activated sludge microorganisms versus time. When the age or SRT is low, the system is dominated by bacteria, free-swimming ciliates, and flagellates, with the relative dominance

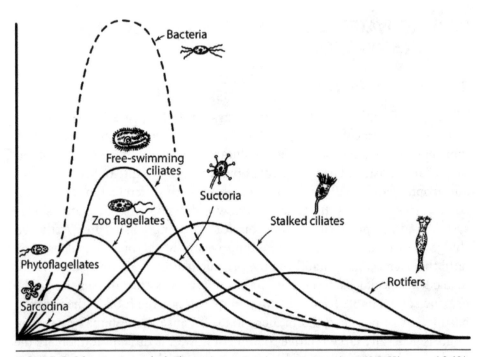

FIGURE 3.36   Activated sludge microorganisms versus time (McKinney, 1962). *Reprinted courtesy of McGraw-Hill*

of each type specific to each facility (if the food is different, the organisms grown are different). As the system gets older, the biology shifts to a stable population of bacteria and a dominance of stalked ciliates and rotifers. The system automatically shifts back and forth in response to changes in both the type and quantity of influent loading. The operator–analyst can shift the system by wasting more sludge to make the system younger, or less sludge to make the system older. In either instance, the shifting may be rapid and highly sensitive to load increases, decreases, over-wasting and under-wasting, and toxic shocks.

By recognizing that the changes are natural, the operator–analyst can use microscopic examination of activated sludge to make process control decisions. Using a microscope, the operator–analyst can accomplish the following:

- Verifying the operating SRT—a system in which rotifers are the dominant metazoan cannot have a low SRT, and a system dominated by free swimmers cannot have a high SRT. This is true no matter what the calculated SRT value indicates. Calculated values are subject to both calculator and testing errors. The biology tends to respond automatically in response to the food available, regardless of the calculations.

- Check sludge settling characteristics—in specific facilities, the dominance of specific types of protozoans and metazoans is associated with good settling and high-quality clarifier effluent TSS. Shifts in the biology, which can occur in less than 30 minutes, are early warning signs of changes in the system and possible problems.

- Check changes in loading and toxic shocks—dead and dying protozoans and metazoans, especially stalked ciliates and rotifers, are indicative of increased loading or toxic shocks. If the organic load increases rapidly, the rapid increase in the bacteria's population drives out the stalks and rotifers, replacing them with free swimmers and flagellates. If the system is shocked with a toxic load, all of the protozoans and metazoans, regardless of type, may disappear from the system.

- Check sludge quality—the presence of filamentous organisms is normal in any activated sludge mixed liquor. However, when sludge bulking occurs, the first reaction is to blame the filaments. This is unfortunate, as the presence of filaments, normal to a specific facility, assists in formation of the sludge blanket and filters out fine particles that would otherwise escape in the clarifier effluent. Filaments are a problem when they dominate the system, excluding other microorganisms. Their problematic presence is readily detected using a microscope at a magnification of 10 to 40 power. Sludge bulking that results from the presence of filaments is confirmed with whole and dilute settleometer data. It is important to note that filaments observed at higher magnifications (i.e., greater than 40 power) may not be filaments and, if they are, may in fact not be a problem. When problem filaments are dominant, their presence at low power (i.e., less than 100) under the microscope is obvious (Figure 3.37).

24.2.1   Indicator Organisms

Microorganisms that are important indicators of the health, stability, and age of the process include the following:

- Amoeboids
- Flagellates
- Ciliates
- Rotifers
- Worms

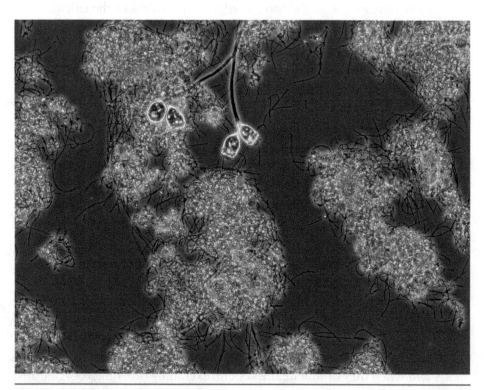

FIGURE 3.37   Filamentous organisms in bulk liquid. *Reprinted with permission from Leach Microbial Consulting*

24.2.1.1 Amoeboids

Amoeboid protozoans, shown in Figure 3.38, are characterized by highly flexible cell membranes. The organisms travel by movements of the protoplasm within the cell. Food is ingested by absorption through the cell membrane.

Amoeboids compete with bacteria for the soluble organics in the system, and thus may predominate in the MLSS startup or when the process is recovering from a process upset. It is important to note that the amoeboid population decreases as the bacteria and flagellate populations increase.

24.2.1.2 Flagellates

Flagellates, which use bacteria for food, are characterized by a tail that extends from a round or elliptical body. The whipping action of the tail allows the flagellates to move in a characteristic corkscrew pattern. In Figure 3.39, a small, oval flagellate can be seen next to the larger floc particle.

Flagellate predominance is generally associated with a low SRT (high F:M), dispersed floc, and a low bacteria population. As the system ages, the floc becomes more dense, the bacteria population increases, and the

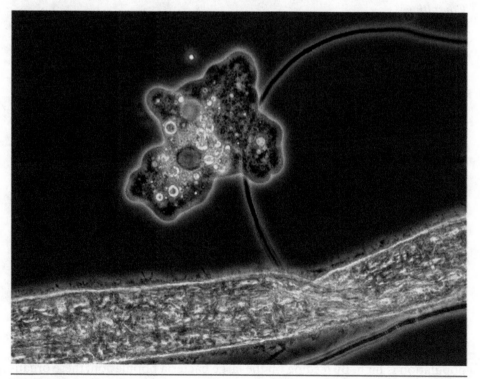

FIGURE 3.38   Amoeba. *Reprinted with permission from Leach Microbial Consulting*

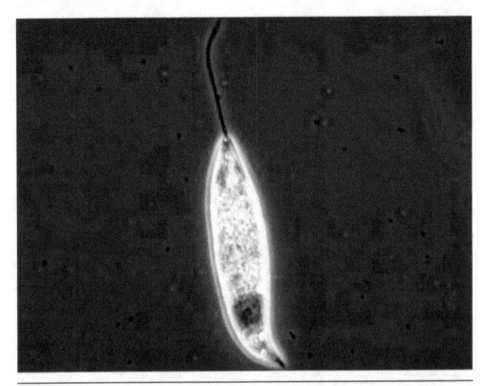

FIGURE 3.39   Flagellate. *Reprinted with permission from Leach Microbial Consulting*

flagellate population decreases to insignificance. Flagellates are less efficient than the stalked ciliates at collecting free bacteria, their preferred food source. As a result, they are washed out of the system.

24.2.1.3 Ciliates

Ciliates are high-energy, bacteria-eating protozoans characterized by moving or rotating hair-like membranes that cover all or part of the cell membrane. The organisms are allowed to move as a result of the motion of the cilia. Cilia around the gullet or mouth are used by organisms for the intake of bacteria. As a group, ciliates are significantly larger than flagellates and, for the purposes of a microscopic exam, are classified as either free swimmers or stalked ciliates.

24.2.1.4 Free-Swimming Ciliates

A free-swimming ciliate is shown in Figure 3.40. These high-energy, fast-moving, free-swimming ciliates generally predominate when the bacteria population is high. By feeding on the large bacteria population, these organisms assist in clarifying the effluent. Therefore, the presence of free swimmers

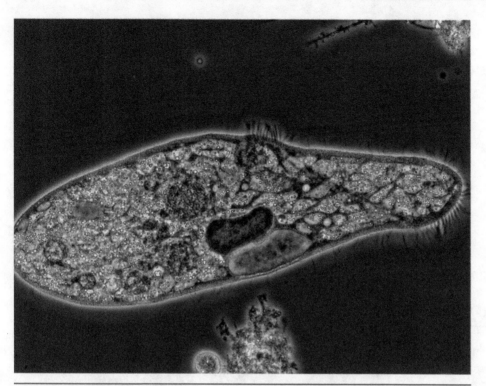

FIGURE 3.40 Free-swimming ciliates. *Reprinted with permission from Leach Microbial Consulting*

is generally indicative of a system that may be approaching or is presently at optimum treatment levels.

The population of free swimmers decreases as the bacteria population (food) decreases. If the decrease in bacteria is the result of a shift to a low SRT, the free swimmers are replaced by flagellates and/or amoeboids. If the decrease is the result of a shift to a higher SRT, the free-swimming ciliates are replaced by stalked ciliates.

24.2.1.5 Stalked Ciliates

Stalked ciliates, shown in Figure 3.41, have lower energy requirements than free-swimming ciliates. The feeding mechanism of the stalked ciliates is to use the cilia to sweep bacteria into the gullet or mouth. Unlike free swimmers, the stalks grow by attaching themselves to the floc or other particles in the mixed liquor. Because they grow in place, less energy is required, and they can successfully compete with free swimmers for available food when the bacteria population drops.

Stalked ciliates grow in MLSS when the bacteria population is declining as a result of an extended SRT (low F:M). Should the influent load increase,

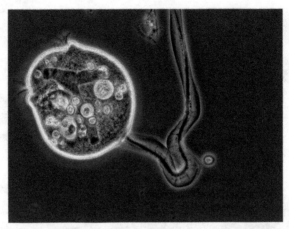

A – single stalked ciliate.

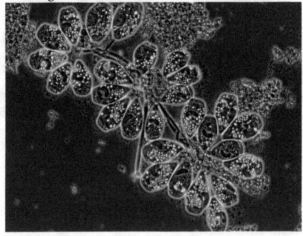

B – colony of stalked ciliates.

FIGURE 3.41   Stalked ciliates. *Reprinted with permission from Leach Microbial Consulting*

thus supplying more soluble food, the bacteria population explodes and the stalks are displaced by free swimmers. It is important to note that the stalked ciliate population decreases as the bacteria population increases and increases as the bacteria population decreases. As the bacteria population continues to decline (the SRT becomes higher), the stalks are replaced by rotifers.

24.2.1.6 Rotifers

Rotifers, shown in Figure 3.42, are indicator metazoans that achieve dominance as the system ages beyond the level of the stalks and appear as the population of the stalks declines. The organisms are vase-shaped filter

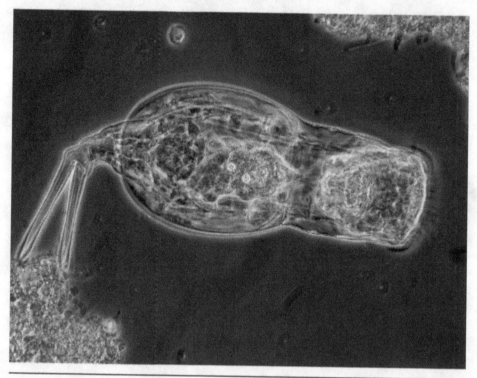

FIGURE 3.42   Rotifer. *Reprinted with permission from Leach Microbial Consulting*

feeders, present when the bacteria population is low, feeding on dead bacteria and particulate matter in the mixed liquor. Dominance by rotifers is an indication of a long SRT. It is important to note that the rotifer population decreases when the bacteria population increases (SRT is shifting lower) and is displaced by stalks. Or, as the SRT becomes longer, rotifers are displaced by nematodes or worms.

24.2.1.7 Nematodes and Worms

Nematodes and sludge worms, shown in Figure 3.43, are true multicelled organisms that may be indicative of an extremely long SRT. Their presence in the MLSS is generally associated with a very dark, fast-settling granular sludge, and may be common in facilities that are underloaded but continue to be operated at high MLSS concentrations. However, because such organisms proliferate in association with scum, rags, or trash, their presence may not be related to the operating SRT, but to poor housekeeping that results in crusted sludge on sidewalls, fouled clarifier weirs, and so on.

It is important to note that worms and nematodes may or may not be a problem with respect to clarifier effluent quality. However, such organisms exiting the clarifier die in effluent filters, gumming up the sand media

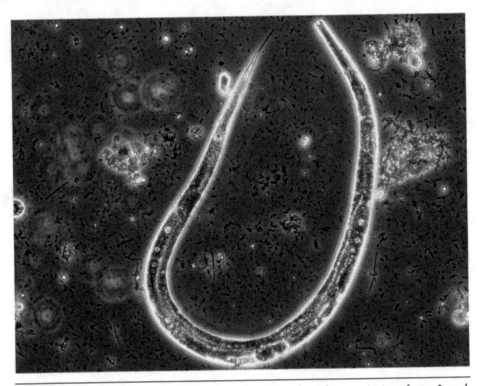

FIGURE 3.43 Nematodes and worms. *Reprinted with permission from Leach Microbial Consulting*

and increasing the required number of filter backwashes. This causes a loss of sand media, which, in turn, allows fine particles to escape and degrade effluent quality. Worms and nematodes, excluding poor housekeeping being a problem, may be associated with dark, scummy, and thick foam.

24.2.1.8 Filaments

- *Beggiatoa* are common during a septic and organically overloaded conditions.
- *Haliscomenobacter hydrossis* is present in a low dissolved oxygen environment, low F:M, and/or nutrient deficiencies.
- Fungi are present at low pH.
- *Microthrix parvicella* is a low F:M organism that loves high grease and fat content with low temperatures. They are easy to spot because they look like a tangle of hair.
- *Nocardia* is common when residence times are long (i.e., greater than 9 days) and at low SRTs (i.e., less than 2 days). The operator–analyst should note that this is really "foam age" versus SRT. It shows up in

warm temperatures when grease, oil, and fat are present. *Nocardia* is a strict aerobe.

- *Sphaerotilus natans* is a strict aerobe. It is capable of growth at dissolved oxygen concentrations as low as 0.01 to 0.03 mg/L, which signals a low dissolved oxygen mixed liquor concentration.

- *Thiothrix I* and *II* are present with septic conditions that have high sulfides and/or organic wastes that are deficient in nitrogen.

- *Type 021N* is a low dissolved oxygen strict aerobe that can be observed over a wide range of F:M values extending down into 0.2 to 0.3.

24.2.2 Relative Dominance

While the succession of indicator organisms is straightforward and occurs as a result of the age of the system and changes in the load, three basic problems exist. All of the indicator organisms can exist in MLSS at the same time. Activated sludge is not a pure culture and operates at a dynamic equilibrium in which the relative numbers of different types of organisms increase and decrease in response to changes in loading and operating conditions. Because each activated sludge process is essentially unique as a result of specific loading and operating conditions, the relative dominance of organisms indicative of high-quality effluent are unique to each facility, and shifts in dominance require unique interpretation by each individual operator–analyst.

Rapid shifts in influent loading or rapid and immediate increases in sludge wasting can result in what may be termed *classic sludge confusion*, in which large populations of organisms that should have a low population are present in association with organisms where the population should be insignificant (e.g., equal portions of free swimmers and rotifers).

What is important to the operator–analyst is the relative dominance of indicator organisms that are associated with excellent treatment in the facility, and the importance and meaning of the relative shifts in dominance. In other words, the operator–analyst needs to ascertain what is going on and why, and whether the shift is transient or indicative of potentially serious problems.

Fortunately, the relative predominance of certain organisms is associated with other indicators that can assist in this process evaluation. This is presented graphically in Figure 3.44, where relative dominance is associated with the type, quantity, and color of foam; MLSS color; floc characteristics; and operating SRT. The relationships presented are intended only as examples and, therefore, should not be considered facts. The operator–analyst's job is to develop data and associations between process indicators that are meaningful to staff and the operation of the facility. Microscopic

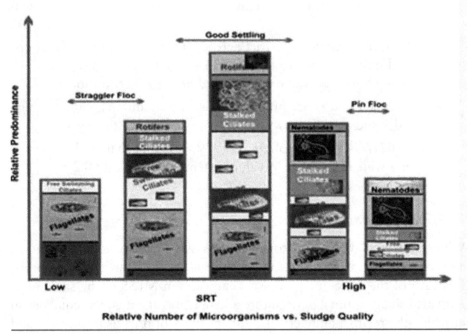

FIGURE 3.44 Organisms distribution versus sludge quality.

examination of the foam may be useful because some filaments will concentrate in the foam, including *Nocardioforms* and *M. parvicella*.

24.3 Interferences

- Dirty or scratched slide or cover slip
- Air bubbles trapped underneath the cover slip

24.4 Apparatus and Materials

- Microscope, preferably phase contrast (see Figure 3.45)
- Mixed liquor sample that has settled
- Slide
- Cover slip
- Dropper

24.5 Sample Collection, Preservation, and Holding Times

Normal samples taken for microscopic examination should include the following:

- Mixed liquor suspended solids—the MLSS sample is used to associate the indicator organisms observed with sludge settling characteristics.

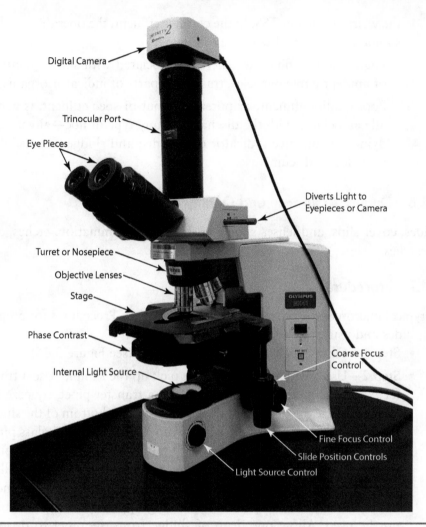

Digital Camera

Trinocular Port

Eye Pieces

Diverts Light to
Eyepieces or Camera

Turret or Nosepiece

Objective Lenses

Stage

Phase Contrast

Coarse Focus
Control

Internal Light Source

Fine Focus Control

Slide Position Controls

Light Source Control

FIGURE 3.45   *Microscope. Courtesy of Indigo Water Group*

The MLSS sample should be taken directly from the sample used in the settleometer or centrifuge spindown tests. Other samples that can be useful in process control include

○ Foam—identify what the foam is composed of (crusty solids, grease, filaments, etc.).

○ Sludge pop-ups—pop-ups can result from a malfunctioning RAS, low dissolved oxygen, and heavily compacted sludge that the RAS cannot pick up. Defining types of organisms present in the pop-ups assists in separating the symptom from the problem.

• Clarifier effluent—with the clarifier effluent sample, the operator–analyst is interested in observing and determining what is going over the weirs as effluent TSS. In the clarifier effluent, the operator–analyst should be concerned with the following:

○ High-quality effluent is generally associated with small quantities of unidentifiable particles, trash, and parts of indicator organisms.

○ Poor-quality effluent, or potentially out-of-spec effluent, is generally associated with the discharge of any type of floc—alive, dead, dying—or inactive indicator organisms; and sludge worms, algal particles, and scum.

24.6   Quality Assurance and Quality Control

Slides, cover slips, and lenses should be free of contamination, etches, and scratches.

24.7   Procedure

A typical microscope setup is pictured in Figure 3.45. Procedures for preparing slides and using the microscope are as follows:

• **Step 1**—Obtain a clean cover slip and slide (see Figure 3.46).

• **Step 2**—Use a wide-mouth pipet to pick up the sample. Use a transfer pipet with bulb or glass pipet. With a transfer pipet, squeeze the bulb, immerse the pipet until the end reaches the bottom of the sludge sample, then release the bulb to draw up a sample. With a glass pipet, put a finger on top of the pipet until the immersed end of the pipet reaches the bottom of the sludge sample. Release your finger to allow sludge into the pipet. Replace your finger on top of pipet and remove the pipet from the sample beaker. A long-tipped eye dropper should not be used. Allow one drop of sludge from the pipet to drop in the middle of the clear area of the glass slide by lifting your finger from the top of the pipet momentarily and then replacing your finger. Figure 3.47 illustrates the release of a single drop of sludge onto a microscope slide.

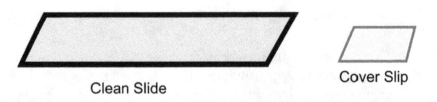

Clean Slide   Cover Slip

FIGURE 3.46   Slide and cover slip.

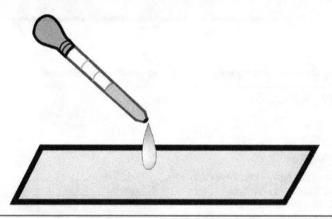

FIGURE 3.47   Release a drop of sample.

- **Step 3**—Pick up the cover slip by two corners. Do not touch the cleaned area. Pull the cover slip along the glass slide toward the drop of sludge, as shown in Figure 3.48.
- **Step 4**—As soon as the cover slip touches the drop of sludge, allow the cover slip to fall onto the glass slide as shown in Figure 3.49.
- **Step 5**—Pick up the glass slide and place it on the microscope stage.
- **Step 6**—Move the stage up to within approximately 3 mm (1/8 in.) of the objective. Look at the glass slide through the eyepiece of the microscope.
- **Step 7**—Use the coarse adjustment on the microscope to bring the sludge into the field of focus.
- **Step 8**—Use fine adjustments to refine to suit your eyes.
- **Step 9**—Identify organisms in the sludge.

A worksheet for use in microscopic examination is presented in Figure 3.50. The organisms presented on the sheet are those that typically occur in activated sludge. In the event that unknown types of organisms that are

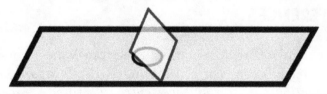

FIGURE 3.48   Pull cover slip along the slide.

FIGURE 3.49   Release cover slip onto slide.

Microorganism Tally Sheet

Count groups in a random sweep across the slides.
Do not search for particular types.
Tally each organism in the space provided.
Calculate percent of total for each organism.
Ignore rare organisms.

	Number	Percent
Amoeboids		
Flagellates	||	
Free-Swimming Ciliates	⑁⑁⑁		
Crawling Ciliates		
Carnivore Ciliates	⑁⑁⑁	
Stalked Ciliates	||	
Suctoria		
Rotifers	||	
Tardigrades		
Nematodes		
Aeleosoma		

FIGURE 3.50   Microorganism tally. *Courtesy of Indigo Water Group*

facility-specific are observed, the operator–analyst should note their presence for later association with other process indicators.

24.8   Calculations

Calculations are not required unless the samples require dilution.

24.9   Reporting

Microscopic examination is performed for activated sludge system process control and is not a regulatory requirement. During the survey of the slide, the organisms should be identified and the number recorded on a tally sheet as shown in Figure 3.50.

25.0   REFERENCES

American Public Health Association, American Water Works Association, & Water Environment Federation. (2017). *Standard methods for the examination of water and wastewater* (23rd ed.). American Public Health Association. Online version: https://www.standardmethods

.org/?gclid=CjwKCAjwh-CVBhB8EiwAjFEPGSxrDNLD_H9lQB64Z-wqCHN9NIH9tqHppnw4Ul5X0DguJDz8vuwh9URoC-qwQAvD_BwE

McKinney, R. E. (1962). *Microbiology for sanitary engineers*. McGraw-Hill.

Metcalf & Eddy, Inc./AECOM. (2014). Wastewater engineering treatment and resource recovery (5th ed.). McGraw-Hill.

U.S. Environmental Protection Agency. (2017). *Federal register*, Part III, 40 CFR Parts 136 and 503. https://www.ecfr.gov/current/title-40/chapter-I/subchapter-D/part-136

Water Environment Federation. (2017). *Operation of water resource recovery facilities* (7th ed.; WEF Manual of Practice No. 11).

Water Environment Federation, American Society of Civil Engineers, & Environmental and Water Resources Institute. (2018). *Design of municipal wastewater treatment plants* (6th ed.; WEF Manual of Practice No. 8/ASCE Manual of Practice and Report on Engineering No. 76). Water Environment Federation.

West, A. W. (1974). *Operational control procedures for the activated sludge process, Part I – Observations, Part II – Control tests* (EPA 330/9-74-001a,b.). U.S. EPA Office of Water Programs, National Waste Treatment Center.

26.0 SUGGESTED READINGS

Jenkins, D., Richard, M. G., & Daigger, G. T. (1993). *Manual on the causes and control of activated sludge bulking, foaming, and other solids separations problems* (3rd ed.). IWA Publishing.

U.S. Environmental Protection Agency. (2007). *Code of federal regulations*, Part III, 40 CFR Part 122, 136, et al.; Guidelines establishing test procedures for the analysis of pollutants under the Clean Water Act; National Primary Drinking Water Regulations; and National Secondary Drinking Water Regulations; Analysis and Sampling Procedures; Final Rule. *Fed. Register* 72(47).

U.S. Environmental Protection Agency. (1994). *A plain English guide to the EPA Part 503 Biosolids Rule* (EPA-832/R-93-003).

4

Spectrophotometry

1.0 INTRODUCTION

In recent years, the spectrophotometer has become an even more important analytical instrument for the examination of wastewater. Results obtained from spectrophotometric methods are generally more versatile, accurate, and reproducible than older, competing methods based on wet chemistry. Because they are not dependent on the visual limitations of an operator–analyst, the results are statistically superior to colorimetric methods for which an operator–analyst must compare colors, for example. Specific methods or spectrophotometers will not be detailed in this chapter; instead, operation theory and tips to keep performance and productivity at a high level will be presented.

2.0 BACKGROUND AND ADVANTAGES

Spectrophotometric methods, as with just about any of the more modern instrument-based methods, have many advantages including minimizing human bias and error, efficient and accurate data handling, and safety. The methods are generally not susceptible to differences in operator–analysts' eyesight or color interpretations and are not even limited to visible colors (i.e., they can be used to measure light that is beyond the ability of humans to detect). Typically, less sample preparation is required than with a conventional or wet method and there is less opportunity to contaminate the test specimen chemically or physically. In addition, modern automation, such as robotics and autosamplers, can allow many samples to be run efficiently. Figure 4.1 shows a typical spectrophotometric method in progress.

3.0 FUNDAMENTALS OF THE TECHNOLOGY

Spectrophotometers use the interaction of light energy with a material to perform an analysis. The graphical data obtained are typically displayed as a spectrum. A spectrum is a plot of the intensity, absorbance, or similar measure of detection versus the wavelength (frequency or wavenumber) of the energy. Interpretation of the spectrum and spectral data can provide more information about the sample than a conventional (wet) method. Atomic and molecular energy levels, molecular geometries, chemical bonds, interactions of molecules, and related processes can sometimes be inferred from a spectrum. Often, spectra are used to identify chemical species (i.e., for qualitative analysis). Spectra may also be used to measure the amount or concentration of material in a sample (i.e., for quantitative analysis). There are currently close to 100 spectrophotometric tests (for water and

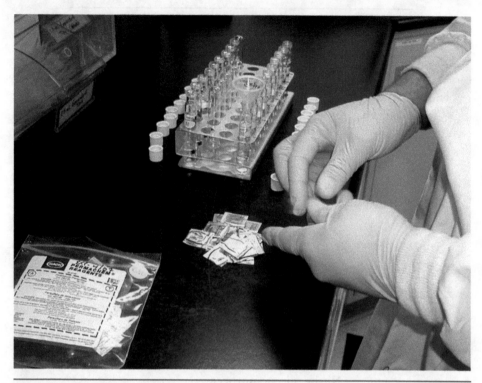

FIGURE 4.1 Typical modern spectrophotometry method in progress. Vials and cuvettes, premeasured reagents, and spectrophotometer are in the background. *Courtesy of Indigo Water Group*

wastewater) commercially available from instrument suppliers, and that number continues to increase. Several dozen are directly applicable and reportable for U.S. Environmental Protection Agency (U.S. EPA) regulatory purposes. Of course, one has to confirm the applicability of any test method with local regulators. Table 4.1 lists spectrophotometer tests currently accepted for U.S. EPA water regulatory monitoring. Figures 4.2 and 4.3 show a few different types of modern spectrophotometers.

The reader should note that, in terms of references involving U.S. EPA, the *Code of Federal Regulations* (CFR) is the source for legal and regulatory information in the test methods, and the most recent edition is the one that should be consulted first. Title 40, Parts 136, 141, and 142, contain the most material on water testing for pollution prevention ("40 CFR 136" is the typical reference format).

3.1   Terminology

A photometer is a device that measures light intensity. A colorimeter is a device that measures color intensity. Color comparators can also be put into this group of devices. When using color comparators, an operator–analyst

TABLE 4.1   U.S. EPA–accepted spectrophotometer-based tests.

Test
Ammonia, nitrogen
Arsenic
Chemical oxygen demand (COD)
Chlorine, free
Chlorine, total
Chlorine dioxide
Chromium, hexavalent
Chromium, total
Copper
Fluoride
Iron, total
Lead
Manganese
Nickel
Nitrate, nitrate-nitrogen
Nitrite, nitrate-nitrogen
Phenols
Phosphorus (phosphates), reactive
Phosphorus, acid hydrolyzable
Phosphorus, total
Sulfate
Sulfide
Total Kjeldahl nitrogen (TKN)
Total organic carbon (TOC)
Zinc

prepares a colored sample and compares (by visual inspection) the sample with tubes or other color standards to obtain a concentration of the analyte. Color comparators are still found in many applications in modern laboratories.

A spectrophotometer uses the concepts of a spectrum and a photometer to measure a specific analyte. A light source in the instrument produces a bright white light composed of many different colors or wavelengths. The light is passed through a device called a monochromineter that splits the light into its separate wavelengths to form a spectrum of colors. The spectrum

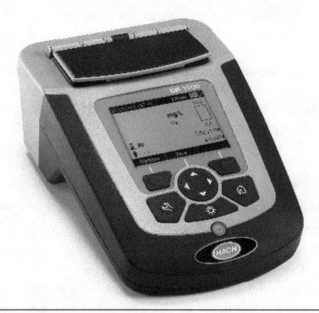

FIGURE 4.2   Typical compact spectrophotometer. *Courtesy of Hach Company*

or color differences are caused by different wavelengths (or frequencies) of
the light. Some wavelengths of light at the ends of the spectrum may not
be visible to the human eye. The monochrometer may be adjusted to pass
a desired range of light wavelengths (colors) through an exit slit and then
through the sample cell. If the analyte of interest is present in the sample,

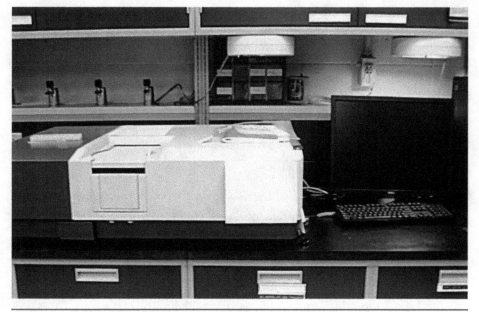

FIGURE 4.3   Typical research laboratory spectrophotometer.

the sample will absorb some of the light as it passes through. The sample cell (cuvette) and other components of the sample may also absorb some light. A detector then measures how much light is transmitted through the sample cell and compares it to the amount of light entering the sample cell to calculate absorption. The part of the molecule that causes the spectral absorbance is sometimes termed the *chromophore,* which is Latin for "color body." Figure 4.4 shows a typical ultraviolet (UV) and visible spectrum.

The following are a few facts about spectra and spectrophotometers:

- Visible white light is the summation of light of all visible colors.
- Colored light has some wavelengths present and some missing.
- Something appears black when there is little or no light being transmitted to one's eye from the object.
- Either a prism or grating in the instrument splits the light into its spectral wavelengths. These disperse the light and are termed *dispersive elements.* (It is important to note that some spectrophotometers use nondispersive technology, such as filters or Fourier-transform technology.)
- Visible light can be described as being made up of the following colors, from long wavelengths to short wavelengths: red, orange, yellow, green, blue, indigo, violet ("ROYGBIV" is a useful abbreviation to remember).
- Infrared is invisible light with a longer wavelength than red.
- Ultraviolet is invisible light with a shorter wavelength than violet.
- Infrared and UV spectrophotometers are fairly common. Instruments for visible wavelengths often incorporate UV and are called *UV and visible spectrophotometers.*

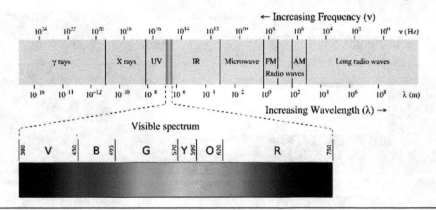

FIGURE 4.4 A typical UV and visible spectrum. Philip Ronan, Gringer, CC BY-SA 3.0 (https://creativecommons.org/licenses/by-sa/3.0), via Wikimedia Commons

Because of the spectrophotometer's design, interferences are minimized. Analytes typically absorb only the wavelength of light being measured in the spectrophotometer and interferences typically do not absorb this wavelength. As such, a spectrophotometric test typically minimizes interferences.

Units of measure in spectrophotometric methods depend on the type of spectrum. Because the spectrum is typically the item to think about in this analysis, the two most common units used are units of intensity at the detector and units of wavelength. Wavelength is a description of a property of the light waves. It is the length of one complete wave of light. Wavelength units in the visible and UV region are typically in nanometers (nm). Frequency can also be used. Frequency units are typically in Hertz (Hz), or waves per second or cycles per second (1 Hz = 1 cycle/second). Frequency (of light) multiplied by its wavelength gives, as an example, length per second, which is a speed (i.e., the speed of light). The amplitude of a wave is its height and is indicative of its intensity. Figure 4.5 shows some features of waves.

In an infrared spectrum, instead of wavelength, wavenumber is often used on the horizontal axis. Wavenumber is the number of waves per unit of length, where wavelength is the length of one wave. Frequency is the number of waves per unit of time. For example, a mid-infrared wavenumber would be about 1000 waves/centimeter. Typically, the "waves" term is left out, so 1000 wavenumbers would appear as 1000 cm^{-1}.

Absorbance is a measure of the ratio of intensity of the sample to the blank intensity. Blank preparation is test method–specific. The blank

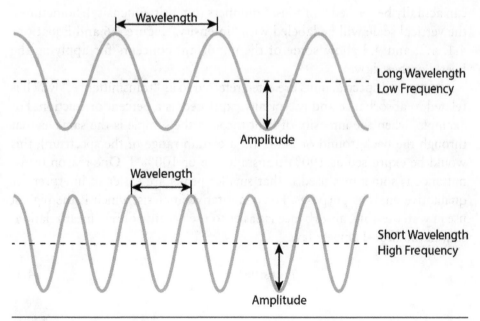

FIGURE 4.5   Wave nature of light showing amplitude and wavelength.

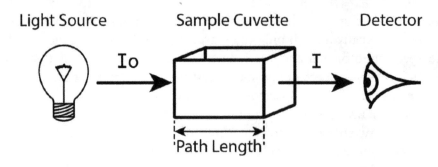

Light Source   Sample Cuvette   Detector

I_o = lamp / source intensity
I = detector intensity

FIGURE 4.6 Typical depiction of Beer–Lambert law. *Courtesy of Indigo Water Group*

may be created in a number of different ways. A blank may simply be a sample cell (cuvette) filled with air or distilled water or the sample itself with no reagents added. A reagent blank may be created by filling a sample cell (cuvette) with laboratory-grade water and then treating it as if it were a sample—adding all reagents in the same order and quantities and duplicating any other steps a regular sample might be subjected to, such as digestion. Typically, the vertical scale of the spectrum is labeled with "absorbance"; these units are abbreviated as "Abs" or "Abs units." These can actually be termed "unit-less" numbers (intensity/intensity). Sometimes, the vertical scale will be labeled with "intensity." Figure 4.6 and Equations 4.1, 4.2, and 4.3 show some of the important concepts for applying the Beer–Lambert law.

In infrared spectra, units are often referred to as "transmittance," which is related to absorbance and is typically expressed as a percent or fraction. For example, when the intensity of light through the sample is the same as that through the background or blank (in a certain range of the spectrum), this would be expressed as 100% transmittance or 100% T. One reason transmittance is sometimes used is that smaller peaks are easier to interpret for qualitative analysis purposes. For quantitative analysis, which is the biggest use in water testing, absorbance is easier to use because of the linear relationship in the Beer–Lambert law.

$$Transmittance = \frac{I}{I_0} \tag{4.1}$$

$$Absorbance = Log\ \frac{I}{I_0} \tag{4.2}$$

$$Absorbance = a \times b \times c \qquad (4.3)$$

Where

a = absorptivity constant
b = path length through sample (e.g., cuvette)
c = concentration

3.2   The Advantage in Water Testing

Water has practically no UV and visible spectrum (i.e., it is transparent in this wavelength [or frequency] range). As such, water generally does not scatter, absorb, or emit radiation in this range, and is an excellent solvent for this type of spectrophotometry. Most routine water testing methods use UV and visible technology. In other spectral ranges, things can be more problematic. With infrared, water absorbs strongly in certain parts of the spectrum. For near-infrared, water does not absorb quite as strongly. In other spectrophotometric methods, such as Raman, fluorescence, light scattering, and turbidimetry, water samples are sometimes amenable. Details should be examined carefully when evaluating a certain type of spectral suitability for environmental testing.

3.2   Calibration and Computer-Based Data

Calibration is essential and is typically linear (over a specified range) in these methods. Beer's (Beer–Lambert) law is the familiar mathematical statement that concentration has a linear relationship to absorption (or absorbance) of light (see Figure 4.6). Absorbance is the log of the ratio of the sample intensity to the blank (or background) intensity. Some spectrophotometers may use (or display) transmission (transmittance or percent transmittance) instead of absorbance on the vertical scale. Transmittance and absorbance can be thought of as "opposites" or complementary ways to display the data, as follows:

$$Absorbance = \log [intensity (sample)/intensity (blank)]$$

This relationship can be used to show that zero absorbance equals 100% transmittance. Figures 4.7 and 4.8 show how concentration and intensity vary in a typical test method.

If a sample has a small or large concentration, and corresponding small or large absorbance, then the linear range of instrument or method operation sometimes will not apply. Outside the linear range, test results will begin to be suspect. This is known as a *Beer's law deviation* and must be scrupulously avoided for most quantitative analytical work (see Figure 4.9).

Modern use of computer-based data presentation and storage allows excellent traceability for quality purposes (e.g., an audit trail). The internal

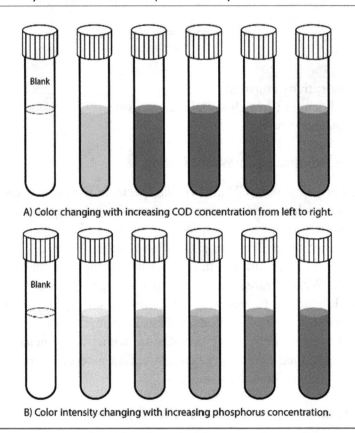

FIGURE 4.7   Variation of visible color (absorbance) with concentration. *Courtesy of Indigo Water Group*

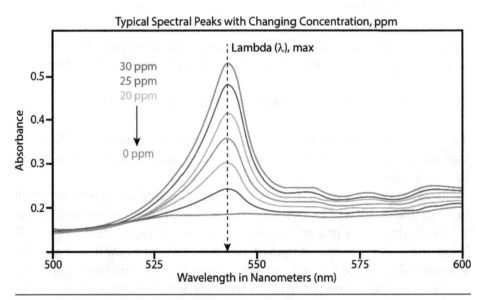

FIGURE 4.8   Variation of peak height (absorbance) with concentration.

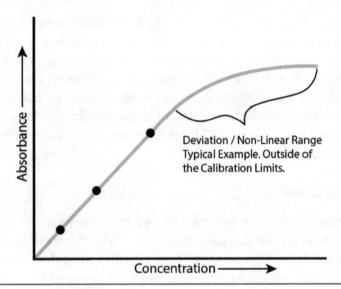

FIGURE 4.9   Beer's law deviation.

workings of the software and calculations can be complex. For absorbance tests, which are the majority, there is an optimum range for the absorbance (to give the most accurate and precise results). This range is about 0.4 to 0.7 absorbance units, but some instruments will perform well outside this optimum range. Near the detection limits, the result will typically be below this range and, therefore, less certain. Even above this range (e.g., above approximately 1 or 2 absorbance units), there will be less certainty due to less light reaching the detector (the analyte may absorb "too much" of the light).

There is an optimum peak or signal to use in a test method. The most consistent performance of a spectrophotometric method will typically be obtained by using the wavelength termed *lambda max*. This is the wavelength of maximum absorbance in the spectrum of the analyte. At the lambda max wavelength, the performance of the spectrophotometer is most consistent; for example, slight imprecision (known as "wavelength jitter") will have a relatively small effect on the final output of the concentration result. This is because, at lambda max, by definition, the measurement is not occurring on the steeply sloping part of the spectral peak. At the peak, a small amount of wavelength jitter will not affect the reading much. In the steeply sloping part of the peak, however, it will.

4.0   QUALITY ASSURANCE AND QUALITY CONTROL

Concepts for quality assurance and quality control (QA/QC) are the same for most other analyses as they are for spectrophotometric methods. Blanks, spikes (matrix spikes), calibration checks, continuing calibration checks, and so on are all important. Blank checks allow a running evaluation of the instrument (e.g., detector sensitivity and source [typically a lamp] intensity). It is important to note that if absorbance is used, which is a ratio of sample intensity to blank or background intensity, this automatically compensates for much detector and source drift. For some methods, a blank is not run (or is not required in each batch, or daily). Essentially, the blank intensity is stored in memory or "assumed" by the instrument software to be constant at some typical value.

4.1   Interferences

Positive interferences are typically caused by more apparent absorbance than should be present because of analyte alone. For example, these can be caused by light absorption or scattering blocking the light by turbid (e.g., suspended) solids. Typically, turbidity is assumed to be eliminated (possibly filtered) for a spectrophotometric measurement. Under the right circumstances, a spectrophotometer can be used to approximate turbidity.

Negative interferences can be caused by several issues and can be thought of as less apparent absorbance than would otherwise occur. This could be attributed to chemical interaction in the sample (destruction and decomposition of the analyte, which gives the "color"). Blank (background) problems can also cause negative interference (i.e., the blank or background absorbs when it typically should not). This will cause the ratio of sample intensity to blank intensity to be larger than normal because of a smaller blank intensity.

4.2   Detection Limits and Quantitation Limits

The concepts of detection and quantitation limits are important in all environmental testing because the desired state of a water sample is typically the smallest possible amount of pollutant present. Sometimes, there are exceptions in water facility applications (e.g., for nutrients) in which an optimum, generally more easily detectable concentration is desired. However, in general, the operator–analyst is working to reliably report data near the limit of quantitation or limit of detection.

4.2.1   Method Detection Limit or Minimum Detection Limit

A method detection limit (MDL) is typically obtained annually and describes the performance of the whole laboratory for the entire analysis. The MDL can encompass more than one instrument or spectrometer and is meant to quantify the smallest concentration that can confidently be measured (above "not detected" or above zero). The reader should note that zero cannot really be detected by definition; as such, below the MDL, the result is termed "not detected." Results that are lower than the MDL are reported as less than. For example, if the detection limit is 1.5 mg/L and a test result is 1.2 mg/L, then the result should be reported as <1.2 mg/L.

4.2.2 Instrument Detection Limit

The instrument detection limit (IDL) is typically reserved for a specific instrument. Often, the IDL only addresses the instrument's performance (i.e., it does not include the whole analysis and method). For example, the IDL describes the "best case" detection limit for a standard tested by the instrument.

4.2.3 Practical Quantitation Limit (Quantitation Limits)

A quantitation limit is generally the amount that can be reported confidently with approximately 10% confidence limits. In other words, a test result with this performance could be reported to about ±10% (relative). Another similar (sometimes identical) term is the minimum quantitation limit or method quantitation limit. The practical quantitation limit (PQL) is often used for the lowest concentration that can be reported for regulatory purposes without qualifiers, flags, or other explanations. For example, many regulators require a flag between the PQL and MDL that describes the number as "an estimate."

Results that are between the MDL and PQL may be reported as a numerical result but flagged with a data qualifier. For example, if the MDL is 1.0 mg/L, the PQL is 3.0 mg/L, and the sample result is 1.8 mg/L, the laboratory may report the sample result as 1.8 mg/L with a "B" data flag. The data flag alerts the person using the data that the result, while numerical, should be treated as an estimate more than an absolute value. Results greater than the PQL are reported as measured without data flags or qualifiers.

Figure 4.10 shows a graphical representation of the aforementioned concepts. This topic has been heavily discussed in recent years and can be complicated. The key idea is that the signal, generally a peak in spectrophotometry, is a certain value when divided by the noise in the baseline. Put simply, the signal-to-noise ratio is the important quantity being estimated.

5.0   OTHER CONSIDERATIONS FOR TESTS

Currently, there are many tests done by spectrophotometry. Some of the most popular are phosphate, amines and ammonia, phenols, and chemical oxygen demand (COD) (APHA et al., 2017). Typical analytes that give visible or even invisible (e.g., UV and infrared) signals can generally

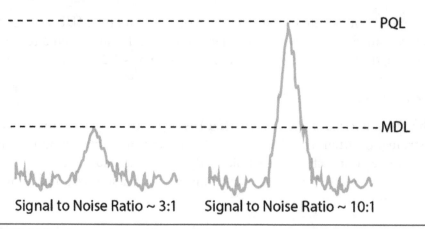

Signal to Noise Ratio ~ 3:1   Signal to Noise Ratio ~ 10:1

FIGURE 4.10 Comparison of limits of detection and limits of quantitation.

be grouped by chemistry. Metals such as iron, chromium, and copper are familiar to most people as having colors such as orange, blue, or green. The colors depend on details of the chemistry and can change during the test. Heavy or transition metals (ions) have complex electron structures with many electrons. These electrons often absorb light energy of specific wavelengths (matching the energy difference between levels of the electron orbitals). This is the physical–chemical process that gives rise to colors. Lighter elements, or those not in the transition element part of the periodic table, do not have this type of electron structure. Therefore, most non-transition metals and elements do not show visible color when dissolved in water solutions. Organic materials with large molecular weights and many double bonds (e.g., azides and other types of dyes) represent another class of chemical compound that have familiar colors. Phenol tests are an example of this type of chemistry.

5.1 Equipment Operation

The basic parts of a spectrophotometer are often called *optical elements*. One of the most familiar pieces of equipment is the sample holder, which is typically a vial or cuvette. *Cuvette* is the name given to the glass (or quartz) sample holder used for most spectrophotometers. Often, these have square corners and flat faces so light will not be affected much when passing through the cuvette and sample within. Some spectrophotometers are designed to use a pump that makes the solution to be tested flow through a cuvette. This can be an advantage in that there is more automation in the method and less cuvette handling. However, it can also make troubleshooting the cuvettes more difficult because they are permanently installed in the spectrophotometer.

Over the years, as cuvette and glassware manufacturing has improved, the quality and reproducibility of cuvettes is such that the point that most cuvettes designed to be identical truly are (i.e., the dimensions of the cuvette are precise and true). If this is an issue, a pair of cuvettes can be designated for a certain test. This is called *matching* the cuvettes. Matching is achieved by putting a pair of cuvettes through various parts of the test, especially the blank and standard checks. If the two cuvettes can be interchanged without detectable change, then they can be termed a *matched* set. Matched cuvettes can remove one variable in the test; however, this is rarely a necessary step today. Figure 4.11 presents examples of cuvettes.

5.2   Tips and Useful Practices

5.2.1   Cuvettes

- Quartz does not absorb UV light; glass often absorbs UV light.
- Handle cuvettes with care and keep them clean and without scratches, chips, and so on.

5.2.2   Internal (Instrument) Windows

- Avoid handling and touching internal windows.
- If necessary, clean internal windows with special cloth materials made for cleaning optics (lenses, windows, etc.).

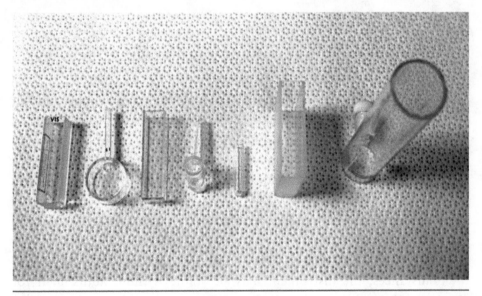

FIGURE 4.11   Typical cuvettes. (Left to right): 1-cm quartz, 1 cm, 1-cm polymer or plastic, 2-cm round face, 1-mm vial shape, 5-cm quartz, and 10 cm. Note that most have flat faces to improve optical performance.

5.2.3   Blanks and Background

- Some familiarity with normal intensity of the blanks or background can be useful.
- Because absorbance methods will compensate for instrument drift, this can be tricky.
- Too much drift cannot be compensated for, and a lamp or source may need replacement.
- Spectral performance of the prism or grating rarely degrades. This can be difficult to recognize and typically requires significant service.
- However, even a small amount of this drift can be handled; again, because of the "absorbance" concept designed into the test, the blank measurement can compensate for this.

5.2.4   Operational Hints

5.2.4.1   New Spectrophotometers

When a new spectrometer is used in the laboratory, it can be valuable to obtain operational and performance characteristics before frequent use. These characteristics can be used in the future to check for degradation in items such as detector sensitivity, source intensity, wavelength reproducibility, and other parameters. This can help avoid wasted time during troubleshooting in the future.

5.2.4.2   Zero and Span Adjustment

- There may be a zero and a "span" adjustment for the instrument.
- This can be either electronic or mechanical.
- For the zero, either a shutter is put in the path of the light to block it completely from the detector, or some electronic adjustment is made inside the instrument to simulate this.
- For the span or 100% adjustment, the opposite is done; that is, the light intensity is measured, typically without anything in the light path.
- Operator–analyst familiarity with these adjustments can be helpful in troubleshooting.
- Caution should be used before changing these adjustments. In some instances, the instrument may be stable without adjustment for months.

5.2.5 Typical Instrument Operations

- Often, spectrometer suppliers suggest a set of standards that check performance.
- Typically, a few different standards at different wavelengths are supplied to check an absorbance at each wavelength.
- Before the standards, a blank (or background or zero) is collected.

5.2.6 Example Steps in a Typical Procedure

- Power on and set the appropriate wavelength. The wavelength setting should be on the wavelength reference standard being used.
- Set the instrument to the same units as the reference standards (e.g., absorbance).
- Insert the blank or zero standard into the unit and zero the instrument. The display should read "0 Abs."
- Remove the blank and insert a reference standard, such as the 0.50 Abs reference standard.
- Read the value for the reference standard. An average of several readings may be used to obtain the reported value. The standard deviation of the replicate "read" values relates to the precision of the instrument.
- Insert the sample and blank and take the reading from the spectrophotometer. It is important to note that there are a few QA/QC steps that are done with each batch.

5.2.7 Instrument Designs

5.2.7.1 Older Designs

- Automation meant that the blank or background could be measured simultaneously with the sample.
- Double-beam designs were built with analog electronics, sophisticated opto-mechanical devices, and chart paper data reporting.

5.2.7.2 Newer Designs

- Computers allowed the use of memory to store background and blank spectra (along with calibration curves and other important data).
- This allowed a simpler mechanical design and single-beam instruments, in many instances.

- Typically, the blank is measured first and then the sample is measured. Computations are done via use of the single-beam blank spectrum comparison to the sample spectrum (as appropriate). Calibration curve computations are also done via computer from data in memory.

6.0   SELECTED EXAMPLES OF COMMONLY USED METHODS

6.1   Chemical Oxygen Demand

Chemical oxygen demand is measured as an amount of oxygen consumed ("demanded") per liter of water and sample in this test, which uses dichromate anion as the oxygen source (oxidizer). The yellow dichromate ion is reduced to a green chromic ion. Because both the reactant dichromate and product chromate are colored and of different colors or wavelengths, either can be measured. Silver is typically added as an oxidation catalyst (i.e., it makes the reaction go faster). Mercury is used to complex chloride anions, which, if present, can interfere with the test. Dichromate can oxidize chloride ions to chlorine (which can cause the interference). The wavelengths typically chosen for these tests are 420 nm for dichromate and 620 nm for chromic cation.

The test involves 2 hours of heating, in addition to some cooling time, to accomplish oxidation. As stated previously, it can be important to know the concentration of chloride expected in the typical sample. In addition, the operator–analyst should be able to estimate the concentration range of expected COD. The details of the test will depend on whether it is above about 150 mg/L or below. Potential QA/QC problems can be encountered if the reagents are not "fresh." This is typically indicated by low (or high) bias of the QA/QC standards. Although these reagents can be useful for many years, experienced operator–analysts report that, in some instances, they do not last for their entire shelf life. It is possible that the oxygen in air and other factors can affect reagent quality if stored for long periods of time.

Current testing chemistry has been reduced in size and volume from what was practiced before 1990. Because of the nature of the test, chemicals still have to be powerful oxidizers and corrosives. In addition, the presence of mercury can be challenging to safe handling and proper disposal.

6.2   Nitrate-Nitrogen

The nitrate anion is relatively stable; as such, this test does not have as many interference or preservation issues as the nitrite test. In a mixed (sulfuric and phosphoric) acid solution, 2,6-dimethylphenol reacts with nitrate, bonding

it to the phenol ring and producing the following colored product: 4-nitro-
2,6- dimethylphenol. The spectral absorbance can be measured between 345
and 370 nm (most methods will specify one of these wavelengths).

6.3   Nitrite-Nitrogen

This test involves detection of a visible and colored complex between a
diazonium salt and another aromatic compound. The product chromophore
is actually a type of compound that is used as an organic dye, and its color
is measured at 515 nm. Handling of the samples and standards can be challenging because nitrite has a relatively strong tendency to decompose to
nitrate simply reacting with common dissolved oxygen in a water sample.
Because acidity speeds up the decomposition reaction, it should be avoided.
Nitrite is a relatively strong oxidizing agent, which, as with many oxidants,
can act as either an oxidizing or a reducing agent (depending on other oxidizing and reducing agents available in the sample).

6.4   Ammonia-Nitrogen

Indophenol is the reaction product between ammonia (as ammonium ions,
even in basic solution at about pH 12 to 13), plus hypochlorite, nitroprus-
side, and salicylate ions. Indophenol is the colored product, whose concentration is proportional to the ammonia or "ammonia-nitrogen." The
spectrophotometer measures the spectral absorbance at 690 nm. As with
many ammonia tests, certain amines (which can be thought of as ammonia
bonded with organic compounds) can interfere.

6.5   "UV 254" Method for Organic Carbon (Dissolved)

This method incorporates the typical absorbance of organic materials at
254 nm in the UV region of the spectrum. Filtration is typically involved to
remove particulate, and this filtration step can be the source of significant
variation in the test (care must be taken to use the proper filter so as not to
contaminate the solution with organic material from the filter). A blank is
measured because of this potential issue, and the operator–analyst should
keep in mind that even lot-to-lot variation in the same kind of filter can show
significant variation in this measurement. While this 254-nm wavelength is
generally good for most situations, certain types of organic materials, such
as long-chain aliphatic materials, will not respond well in this test (e.g., some
hydrocarbons such as gasoline-related or mineral and petroleum oil–related
compounds). Even this statement has exceptions because these materials
are often accompanied by aromatic and ring-based organic materials that
will typically absorb at the 254-nm wavelength. Although this test does not

TABLE 4.2 Typical benchsheet for spectrophotometric analysis.

Facility name

Spectrophotometric analysis

Analyte _____   Method used _____

Wavelength _____ nm   Analysis date _____

Analyst _____   Check standard recovery _____%

Sample/ standard date	Sample identification	Sample volume (mL)	Final volume (mL)	Abs ☐ Trans ☐	Initial concentration (mg/L)	Dilution factor	Final concentration (mg/L)
	Cal. blank					1	
	Standard 1					1	
	Standard 2					1	
	Standard 3					1	
	Standard 4					1	
	Standard 5					1	
	Check standard						
	Reagent blank					1	
	Duplicate						
	Duplicate						
	Spike						

$$\text{Spike added (mg/L)} = \frac{\text{Concentration of spike added (mg/L)} \times \text{volume of spike added (mL)}}{\text{Final volume (mL) of spiked sample used for colorimetric determination}}$$

$$\% \text{ Spike recovery} = \frac{[\text{Spiked sample initial concentration (mg/L)} - \text{Sample initial concentration (mg/L)}] \times 100}{\text{Spike added (mg/L)}}$$

have a typical, primary standard, potassium hydrogen phthalate is often used because of its general availability and its absorbance at 254 nm (it has a ring structure).

6.6   Phosphate Phosphorus

Phosphate analysis is generally comprised of two steps. The first step is the conversion of the phosphate to dissolved orthophosphate. This is accomplished by a heat digestion in which molybdate reacts in the acid medium with orthophosphate to form phosphomolybdic acid. The second step involves the colorimetric determination of the dissolved phosphate. The addition of ascorbic acid reduces the phosphomolybdic complex, giving an intense molybdenum blue color that can be read as concentration on the spectrophotometer at 890 nm. Table 4.2 provides a typical benchsheet for spectrophotometric analysis.

7.0   REFERENCE

American Public Health Association, American Water Works Association, & Water Environment Federation. (2017). *Standard methods for the examination of water and wastewater* (23rd ed.). American Public Health Association.

5

Microbiological Examinations

1.0   INTRODUCTION

The bacterial requirement of final effluent from a water resource recovery facility must be in compliance with the National Pollutant Discharge Elimination System (NPDES) permit prior to discharge into a body of water (rivers, lakes, and oceans). Available test methods and organisms are summarized in Table 5.1. Before 2007, the bacteriological indicators were

TABLE 5.1   U.S. EPA–approved test methods for wastewater.

Method	Method	Total coliforms	E. coli	Fecal coliforms	Enterococci	Membrane filtration	MPN	Confirmation step
MTF[1]	Standard Method 9221	Yes	Yes	Yes			Yes	Yes, required
m-Endo	Standard Method 9222B	Yes				Yes		Yes, recommended
m-FC*2	Standard Method 9222D			Yes		Yes		Yes, recommended
Modified m-TEC	U.S. EPA 1603		Yes			Yes		
m-ColiBlue®24	Hach Company 10029	Yes	Yes			Yes		No
Colilert®	Standard Method 9223B	Yes	Yes				Yes	No
Colilert®-18	Standard Method 9223B	Yes	Yes	Yes			Yes	No
m-EI3	U.S. EPA 1600				Yes	Yes		
Enterolert®					Yes		Yes	

*May be used to test fecal coliform for sludge samples.
[1]MTF = multiple tube fermentation, [2]FC = fecal coliform, [3]EI = enterococci

either total coliforms or fecal coliforms (now classified as thermotolerant bacteria). In 2005, the U.S. Environmental Protection Agency (U.S. EPA) proposed to use other indicators that were more specific of true fecal coliforms. This included *Escherichia coli* (*E. coli*) and enterococci. In 2007, U.S. EPA promulgated these indicators with the approval of several new methods for detection in addition to existing methods for total coliforms and fecal coliforms.

The *E. coli* indicator was approved for use when discharging into fresh waters and enterococci for marine waters (oceans). The new methods for *E. coli* include two membrane filtration methods, modified membrane thermotolerant *E. coli* (m-TEC) and m-ColiBlue® and, for the most probable number (MPN), Colilert®, and Colilert®-18 with the Quanti-Tray® system. The new methods for enterococci are membrane enterococcus indoxyl 1-D glucoside agar (m-EI) (U.S. EPA Method 1600 [U.S. EPA, 2002a]) for membrane filtration and, for MPN, Enterolert®, and the Quanti-Tray® system. The 15-tube multiple tube fermentation (MTF) procedure and the m-Endo membrane filtration procedure can be used for either total coliforms, fecal coliforms, and/or for *E. coli.* Membrane fecal coliform (m-FC) can also be used for testing fecal coliforms.

E. coli and enterococci are the predominant bacteria found in the feces of humans and warm-blooded animals.

2.0   INDICATOR ORGANISMS

Pathogens are present in wastewater as viruses, bacteria, and microorganisms. It is nearly impossible to test for every possible pathogen. Many pathogens are difficult to test for because they are difficult to isolate, are present in small numbers, or simply because an approved test method does not exist. Instead, water and wastewater methods target indicator organisms. Indicator organisms are generally not pathogenic themselves but are closely associated with the pathogens, so the presence of the indicator organism indicates that the pathogen might also be present. Characteristics of an ideal indicator organism include

- Always present when the pathogen is present
- Not ubiquitous
- Present in larger numbers than the pathogen
- Easy to analyze for and approved test methods exist
- More difficult to kill than the target organism

Indicator organisms used in water and wastewater testing include total coliforms, fecal coliforms, *E. coli,* and enterococci.

2.1   The Coliform Group

Total coliform bacteria (including thermotolerant coliforms and *E. coli*) belong to the family *Enterobacteriaceae;* some of the genera are *Escherichia, Klebsiella, Enterobacter,* and *Citrobacter.* In general, they are gram-negative, aerobic, and facultative anaerobic rod-shaped bacteria, and are further defined by the following methods used to test for them:

- Multiple tube fermentation—ferments lactose to produce lactic acid, turbidity, and/or gas (CO_2) at 35 °C within 24 to 48 hours—presumptive requiring confirmation for total coliforms and fecals or *E. coli* (gas and turbidity);

- Membrane filtration—produces metallic sheen to red colonies on m-Endo media within 24 hours, which is a presumptive test requiring confirmation as indicated;

- Enzymatic—the *total coliform group* is defined as all bacteria possessing the enzyme 1-D galactosidase, which cleaves the chromogenic substrate resulting in the release of the chromogen. *E. coli* is defined as possessing the enzyme 1-D glucuronidase, which cleaves a fluorogenic or chromogenic substrate resulting in the release of the fluorogen or the chromogen (Colilert® and Colilert®-18, modified m-TEC [*E. coli* only] and m-ColiBlue®); and

- Thermotolerant coliforms—the ability to grow at 44.5 °C that
  - Produces gas and turbidity in *E. coli* medium with MTF,
  - Produces blue colonies in m-FC media with membrane filtration, and
  - Produces yellow color wells with Colilert®-18.

2.2 The Enterococci Group

Enterococci bacteria belong to the streptococci family and are gram-positive bacteria that can grow in 6.5% saline. Enzymatically, they possess the enzyme 1-D-glucoside, which cleaves the fluorogenic or chromogenic substrate resulting in the release of a fluorogen or chromogen (Enterolert® and m-EI).

3.0 METHODS

3.1 General Description

The procedures and techniques required to test for total coliforms, fecal coliforms, *E. coli*, and enterococci are based either on membrane filtration or MPN methods. They are cultural-based methods and require incubation. The time and temperature of incubation is dependent on the method. All methods will be reviewed in this chapter.

3.2 General Laboratory Equipment and Materials

- pH meter
- Balance (0.01 to 0.1 g accuracy)

- Sterilizers (autoclave or other steam-pressure sterilizers), oven sterilizer, and ultraviolet (UV) sterilizer
- Hot plate, magnetic stirrer, and stir bars
- Incubators
- Water bath
- Thermometers enclosed in liquid to be on the shelf of the incubator
- Assorted glassware such as beakers (50 mL, 100 mL, and 1 L), Erlenmeyer flasks (250 and 1000 mL), and graduated cylinders (100, 500, and 1000 mL)
- Graduated glass or sterile disposable plastic pipets (1, 10, 25, and 50 mL) or equivalent
- Sample bottles, 120-mL glass or sterile disposable plastic (with sodium thiosulfate for chlorinated effluent); 0.1 mL of a 10% sodium thiosulfate solution is added to the vessel for neutralization of the chlorine or chloramines to neutralize up to 15 mg/L of chlorine
- Microscope
- Incubation tubes and Petri plates

3.3 Reagents

Reagents must be sterilized before use (follow the directions given in Section 3.4). Preparation of reagents includes the following:

- Sodium thiosulfate solution (10%): Sodium thiosulfate solution is added to sample containers before autoclaving. It neutralizes any residual chlorine up to 15 mg/L that may be present in the sample. This ensures that disinfection is stopped and that the number of organisms measured during the test is representative. The solution is prepared as follows:
  1. Weigh 10 g of sodium thiosulfate in a weighing dish.
  2. Add this to clean beaker or flask (sufficient to hold 100 mL).
  3. Add 75 mL of deionized water or distilled water and stir until thoroughly dissolved.
  4. Dilute to 100 mL with deionized water or an equivalent and stir.
  5. Pour contents into a clean dilution bottle (or 125-mL Erlenmeyer screw-cap flask).
  6. Label the container with the name of the chemical, date prepared, manufacturer, lot number, expiration date, and operator–analyst's initials.

- Phosphate buffered water: This buffer is used to dilute and rinse samples between filtration (unless otherwise noted). This solution must be sterile to prevent contamination of the sample. The solution is prepared as follows:

  1. Stock solution I: Dissolve 34.0 g of KH_2PO_4 in 500 mL of deionized water or distilled water using a 1-L volumetric flask or equivalent. Verify the pH is 7.2 ± 0.5. If it is not, adjust the pH with 1-N NaOH. Dilute with deionized water or distilled water to produce 1 L of stock buffer solution. Sterilize by autoclaving or sterile filtration and refrigerate the stock buffer. Discard it if it becomes turbid.

  2. Stock solution II: Dissolve 38 g of magnesium chloride ($MgCl_2$) or 81.4 g of MgCl2·6H2O in 1 L of deionized water or distilled water. Sterilize by autoclaving or sterile filtration and refrigerate the stock buffer. Discard it if it becomes turbid.

  3. Working solution: Add together 1.25 mL of stock solution I and 5.0 mL of stock solution II and dilute to 1 L with deionized water or distilled water. Mix completely and sterilize by autoclaving or sterile filtration.

- Amphyl solution (0.5%): Dilute 5 mL of concentrated amphyl to a final volume of 1000 mL with deionized water. Store in a glass bottle at room temperature.

3.4   Sterilization of Equipment and Materials

When sterile disposable equipment and materials are not used, the filter holders, media, buffer water, sample bottles, dilution bottles, and pipet must be sterilized before use. Depending on the units used, the methods available are pressurized steam, dry heat, UV irradiation, and membrane filtration.

- Autoclave technique: Buffered water should be kept in a 1-L borosilicate container or dispensed into 100-mL bottles. Loosely cap or cover with foil or kraft paper and autoclave at 121 °C, 103 kPa (15 psi), for 15 minutes (100 mL or less) or 30 minutes (500 mL to 1 L). For dilution bottles, dispense in amounts that will provide 99 ± 2.0 mL or 9.0 ± 0.2 mL. Loosely cap or cover with foil or kraft paper and autoclave at 121 °C, 103 kPa (15 psi), for 15 minutes.

- Membrane filter technique: If an autoclave is unavailable or if efficiency and convenience are essential, remove particles by filtering buffer with a filter holder fitted with a 47-mm diameter, 4-µm membrane filter and a 42-mm microfiber glass prefilter. Pour filtrate into a sterilized container, then dispense as required (at the time of sample filtration) directly into the filter funnel for the rinse step or into dilution bottles using a 50-mL syringe filled with a disposable, presterilized filter unit of 0.45- or 0.22-μm pore size.

3.5   Sample and Dilution Bottles

The following recommendation applies to borosilicate glass or polypropylene sample bottles (120-mL capacity). If the water to be sampled contains chlorine or chloramines, add 0.1 mL of a freshly prepared 10% solution of sodium thiosulfate to the bottle before sterilization. Wrap the cap and neck of the bottle with foil or kraft paper and autoclave at 121 °C, 103 kPa (15 psi), for 15 minutes. For empty glass bottles or a glass pipet, either autoclave at 121 °C, 103 kPa (15 psi), for 15 minutes or apply dry heat at 170 °C for 1 hour (2 hours for a glass pipet if placed in metal containers). This procedure is not required if sterilized disposable plastic containers with sodium thiosulfate are used.

3.6   Filter Holders

All nondisposable glass, plastic, or stainless steel filter holders can be sterilized by autoclaving, UV sterilization, or dry heat. Sterilization methods include

- Autoclaving technique: Wrap the funnel and base with foil or kraft or other autoclave paper and autoclave at 121 °C, 103 kPa (15 psi), for 15 to 20 minutes. The units will remain sterile until opened.
- Ultraviolet light: Where an autoclave or dry heat oven are not available to the laboratory, UV sterilization for 2 to 3 minutes has been found to be effective for filter holders.
- Dry heat: This is only recommended for glass or stainless steel holders. Remove any rubber stoppers, wrap the filter holders in aluminum foil, and heat for 1 hour at 170 °C. The units will remain sterile until opened.

3.7   Alternate Sterilization Procedures

For field applications or where conditions preclude use of the aforementioned methods, one of the following two alternate methods may be applied:

- Boiling: Any glass or polypropylene sample bottles or holders may be sanitized just before use by immersion in boiling water for 10 minutes. Remove with metal tongs and wrap openings and the base surface with flamed aluminum foil. Use promptly after cooling.

- Alcohol: A 70% nondenatured ethanol is preferred or, if not available, 80% to full-strength isopropyl alcohol is acceptable. Immerse units in alcohol for 5 to 10 minutes, allow them to dry, and use promptly.

4.0 QUALITY ASSURANCE AND QUALITY CONTROL REQUIREMENTS FOR MICROBIOLOGICAL METHODS

4.1 Sampling

Sampling is a critical part of sanitary water testing and represents the body of water being evaluated. It is critical to collect a sample without contaminating it. Plastic gloves should be used for collecting the sample; in addition, an aseptic technique should be used. The sample should be filled to the milliliter line (tolerance of ± 2.5 mL). There should be sufficient head space to allow proper mixing of the sample (approximately 2.5 cm).

4.2 Dechlorination

Samples must be dechlorinated prior to testing. This applies to the use of chlorine, chloramines, or bromine. The samples are neutralized by adding 0.1 mL of a 10% sodium thiosulfate solution to the 100-mL sample or by using containers that contain sodium thiosulfate sufficient to neutralize the oxidant (up to 15 mg/L). The samples should be tested to verify the oxidant is neutralized using N, N-diethyl-p-phenylenediamine or an equivalent method.

4.3 Storage and Holding Times

All samples should be tested as soon as possible. If this is not possible, the samples must be stored at <10 °C (the samples should not be allowed to freeze). The required hold time of samples is 6 hours from collection to initiation of testing (Federal Register, 2007).

4.4 Sample Containers

Containers must be of sufficient size to hold 100 ± 2.5 mL with sufficient head space for mixing. The containers are typically 120 mL and can be either sterile borosilicate glass or sterile plastic containers such as polypropylene or polystyrene. There should be sufficient space (2.5 cm) in the container to allow mixing of the sample. It is important to avoid contaminating the opening of the container.

4.5   Sample Volume and Dilutions

The bacterial counts in surface water, other naturally occurring nontreated and non-potable water, and wastewater may range from a few hundred to several million per 100 mL. Therefore, smaller sample volumes will be required for analysis. When a reasonably reliable history of the water source is known, the appropriate selection of sample volume can be ascertained for the test procedure chosen. If this information is unavailable, suggested guidelines are outlined in Table 5.2.

Using a small volume of sample may result in either (a) overdiluting of the sample or (b) erroneously overestimating the final results because of a low number. Samples less than 1 mL in volume must be diluted appropriately.

4.5.1   Dilutions for Membrane Filtration

When the sample size is between 1 and 100 mL, it is frequently more convenient, efficient, and less expensive to introduce the sample directly into the filtration unit funnel. Samples greater than 30 mL can be poured directly into the funnel and filtered by a rinse with phosphate buffer water. For samples less than 30 mL, first pour 20 to 30 mL of phosphate buffered water into the funnel and dispense the sample into the buffer phosphate water while swirling the funnel. After filtering this, rinse the funnel with 20 to 30 mL of the phosphate buffered water. Complete the procedure and record the sample volume used.

4.5.2   Dilutions for Membrane Filtration and Most Probable Number (Quanti-Tray® or Equivalent)

Accurate measurement of sample sizes smaller than 1 mL requires dilution in sterile buffered phosphate water or the recommended diluent for the

TABLE 5.2   Suggested sample volumes.

Types of water	Sample volume, mL	
	Total coliform	Fecal coliform
Disinfected wastewater effluent (primary)	0.003, 0.01, 0.02, 0.08	0.01, 0.03, 0.1, 0.3, 1, 3, 10
Raw wastewater	0.0001, 0.0003, 0.0001, 0.003, 0.01	0.0001, 0.0003, 0.003, 0.01, 0.03
Disinfected secondary effluent	0.01, 0.1, 1	0.1, 1, 10, 20, 50
Surface water, river water, bathing, beaches	0.001, 0.01, 0.1, 1	0.01, 0.1, 1, 10, 50

method. The *diluents,* which are defined as the tube or vessel used for dilution of the sample, must be sterile. The most convenient method is to sterilize aliquots of buffer water in dilution bottles or test tubes. Sterile buffer water may also be purchased pre-aliquoted in disposable plastic bottles.

4.5.2.1   1:10 Dilution

Pipet 1.0 mL of a well-mixed sample into a tube containing 9.0 mL of the sterile diluent (see Figure 5.1) and mix thoroughly. This represents one-tenth of the concentration of the original sample.

4.5.2.2   1:100 Dilution

Aseptically pipet 1.0 mL of the well-mixed sample into a sterile vessel containing 99 mL of sterile diluent and mix thoroughly (1:100-mL sample) (see Figure 5.2). This represents one-hundredth of the concentration of the original sample.

Other dilutions can be made, such as a 1:20 by pipeting 5.0 mL of a well-mixed sample into 95 mL of sterile diluent. Any dilution is possible using this basic technique.

It is important to note that it is always advisable to run a diluted sample as soon as possible after preparation of the dilution (preferably within 30 minutes); otherwise, a die-off or an increase of the sample population may occur.

4.5.2.3   Serial Dilutions

Serial dilutions are a series of dilutions that are made by creating each new dilution from the previous dilution. Serial dilutions are convenient for the

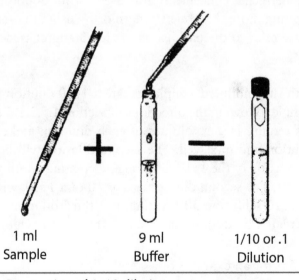

| 1 ml | 9 ml | 1/10 or .1 |
| Sample | Buffer | Dilution |

FIGURE 5.1   Preparation of 1:10 dilution.

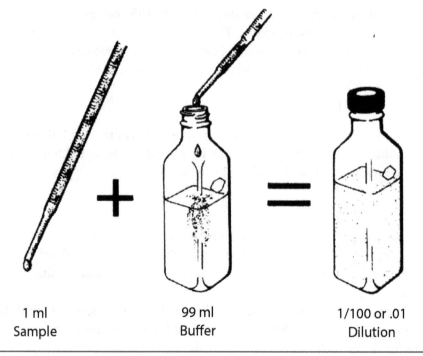

| 1 ml | 99 ml | 1/100 or .01 |
| Sample | Buffer | Dilution |

FIGURE 5.2 Preparation of 1:100 dilution.

MTF methods because each dilution used must be an increasing decimal dilution for the MPN tables to be valid. An example of an increasing decimal dilution would be aliquots of 10, 1.0, and 0.1 mL. Each dilution moves the decimal point one place further to the right.

Serial dilutions may be necessary when using prepurchased MTF tubes or when the same media concentration is used for all dilutions. This may be done by creating more than one strength or by using the serial dilution technique. Please refer to the package insert for prepurchased MTF tubes, or as follows:

- Start with the undiluted sample. Create a 1:10 dilution of the undiluted sample following the directions in Section 4.5.2.1. Because MTF methods require larger volumes of each dilution, make 100 mL of each dilution and mix well. The second dilution will be created by taking a portion of the 1:10 dilution and creating another 1:10 dilution from it. The second dilution is a 1:10 of a 1:10, which ends up being a 1:100 dilution all together. If a third dilution is desired, it will be created by diluting the second dilution to achieve a 1:1000 dilution all together.

See Figure 5.3 for an example of a serial dilution.

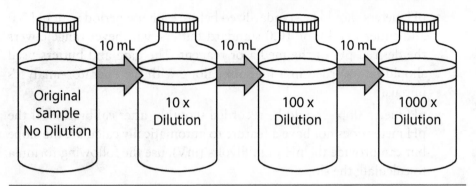

FIGURE 5.3   Preparation of serial dilution. *Courtesy of Indigo Water Group*

4.6   Positive and Negative Controls and Blanks

Before use, each new lot of dehydrated or prepared commercial medium and each batch of laboratory-prepared medium should be checked for sterility using positive and negative culture controls. Those controls can be obtained from proficiency testing providers, scientific supply vendors, or through the use of American type culture collection (ATCC) strains or equivalent. Examples of positive strains that can be used are (a) total coliforms is *Klebsiella pneumoniae* (thermotolerant) ATCC 13883 or equivalent; (b) *E. coli* /fecal coliforms is *Escherichia coli* ATCC 25992 or ATCC 11775 or equivalent; and (c) a negative strain for total coliforms is *Pseudomonas aeruginosa* ATCC 27853. For enterococci, a positive strain is (a) *Enterococcus faecalis* ATCC 11700 or equivalent, and a negative control is (b) *Staphylococcus aureus* ATCC 6538, *Streptococci bovis,* or equivalent. A blank should be performed by substituting sterile deionized water or distilled water following the procedure for the specific method. The blank should not yield any positive reaction for the specific method.

4.7   Duplicates

Duplicate analyses should be performed on 10% of samples and on at least one sample per test run. If the laboratory performs fewer than 10 tests per week, a duplicate analysis should be performed on at least one sample per week.

4.8   pH Meter and Buffers

- Accuracy and scale graduations should be within ±0.1 units.
- pH buffer aliquots should be used only once.
- Electrodes should be maintained according to the manufacturer's recommendations.

- pH meters should be standardized before each use period with pH 7.0 and either pH 4.0 or 10.0 standard buffers, whichever range covers the desired pH of the media or reagent. The date and buffers used should be recorded in a logbook, along with the operator–analyst's initials.

- pH meter slope should be recorded monthly, after calibration. If the pH meter does not have a feature to automatically calculate the slope but can provide the pH in millivolts (mV), use the following formula to calculate the slope:

$$\text{Slope (as \%)} = [\text{mV at pH 7} - \text{mV at pH 4}] \times 100/177 \quad (5.1)$$

- If the slope is below 95% or above 105%, the electrode or meter may need maintenance. Follow manufacturer's instructions for electrode maintenance and general cleaning.

- Commercial buffer solution containers should be dated upon receipt and when opened. Buffers should be discarded by the expiration date.

4.9   Temperature Monitoring

- Glass or electronic thermometers must be graduated in 0.5 °C increments or less (0.2 °C increments for tests that are incubated at 44.5 °C). The fluid column in glass thermometers should not be separated. The calibration of glass and electronic thermometers should be checked annually, at the temperature used, and against a National Institute of Standards and Technology (NIST)-traceable reference thermometer or one that meets the requirements of National Bureau of Standards Monograph SP 250-23. The calibration factor and date of calibration should be indicated on the thermometer.

- The laboratory should record in a quality control record book the following information: serial number of laboratory thermometer if applicable, serial number of NIST-traceable thermometer (or other reference thermometer), temperature of laboratory thermometer, temperature of NIST-traceable thermometer (or other reference thermometer), correction (or calibration) factor, date of check, and the operator–analyst's initials.

- If a thermometer differs by more than 1 °C from the reference thermometer, it should be discarded.

- Reference thermometers should be recalibrated at least every 5 years. Reference thermometer calibration documentation should be maintained.

- Continuous recording devices that are used to monitor incubator temperature should be recalibrated at least annually. A reference thermometer that meets the specifications described in the aforementioned paragraph should be used for calibration.

- Incubator units must have an internal temperature monitoring device and maintain the temperature specified by the method used, typically 35 ± 0.5 °C and 44.5 ± 0.2 °C. For nonportable incubators, thermometers should be placed on the top and bottom shelves of the use area and immersed in liquid as directed by the manufacturer (except for electronic thermometers). If an aluminum block incubator is used, culture dishes and tubes should fit snugly.

- Calibration-corrected temperature should be recorded at least twice per day for each thermometer being used during each day the incubator is in use, with readings separated by at least 4 hours. Documentation should include the date and time of reading, temperature, and the technician's initials.

5.0   TEST METHODS

5.1   Multiple Tube Fermentation

5.1.1   Introduction

Method 9221 from *Standard Methods for the Examination of Water and Wastewater* (APHA et al., 2017) is based on lactose fermentation to lactic acid within 48 hours when incubated at 35 ± 0.5 °C. Formation of gas and/or turbidity in any quantity in the inverted Durham tube is a positive reaction. Three dilutions are performed on the sample, typically 10, 1, and 0.1 mL. Additional dilutions may be made depending on the estimated bacterial density. Five tubes are used for each dilution for a minimum total of 15 tubes. Confirmation is required. There are three procedures for this method: presumptive test, confirmation for total coliforms, and confirmation for fecal coliforms and/or *E. coli*. Densities are calculated from the MPN table on the basis of the positive confirmation step. The procedures are as follows:

- Presumptive test—lauryl tryptose broth (LTB) provides a preliminary estimate of bacterial density. The tubes are incubated up to 48 hours at 35 ± 0.5 °C. At 24 hours, the tubes are removed from the incubator and checked for turbidity and/or gas production. Any tubes that do not show turbidity and/or gas should be incubated for an additional 24 hours. All tubes that are either turbid or produce gas are then confirmed for either total coliforms, fecal coliform, or *E. coli*. A sterile loop or an applicator stick is used to remove a sample from the tube and is placed into a tube of brilliant green lactose broth (BGLB) for total coliform confirmation. At the same time, another loop is used to place a sample into a tube of *E. coli* medium or *E. coli* medium and 4-methylumbelliferyl-1-D-glucuronide (MUG) for confirmation of fecal coliform or *E. coli*.

- Confirmation for total coliforms—BGLB is a broth that will produce turbidity and gas within 48 hours, confirming total coliforms. The tubes are incubated at 35 ± 0.5 °C. At 24 hours, the tubes are removed from the incubator and checked for turbidity and gas. Any tubes that do not show turbidity and/or gas should be incubated an additional 24 hours and checked again for gas and turbidity. The number of tubes confirmed for each dilution should be indicated.

- Confirmation for fecal coliforms or *E. coli*—*E. coli* medium is incubated at 44.5 ± 0.2 °C in either a water bath or air incubator that can maintain the ± 0.2 °C requirement. The samples are incubated for 24 hours. Turbidity and gas at 24 hours indicate the presence of fecal coliforms. If the *E. coli* medium contains MUG, then the operator–analyst should check the sample for fluorescence using a 6-W, 365-nm UV light source. Any tubes that fluoresce are positive for *E. coli*.

5.1.2 Apparatus and Materials

- Air incubator or circulating water baths set at 35 ± 0.5 °C and 44.5 ± 0.2 °C
- Pipets, graduated glass or sterile plastic (10 mL, 1 mL), or equivalent (micropipets)
- Inoculation loops, 3-mm diameter or disposable applicator tips; Pyrex culture test tubes with closures (metal caps or equivalent), 150×20 mm containing inverted Durham tube 50 mm $\times$ 16 (or equivalent)
- 6-W, 365-nm UV light source

5.1.3 Reagents

It is strongly recommended that media be purchased rather than purchasing individual ingredients to make the media in the laboratory. Reagents required for testing include the following:

- Phosphate buffer (refer to Section 3.3 for recipe)
- Sodium thiosulfate solution (refer to Section 3.3 for recipe)
- Lauryl tryptose broth for presumptive test:

○ Tryptose   20.0 g

○ Lactose   5.0 g

○ Dipotassium hydrogen phosphate, K_2HPO_4 2.75 g

○ Potassium dihydrogen phosphate, KH_2PO_4 2.75 g

○ Sodium chloride, NaCl   5.0 g

○ Sodium lauryl sulfate   0.1 g

○ Deionized or distilled water   1 L

1. If using 1 mL or less of sample volume, add the dehydrated ingredients previously listed (total weight of 35.6 g) to 1 L of deionized water.

2. If using 10 mL of sample volume, double the quantities of the ingredients previously listed (total weight of 76.2 g) and add to 1 L of deionized water. Table 5.3 lists the volume of dry ingredients needed for different sample volumes.

3. Mix thoroughly with a stir bar and heat to dissolve. Heat is necessary to completely dissolve the ingredients. Remember to never leave media to warm on a hot plate unattended.

4. Dispense 10 mL of broth into each fermentation tube.

5. Place an inverted Durham tube into each fermentation tube. The LTB should cover the tube at least one-half to two-thirds after sterilization.

6. Close the tubes with metal or heat-resistant plastic caps. Leave caps loose to allow for liquid expansion and for gases to escape during sterilization.

7. Autoclave the media for 12 to 15 minutes at 121 °C (U.S. EPA, 1978).

8. As soon as the autoclave pressure has fallen to zero, the sterilized media should be removed from the autoclave for cooling before use or storage.

9. After media has cooled to room temperature, tighten lids if using caps.

10. The pH of the finished media should be 6.8 ± 0.2. Make minor adjustments in pH (<0.5 pH units) with 1.0N NaOH or 1.0N HCl solution to the pH specified in formulation. If the pH difference is larger than 0.5 units, discard the batch and check preparation instructions and pH of reagent water to resolve the problem. Incorrect pH values may be caused by reagent water quality, medium deterioration, or improper preparation.

11. Verify that the Durham tubes do not contain any air bubbles. Invert the tubes as needed to dislodge air bubbles from the Durham tubes.

12. Incubate the media at 20 °C overnight if refrigerated after sterilization. Discard any tubes showing bubbles. The remaining tubes may be used for testing.

13. Store prepared tubes at 10 °C or less until they are needed. Do not freeze them.

- Brilliant green lactose broth for confirmation test for total coliforms:
  - Peptone   10.0 g
  - Lactose   10.0 g
  - Oxgall   20.0 g
  - Brilliant green   0.0133 g
  - Deionized or distilled water   1 L

    1. Add the dehydrated ingredients previously listed above (total weight of 40.01 g) to 1 L of deionized water.

    2. Follow steps 3 through 9 under the LTB media.

    3. The pH of the finished media should be 7.2 ± 0.2 s.u. Make minor adjustments in pH (<0.5 pH units) with 1.0N NaOH or 1.0N HCl solution to the pH specified in formulation. If the pH difference is larger than 0.5 units, discard the batch and check preparation instructions and pH of reagent water to resolve the problem. Incorrect pH values may be caused by reagent water quality, medium deterioration, or improper preparation.

    4. Follow steps 11 and 12 under the LTB media.

- *E. coli* broth (+MUG) for confirmation test for fecal coliforms or *E. coli:*
  - Tryptose or trypticase   20.0 g
  - Lactose   5.0 g
  - Bile salts mixture or bile salts no. 3   1.5 g
  - Dipotassium hydrogen phosphate, K_2HPO_4 4.0 g
  - Potassium dihydrogen phosphate, KH_2PO_4 1.4 g
  - Sodium chloride, NaCl   5.0 g
  - 4-methylumbelliferyl-1-D-glucuronide (optional)   0.1 g
  - Deionized or distilled water   1 L

1. Add the dehydrated ingredients listed above (total weight of 41.9 g without MUG) to 1 L of deionized water.

2. Follow steps 3 through 9 under the LTB media.

3. The pH of the finished media should be 6.9 ± 0.2 s.u. Make minor adjustments in pH (<0.5 pH units) with 1.0N NaOH or 1.0N HCl solution to the pH specified in formulation. If the pH difference is larger than 0.5 units, discard the batch and check the preparation instructions and pH of reagent water to resolve the problem. Incorrect pH values may be caused by reagent water quality, medium deterioration, or improper preparation.

4. Follow steps 11 and 12 under the LTB media. It is important to note that the addition of MUG is required for confirming the presence of *E. coli*. If only fecal coliform quantification is needed, then MUG may be omitted from the media.

5.1.4 Quality Control

Media prepared in the laboratory or purchased media from a vendor should be tested with both a positive and negative control and a blank prior to testing actual samples (see Section 4.6 for information on how to run positive and negative controls and blanks). Premade media from an outside vendor should come with a certificate showing that it has been tested by the manufacturer. These certificates should be kept on file. The post-sterilization pH must be within the specified range for each media type. Duplicates should be prepared in accordance with Section 4.7.

TABLE 5.3 Preparation of lauryl tryptose broth (LTB).

	Dehydrated LTB broth	
mL of inoculums	mL of medium	Ingredients required, g/L
1 or 0.1	10	35.6
10	10	71.2
10	20	53.4
20	10	106.8
100	50	106.8
100	35	137.1
100	20	213.6

Source: APHA et al., 2017.

5.1.5 Procedure

The first step of the MPN procedure is called the *presumptive test*. Inoculated fermentation tubes are incubated for 24 to 48 hours. Any tube showing turbidity and/or gas production during the presumptive test indicates the possible presence of coliform group bacteria and is recorded as a positive presumptive tube on the benchsheet (see Section 5.1.1). Before beginning the procedure, the operator–analyst should refer to Figure 5.4, which shows a flow chart illustrating the various steps and confirmation steps required to confirm the presence of either total coliforms or fecal coliforms and/or *E. coli*. In addition, a sample benchsheet is included in Table 5.4, which may also help to clarify the various stages of presumptive and confirmatory testing. This table is for *E. coli* only and not for total coliforms and is an example that can be modified to fit specific requirements.

The procedure for testing is as follows:

- **Step 1**—Setting up the presumptive phase test: 15-tube MTF:
    1. Wash hands thoroughly with soap and water.
    2. It is important to remember to shake all samples vigorously before aliquoting.

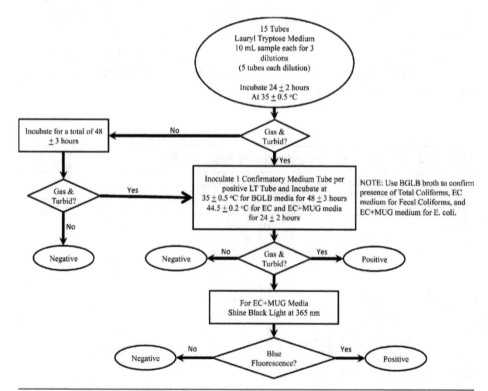

FIGURE 5.4   Multiple tube fermentation procedure flow chart.

TABLE 5.4 Sample benchsheet for MPN analysis.

Sample identification:
Loc. code:

LT media log no./expiration date: EC media I no./ expiration date:

Sample setup: operator-analyst/ date/time: Sample set up after 24 hours:
Sample read back at 24 hours: Sample read back at 24 hours:
Sample readback at 48 hours: Sample set up after 48 hours:
                                                                   Sample read back at 48 hours:

Incubator for 35.0 °C ± 0.5 °C: Incubator for 44.5 °C ± 0.2 °C:

| | | LT media at 35.0 °C ± 0.5 °C | | | | EC media at 44.5 °C ± 0.2 °C | | | |
Dilution	Tube	Tray position	24 hours	48 hours		Tray position	24 hours	48 hours	MPN
1	1								
	2								
	3								
	4								
	5								
10	1								
	2								
	3								
	4								
	5								
100	1								
	2								
	3								
	4								
	5								

Courtesy of Littleton/Englewood Wastewater Treatment Plant Laboratory, Englewood, Colorado

3. Record the lot number of LTB media on the benchsheet.

4. Arrange fermentation tubes in three rows of five tubes each in a test tube rack. One row of five tubes will be used for each dilution and/or the undiluted sample. The operator–analyst may elect to use more than three rows; however, a minimum of three must be used.

5. Unscrew the lids or remove metal caps of the first row of five presumptive LTB tubes.

6. Shake sample and dilutions vigorously about 25 times.

7. Using a sterilized, 10-mL wide-mouth pipet, pipet 10 mL of the well-mixed sample into each of the five LTB tubes containing double-strength LTB in the first row. Shake the bottle before each withdrawal.

8. Tighten the caps on the tubes and invert to mix. Allow all air bubbles to escape from the Durham tube (the inner tube), then carefully upright the test tube and place it in the test tube rack.

9. Repeat steps 5 through 8 with each sample dilution. Using a new, sterile pipet for each dilution, pipet 1 mL of the well-mixed

sample into the second set of tubes and either 0.1 mL of the well-mixed sample or 1 mL of a 1:10 dilution of the sample into the third set of tubes.

10. Incubate inoculated tubes at 35 ± 0.5 °C.

11. Allow the samples to incubate for 24 ± 2 hours at 35 ± 0.5 °C.

12. After 24 ± 2 hours, swirl each tube or bottle gently and examine each one for growth, gas, and/or turbidity. Gas present in the Durham tube and/or turbidity or both in any amount is considered a positive result, regardless of whether the broth is cloudy or clear.

13. Record the presence or absence of growth, gas, and acid production for each LTB tube. Figure 5.5 shows an example of gas within a Durham tube. It is important to note that turbidity may make it difficult to see the gas bubble. Holding each vial up to the light can make viewing easier.

14. If no gas or turbidity is evident, those LTB tubes should be returned to the incubator for re-incubation at 35 °C; the tubes should be reexamined at the end of 48 ± 3 hours from the original sample setup time (24 hours after first read back).

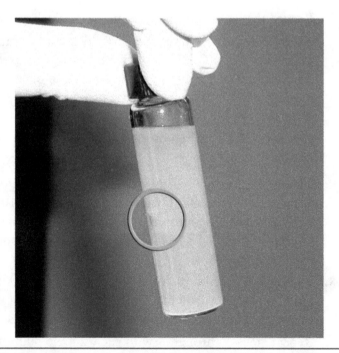

FIGURE 5.5 Gas bubble in Durham tube. *Courtesy of Indigo Water Group*

15. The presumptive LTB tubes that contain gas and or turbidity or both should now be tested using the confirmed media. Submit tubes with a positive presumptive reaction to BGLB for total coliforms and *E. coli* broth or EC–G for fecal coliforms or *E. coli*. The absence of gas formation and turbidity at the end of 48 ± 3 hours of incubation constitutes a negative.

16. At the end of 48 hours, submit all additional positive LTB tubes to one of the confirmation steps.

17. Discard any negative LTB tubes that remain using appropriate safety precautions. It is important to note that only step 2 or 3 is typically performed. These steps may be done at the same time if confirmation of both total coliforms and fecal coliforms and/or *E. coli* is desired. Steps 2 and 3 are not sequential.

- **Step 2**—Setting up the confirmation step for total coliforms with BGLB media:

  1. Submit all presumptive tubes showing growth, any amount of gas, or turbidity within 24 to 48 hours (or earlier) to tubes of BGLB to confirm the presence of total coliforms.

  2. Place one BGLB media tube into a new test tube rack for each of the presumptive tubes. Tubes should be placed into the same location in the new rack that the presumptive tubes currently occupy in the LTB rack. Placing the tubes this way will make it easier for the operator–analyst to track which tubes correspond to each dilution.

  3. Loosen the caps on the BGLB media tubes.

  4. Flame a calibrated wire loop (3.0 to 3.5 mm in diameter) by holding it in the flame of a Bunsen burner or similar flame source until it glows red. Allow it to air cool. Do not set the loop down on the laboratory bench or other surface.

  5. Gently shake or rotate presumptive tubes gas or acidic growth to resuspend the organisms.

  6. With a sterile loop 3.0 to 3.5 mm in diameter, transfer one or more loopfuls of culture from the first positive LTB tube into a fermentation tube containing BGLB as shown in Figure 5.6:

     a) Insert the wire loop into the LTB media.

     b) Swirl gently.

     c) Remove the loop from the LTB media.

     d) Insert the loop into the corresponding BGLB tube.

     e) Swirl gently.

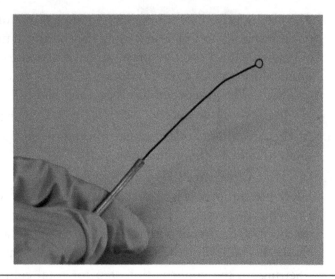

FIGURE 5.6   Wire loops. Jeffrey M. Vinocur, CC BY-SA 3.0 (http://creativecommons. org/licenses/by-sa/3.0/), via Wikimedia Commons (https://commons.wikimedia.org/ wiki/File:Inoculation_loop.JPG).

    f) Repeat these steps for each positive LTB tube. Be sure to flame the loop as described in step 4 before each new LTB tube is transferred.

7. Sterile wooden applicators or sterile plastic loops may be used in place of the calibrated loop. In this instance, insert a sterile wooden applicator at least 2.5 cm into the LTB broth culture, promptly remove, and plunge the applicator to the bottom of the fermentation tube containing BGLB. Remove and discard the applicator. Repeat for all other positive presumptive tubes.

8. Incubate the inoculated BGLB tube at 35 ± 0.5 °C for 24 to 48 hours. Formation of gas in any amount in the inverted tube (Durham) of the BGLB fermentation tube at any time (e.g., 6 ± 1 hours or 24 ± 2 hours) within 48 ± 3 hours constitutes a positive test for total coliforms.

9. Add up the total number of positive tubes for the confirmed test for each dilution. If all five tubes in the first dilution row are positive, then the number of positive tubes is five. If only three tubes in the second dilution have gas bubbles in their Durham tubes, then the number of positive tubes is 3; and

10. Determine the MPN value from the number of positive tubes as indicated in Section 5.1.6.

- **Step 3**—Setting up the confirmation test for fecal coliforms and/or *E. coli* with *E. coli* broth or EC–MUG broth:

  1. Submit all presumptive tubes showing growth, any amount of gas, or turbidity within 24 to 48 hours (or earlier) to tubes of *E. coli* medium or EC–MUG medium to confirm the presence of fecal coliforms and/or *E. coli*.

  2. Follow steps 2 through 7 under Step 2; however, substitute *E. coli* media or *E. coli* media-MUG in place of BGLB media.

  3. Incubate the inoculated *E. coli* broth or EC–MUG broth tubes at 44.5 ±0.2 °C for 24 hours in a circulating water bath or dry incubator that can maintain 44.5 ± 0.2 °C.

     a) Formation of turbidity and gas in any amount in the fermentation tube at any time within 24 ± 3 hours constitutes a positive test for fecal coliforms.

     b) For EC–MUG, shine a 6-W, 365-nm UV lamp on the tube. Any tube with a blue fluorescence is positive for *E. coli*.

  4. Add up the total number of positive tubes for the confirmed test for each dilution.

  5. Determine the MPN value from the number of positive tubes as indicated in Section 5.1.6.

5.1.6 Calculations

The calculated density of the confirmed test may be obtained from the MPN table (see Table 5.5) or *Standard Methods* (APHA et al., 2017), Section 9221, based on the number of positive tubes and reactions in each dilution.

Three dilutions are necessary to create the MPN code. The operator–analyst should note that the top of the MPN table is labeled for dilutions of 10, 1, and 0.1 mL. For example, if all five tubes of the 10-mL dilution are positive, then the number of positive tubes is five. If only three tubes in the 1-mL dilution have gas bubbles in their Durham tubes, then the number of positive tubes is three. In the last dilution, 0.1 mL, only one tube was a confirmed positive result. The operator–analyst will record "5-3-1" as the MPN code and will look this value up in the MPN table. In this example, the corresponding result from the MPN table is 110 with a range of 40 to 300, which means that the most likely number of target organisms in the original sample (total or fecal coliforms or *E. coli*) is 110 organisms per 100 mL of sample volume.

TABLE 5.5   Most probable number table.

Combination of Positives	MPN Index/ 100 mL	Confidence Limits Low	Confidence Limits High	Combination of Positives	MPN Index/ 100 mL	Confidence Limits Low	Confidence Limits High
0-0-0	<1.8	—	6.8	4-0-3	25	9.8	70
0-0-1	1.8	0.090	6.8	4-1-0	17	6.0	40
0-1-0	1.8	0.090	6.9	4-1-1	21	6.8	42
0-1-1	3.6	0.70	10	4-1-2	26	9.8	70
0-2-0	3.7	0.70	10	4-1-3	31	10	70
0-2-1	5.5	1.8	15	4-2-0	22	6.8	50
0-3-0	5.6	1.8	15	4-2-1	26	9.8	70
1-0-0	2.0	0.10	10	4-2-2	32	10	70
1-0-1	4.0	0.70	10	4-2-3	38	14	100
1-0-2	6.0	1.8	15	4-3-0	27	9.9	70
1-1-0	4.0	0.71	12	4-3-1	33	10	70
1-1-1	6.1	1.8	15	4-3-2	39	14	100
1-1-2	8.1	3.4	22	4-4-0	34	14	100
1-2-0	6.1	1.8	15	4-4-1	40	14	100
1-2-1	8.2	3.4	22	4-4-2	47	15	120
1-3-0	8.3	3.4	22	4-5-0	41	14	100
1-3-1	10	3.5	22	4-5-1	48	15	120
1-4-0	10	3.5	22	5-0-0	23	6.8	70
2-0-0	4.5	0.79	15	5-0-1	31	10	70
2-0-1	6.8	1.8	15	5-0-2	43	14	100
2-0-2	9.1	3.4	22	5-0-3	58	22	150
2-1-0	6.8	1.8	17	5-1-0	33	10	100
2-1-1	9.2	3.4	22	5-1-1	46	14	120
2-1-2	12	4.1	26	5-1-2	63	22	150
2-2-0	9.3	3.4	22	5-1-3	84	34	220
2-2-1	12	4.1	26	5-2-0	49	15	150
2-2-2	14	5.9	36	5-2-1	70	22	170
2-3-0	12	4.1	26	5-2-2	94	34	230
2-3-1	14	5.9	36	5-2-3	120	36	250
2-4-0	15	5.9	36	5-2-4	150	58	400
3-0-0	7.8	2.1	22	5-3-0	79	22	220
3-0-1	11	3.5	23	5-3-1	110	34	250
3-0-2	13	5.6	35	5-3-2	140	52	400
3-1-0	11	3.5	26	5-3-3	170	70	400
3-1-1	14	5.6	36	5-3-4	210	70	400
3-1-2	17	6.0	36	5-4-0	130	36	400
3-2-0	14	5.7	36	5-4-1	170	58	400
3-2-1	17	6.8	40	5-4-2	220	70	440
3-2-2	20	6.8	40	5-4-3	280	100	710
3-3-0	17	6.8	40	5-4-4	350	100	710
3-3-1	21	6.8	40	5-4-5	430	150	1100
3-3-2	24	9.8	70	5-5-0	240	70	710
3-4-0	21	6.8	40	5-5-1	350	100	1100
3-4-1	24	9.8	70	5-5-2	540	150	1700
3-5-0	25	9.8	70	5-5-3	920	220	2600
4-0-0	13	4.1	35	5-5-4	1600	400	4600
4-0-1	17	5.9	36	5-5-5	>1600	700	—
4-0-2	21	6.8	40				

Source: APHA et al., 2017.

When the series of decimal dilutions is other than 10, 1, and 0.1 mL, select the MPN value from the table and then calculate according to the following formula:

$$\text{MPN }(\textit{from table}) \times \frac{10}{\text{Sample portion selected at the lowest dilution}} \quad (5.2)$$
$$= \text{MPN/100 mL}$$

For additional information and guidance for unusual dilutions and results, the operator–analyst should refer to *Standard Methods* (APHA et al., 2017) and/or the 1978 U.S. U.S. EPA manual, *Microbiological Methods for Monitoring the Environment: Water and Wastes* (U.S. EPA, 1978). This extensive and detailed manual is available as a free download from U.S. EPA at http://nepis.epa.gov. The manual also contains detailed instructions for making different types of media and information on test interferences.

5.2   General Procedure for Membrane Filtration Methods

5.2.1 Introduction

There are several variations of the membrane filtration method, with each using a different type of growth media. However, the basic procedural steps for each method are similar and are given in Section 5.2.5. Incubation times and temperatures and type of media used vary depending on the method and target organism. For methods that use proprietary growth media such as m-ColiBlue® (Hach Company, Loveland, Colorado [Hach Company, www.hach.com]), the operator–analyst should refer to the manufacturer's instructions.

It is important to note that because the membrane filtration procedure can yield low or highly variable results for chlorinated wastewater, U.S. EPA requires verification of these results using the MPN procedure to resolve any controversies. Even without controversies, the operator–analyst should consider either regularly running samples by both methods or sending splits to an outside laboratory.

A 0.45-μm filter is placed onto the bottom of a funnel. A specific volume of sample is filtered through a funnel where a membrane filter of a specific porosity size (typically 0.45 μm) is placed to capture bacteria from the sample. Vacuum is applied to allow the liquid to pass through the filter. The filter is removed aseptically to a plate containing either a pad saturated with the media or agar containing the media. The plates are incubated at a specific temperature for a specific period of time, and growth of colonies occurs if bacteria are present. Following incubation, the number of colonies on each plate is counted. Results are multiplied by the appropriate dilution factor (if required) to get the final result.

5.2.2 Apparatus and Materials

- Air incubator or circulating water bath set at 35 ± 0.5 °C and 44.5 ± 0.2 °C
- Colony counter
- Vacuum pump or house vacuum
- Vacuum tubing (sufficient thickness to allow proper vacuum without collapsing)
- Manifold for multiple filtrations
- Vacuum flask (1 L or larger)
- Wide-mouth, 10-mL (disposable or autoclavable) pipets
- Pipets (1-mL disposable or autoclavable)
- Autoclavable volumetric cylinders of various sizes
- Bunsen burner or alcohol lamp
- Membrane filtration funnel and glass, metal, plastic, or sterile disposable filter funnel
- Forceps with smooth tips
- Stereoscopic microscope with a fluorescent illuminator or equivalent
- Sterile Petri dishes (50-mm diameter)
- Sterile membrane filters (47-mm diameter with a pore size of 0.45 μm)
- Absorbent pads for liquid media (47 mm)
- Sterile deionized water or sterile buffered water
- Waterproof plastic containers such as Whirl-Pak® bags (manufactured by NASCO, California) or an equivalent

5.2.3 Reagents

- Phosphate buffer (refer to Section 3.3 for recipe)
- Sodium thiosulfate solution (refer to Section 3.3 for recipe)
- Growth media required by test method; recipes for each type of growth media are given within each method

5.2.4 Quality Assurance and Quality Control

Insert a sterile dilution/rinse water sample (100-mL buffer water) after filtration of a series of 10 samples, or once per batch, whichever is more frequent, to check for possible cross-contamination and/or contaminated rinse water. Incubate the rinse water control membrane culture under the same conditions as the sample. Alternatively, use an additional buffer rinse of the filter unit after the filter is removed to prevent carryover between samples.

With each test method, the growth media prepared, including premade media from a vendor, should be tested with a positive and negative control and a blank prior to testing actual samples. Premade media from an outside vendor should come with a certificate showing that it has been tested by the manufacturer. These certificates should be kept on file. The post-sterilization pH must be within the specified range for each media type. (See Section 4.6 for controls and blanks and see the procedure under each method—Section 5.3—for testing.) Lastly, duplicates should be prepared in accordance with Section 4.7.

5.2.5 General Procedure

- **Step 1**—Clean the countertop with 0.5% amphyl solution or equivalent disinfectant.
- **Step 2**—Prepare the desired number of media plates. The type of media used will vary depending on the method and target organism. The methods and media types are summarized in Table 5.6. Discard plates containing growth on the media (mold or bacteria) and plates with incomplete or uneven media coverage.
  1. Solid agar media plates should be prepared ahead of time according to the recipes within the individual test methods. They should be tested with a positive and negative control and a blank prior to their first use.
  2. If using a pad and filter configuration instead of solid media, open a Petri dish, remove an absorbent pad from the envelope (containing

TABLE 5.6 Summary of media types by method.

Method	Target organism	Media type
m-Endo	Total coliforms	Les Endo agar or
		m-Endo medium
m-FC	Fecal coliforms	m-FC medium
Modified m-TEC	*E. coli*	Modified m-TEC Agar
m-ColiBlue®	Total coliforms	Hach m-ColiBlue®
	E. coli	24 ampules
m-EI	Enterococci	m-EI agar

filters and pads) or dispenser-cartridge pack, and place the pad into the Petri dish. When handling the pad, be sure to use smooth-tipped tweezers that have been sterilized by being dipped in alcohol and flaming. With a pad-loaded Petri dish, proceed to the Step 2(3).

3. Media may be contained in a glass ampoule or plastic ampoule with screw cap. If using media contained in a glass ampoule, flick the top of the ampoule with a finger to force the media down into the body below the score line. Then, use a commercially available ampoule breaker to break off the top portion above the score. If an ampoule breaker is not available, place the ampoule between several folds of a clean paper towel and break the tip by forcing it off with thumb pressure.

4. Aseptically dispense the ampoule contents over the absorbent pad. Be sure to cover the entire pad surface.

5. If stock media have been prepared instead of purchasing ampoules, pipet 2 mL of this solution into the Petri dish to saturate the pad.

6. For both ampoules and stock media, carefully remove any surplus liquid from the culture dish.

7. Close the dish.

- **Step 3**—Label media plates with the sample identification, date, and volume. Be sure to mark the plates on the media side rather than the lid. This will prevent confusion if the lids become separated from the media side of the plate.

- **Step 4**—Prepare the filter funnel or filter manifold by assembling the sterilized pieces as shown in Figure 5.7. It is important to note that many operator–analysts find it convenient to leave the filter funnels, graduated cylinders, and other glassware in the autoclave after the autoclave cycle is complete. Alternatively, items can be wrapped in aluminum foil and be stored on shelves or in drawers after sterilization.

- **Step 5**—Sterilize the forceps by dipping them into methanol and passing them through a flame. Let them air cool for a moment.

- **Step 6**—Place a sterile membrane filter on the filter base, grid side up, and attach the funnel to the base so that the membrane filter is held between the funnel and the base. It is important to note that the membranes are supported in their packages by a more rigid backing. Only the membrane should be placed on the filter base.

- **Step 7**—Shake the sample bottle vigorously at least 25 times to distribute the bacteria uniformly. Immediately measure the desired volume of sample or dilution into the funnel using either a pipet or graduated cylinder.

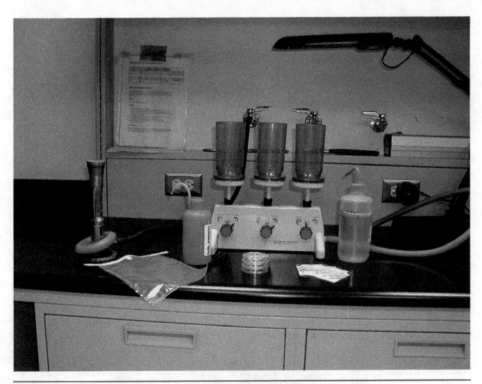

FIGURE 5.7 Filter manifold and equipment. *Courtesy of Indigo Water Group*

1. Select sample volumes based on previous knowledge of the pollution level to produce 20 to 80 colonies for total coliforms or *E. coli*, and 20 to 60 colonies for fecal coliforms or enterococci. Sample volumes of 1 to 100 mL are typically tested at half-log intervals (e.g., 100, 30, 10, and 3 mL).

2. When less than 20 mL of sample (diluted or undiluted) is to be filtered, add approximately 10 mL of sterile buffer water or the equivalent to the funnel before filtration. The membrane filter should be completely submerged. Adding buffer water helps to distribute the bacteria present in the sample over the filter surface. Alternatively, the operator–analyst may pipet the sample volume into a bottle of sterile dilution water, mix well, and add the entire dilution volume to the filter funnel. It is important to note that smaller sample sizes or sample dilutions can be used to minimize the interference of turbidity or for high bacterial densities. Multiple volumes of the same sample or sample dilutions may be filtered, and the results may be combined by averaging.

- **Step 8**—Turn on the vacuum system and filter the samples to dryness.

- **Step 9**—After the samples have completely filtered, rinse the sides of the funnel at least twice with 20 to 30 mL of sterile buffered rinse water.

- **Step 10**—While keeping the membrane filters under vacuum, remove the funnel(s) from the filter base.

- **Step 11**—Use sterile forceps to aseptically remove the membrane filter(s) from the filter base(s) and roll them onto the appropriately labeled media to avoid the formation of bubbles between the membrane and the agar surface (see Figure 5.8 for an illustration). NOTE: Sterilize forceps by dipping them in 70% ethyl or isopropyl alcohol. Just before use, burn off the alcohol by passing the forceps through the open flame of a Bunsen burner. Allow the forceps to cool before touching the membrane. If the forceps are too warm, the membrane may catch on fire. This procedure should be followed every time the forceps are used to touch anything associated with the sample to prevent contamination.

1. Reseat the membrane filter if bubbles occur.

2. Run the forceps around the edge of the filter to be sure that the filter is properly seated on the agar.

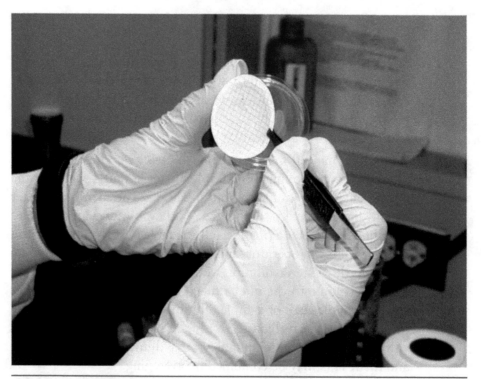

FIGURE 5.8   Placing filter onto media. *Courtesy of Indigo Water Group*

- **Step 12**—Close the dish and invert so the media side of the plate faces up. This step is critical to prevent the bacteria from drowning in their own byproducts and to keep bacteria from being carried from one area on the plate to another.
- **Step 13**—Repeat Steps 5 through 12 for all remaining sample aliquots. When finished, turn off the vacuum pump.
- **Step 14**—Transfer the finished, labeled plates to a Whirl-Pak® bag or other waterproof container. Remove all the air from the bag before closing the bag, as shown in Figure 5.9.
- **Step 15**—Place the bag with the plate(s) inverted between the upper and lower horizontal pieces of a test tube rack. This will keep the plates horizontal and prevent them from sliding around. Place the rack in a water bath incubator for the required incubation time. Place a lead ring or equivalent weight (no brick or bottled water) on top of the test tube rack to hold it in place. Ensure that the bag with the plates in it is completely submerged. Place all prepared cultures in the water bath within 30 minutes after filtration. The desired water bath temperature is specific to the method and the target organism. Table 5.7 lists the different methods, temperatures, and target

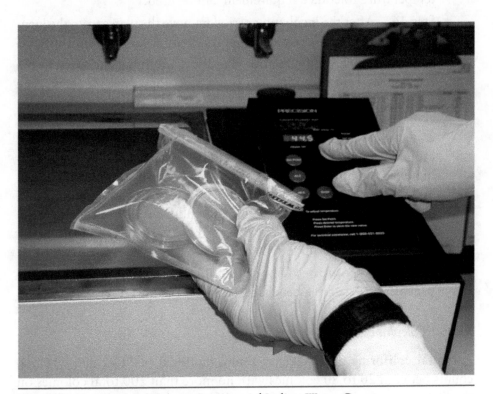

FIGURE 5.9 Whirl-Pak® bag. *Courtesy of Indigo Water Group*

TABLE 5.7   Incubation temperature reference guide.

Method	Target organism	Incubation temperature	Colony color
m-Endo	Total coliforms	35.0 ± 0.5 °C for 24 hours	Pink to dark red with a metallic surface sheen
m-FC	Fecal coliforms	44.5 ± 0.2 °C for 24 hours	Various shades of blue
Modified m-TEC	*E. coli*	35.0 ± 0.5 °C for 2 hours, then 44.5 ± 0.2 °C for 22 hours	Red or magenta
m-ColiBlue®	Total coliforms *E. coli*	35.0 ± 0.5 °C	Total coliforms will be red or blue. *E. coli* colonies will be blue to purple.
m-EI	Enterococci	41.0 ± 0.5 °C	All colonies (regardless of color) with a blue halo are recorded as enterococci colonies.

organisms as a quick reference guide. Alternatively, an appropriate, accurate, solid heat sink or equivalent incubator that can hold the temperature tolerance requirement can be used.

- **Step 16**—After the incubation period, remove the plates from the incubator and count the desired colonies. Colony color is method-specific.
- **Step 17**—Place the bottom of the dish containing the filter in the microscope field. Examine and count all colonies per the membrane filtration method used (see Section 5.3 for the specific methods). To count colonies, use a low-power (5 to 20 times magnification) stereomicroscope with a daylight fluorescent light source, either fitted around the nosepiece or positioned in the same place as the filter, shining onto the filter. Colonies shown in every grid should be counted. To facilitate this counting procedure, proceed from top to bottom and left to right as shown in Figure 5.10, and have a consistent method for counting colonies that straddle grid lines as shown in Figure 5.11 to prevent over- or undercounting.
- **Step 18**—Record the number of colonies for each aliquot on the benchsheet. A sample benchsheet is shown in Table 5.8.

5.2.6   Calculations

For total coliforms or *E. coli* (m-Endo, modified m-TEC, m-ColiBlue), counts between 20 to 80 colonies with no more than 200 total colonies on the filter must be obtained. For fecal coliforms (m-FC) where the colonies

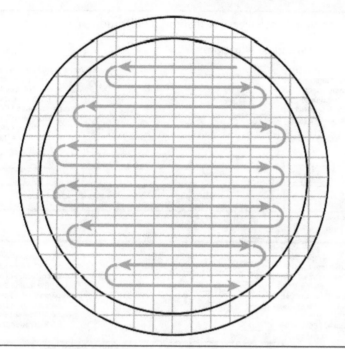

FIGURE 5.10   Counting procedure for coliform colonies. U.S. EPA (1978), Figure II-C-8

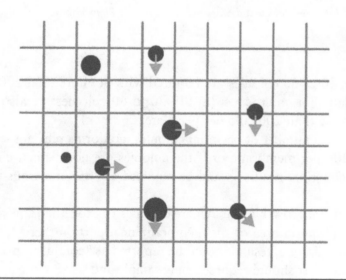

FIGURE 5.11   Locating borderline colonies (indicated by arrows pointing to various squares). U.S. EPA (1978), Figure II-C-9

TABLE 5.8   Sample benchsheet for membrane filtration analysis.

Fecal Coliforms, Membrane Filtration   Standard M

Sample Date:	_____	Bottle Sterilization Batch: ___
Set-Up Date/Time/Analyst:	_____	Filter lot #: ___
In 44.5° Incubator Date/Time:	_____	E. coli Std. SLOG#: ___
Incubator ID #:	_____	Expiration Date: ___
Read-Back Date/Time/Analyst:	_____	Prepared Media LOG#: ___
Reviewed by/Date:	_____	Expiration Date: ___
		Buffered Water LOG#: ___
QA Batch #: FECAL-MF-	_____	Expiration Date: ___

Loc. Code:_____	E. coli Std. (+ or –)	Rinse Blank	Sample Aliquot (ml)					
Sample ID:_____								
Colony Count								
Colonies/100 ml	n/a							

Comments: * Colonies present that are not blue in color

Sample Date:	_____	Bottle Sterilization Batch: ___
Set-Up Date/Time/Analyst:	_____	Filter lot #: ___
In 44.5° Incubator Date/Time:	_____	E. coli Std. SLOG#: ___
Incubator ID #:	_____	Expiration Date: ___
Read-Back Date/Time/Analyst:	_____	Prepared Media LOG#: ___
Reviewed by/Date:	_____	Expiration Date: ___
		Buffered Water LOG#: ___
QA Batch #: FECAL-MF-	_____	Expiration Date: ___

Loc. Code:_____	E. coli Std. (+ or –)	Rinse Blank	Sample Aliquot (ml)					
Sample ID:_____								
Colony Count								
Colonies/100 ml	n/a							

Comments: *Colonies present that are not blue in color

Courtesy of Littleton/Englewood Wastewater Treatment Plant Laboratory, Englewood, Colorado

are larger, 20 to 60 colonies are required with no more than 200 colonies on the plate. For enterococci (m-EI), 20 to 60 colonies are also required with no more than 200 colonies per plate.

If the total number of bacterial colonies (coliforms plus non-coliforms) exceeds 200 per membrane or if the colonies are not distinct enough for accurate counting, results should be reported as "too numerous to count" (TNTC).

If confluent growth occurs where colonies run together, covering either the entire filtration area of the membrane or a portion thereof, and colonies are not discrete, results should be reported as "confluent growth." The operator–analyst should request a new sample and select more appropriate volume(s) for filtration. Typically, three dilutions are performed on each sample to achieve this requirement.

Count the number of colonies on each plate. Use the dilution or dilutions that contain between 20 and 80 (or 20 and 60 for fecal coliforms and enterococci) colonies. Do not use or count dilutions that contain more than 200 colonies of all types per plate. If counts do not fall in the range, then the counts from the plates will have to be calculated as shown in the following formula or refer to Section 9222 of *Standard Methods* (APHA et al., 2017). Results are reported as the number of colony forming units (CFU) per 100 mL of sample. The term CFU is used because two bacteria landing on the membrane filter close together may produce a single colony, and nonviable bacteria that land on the membrane filter will not produce a colony at all. General formula:

$$\frac{Colonies}{100\ mL} = \frac{Colonies\ counted \times 100}{Sample\ filtered,\ mL} \quad\quad (5.3)$$

Example 1: When one plate has 23 colonies.

Colonies/100 mL = 23 × 100 ÷ 100 mL = 23 CFU/100 mL

Example 2: When none of the dilutions has a colony count falling within the desired range, count the colonies on all filters and report as number per 100 mL. For example, if duplicate 50-mL portions of the same sample were examined and the two membranes had counts of five and three colonies, then calculate the colonies per 100 mL as

$$\frac{[(5+3) \times 100]}{(50+50)} = 8\ colonies\ per\ 100\ mL$$

Example 3: If the total number of colonies is fewer than or greater than 20 to 80 or 20 to 60, total the counts on all countable filters. When three plates with dilutions of 50, 25, and 10 mL have 15, 6, and <1 colonies, calculate the colonies/100 mL as

$$\frac{[(15+6+0) \times 100]}{(50+25+10)} = 25\ colonies\ per\ 100\ mL$$

5.3 Specific Membrane Filtration Methods

5.3.1 m-Endo Method

5.3.1.1 Introduction

Method 9222(B) from *Standard Methods* (APHA et al., 2017) is based on lactose fermentation. Bacteria such as *E. coli* and *K. pneumoniae* have sufficient acid to produce golden sheen colonies. Other coliform bacteria such as *Enterobacter* spp. produce only red colonies. The m-Endo method is a presumptive test and confirmation is required to verify total coliforms, fecals, or *E. coli.*

5.3.1.2 Method-Specific Reagents

Prepare the m-Endo media as described in this section. Verify the finished media pH and run a positive and negative control and blank prior to use as described in Section 4.6. The operator–analyst should note that m-Endo broth and agar can be purchased or can be prepared from purchased media in the laboratory. However, it is strongly recommended that m-Endo broth and agar not be prepared from individual ingredients.

- m-Endo agar:
  - Tryptose or polypeptone   10.0 g
  - Thiopeptone or thiotone   5.0 g
  - Casitone or trypticase   5.0 g
  - Yeast extract   1.5 g
  - Lactose   12.5 g
  - Sodium chloride, NaCl   5.0 g
  - Dipotassium hydrogen phosphate, K_2HPO_4 4.375 g
  - Potassium dihydrogen phosphate, KH_2PO_4 1.375 g
  - Sodium lauryl sulfate   0.05 g
  - Sodium desoxycholate   0.10 g
  - Sodium sulfite, Na_2SO_3 2.10 g
  - Basic fuchsin   1.05 g
  - Agar   15.0 g
  - Deionized or distilled water   1 L

    1. Rehydrate m-Endo media in 1 L of water containing 20 mL of 95% ethanol.

    2. Use a magnetic stir bar and hot plate. Heat to near boiling to dissolve agar (but do not allow to boil), then promptly remove from heat and cool to between 45 and 50 °C.

    3. Check the pH of the medium and, if necessary, adjust the pH to between 7.1 and 7.3 with 1.0 N NaOH or 1.0 N HCl. The pH of reconstituted dehydrated media seldom will require adjustment.

    4. Dispense 5 to 7 mL quantities into 60-mm sterile glass or plastic Petri dishes or 3 to 5 mL into 47-mm dishes.

5. Do not sterilize by autoclaving. A precipitate is normal in Endo-type media. Refrigerate finished medium in the dark and discard unused agar after 2 weeks.

- m-Endo broth:
  1. Prepare as m-Endo agar as previously described but omit agar from the recipe.
  2. Refrigerated broth may be stored for up to 4 days. The broth may have a precipitate, but this does not interfere with medium performance.
  3. Immediately before use, dispense liquid medium (between 1.8 and 2.2 mL per plate) onto absorbent pads. Pads must be certified free of sulfite and other toxic agents at a concentration that could inhibit bacterial growth.
  4. Carefully remove excess medium by decanting the plate.

5.3.1.3 Procedure

Follow the general membrane filtration method given in Section 5.2.5 using m-Endo agar or m-Endo broth plates. Incubate at 35.0 ± 0.5 °C for 24 hours.

5.3.1.4 Interpretation

After 22 to 24 hours, remove the plate from the water bath. Count and record the number of metallic sheen colonies and atypical red colonies with the aid of an illuminated lens with a 10 to 20 times magnification or a stereoscopic microscope.

The sheen area may vary in size from a small pinhead to complete coverage of the colony surface. Atypical coliform colonies can be dark red, mucoid, or nucleated without sheen. Generally, pink, blue, white, or colorless colonies lacking sheen are considered non-coliforms. Verification of both colony types (verify at least one colony from each typical and atypical colony type and verify 10% of positive samples) is advisable, but not required.

5.3.1.5 Verification

Standard Methods (APHA et al., 2017) recommends that all typical and atypical colonies included in the direct count be verified by submitting them to confirmation tests. The procedures for confirmation testing are given in Section 5.1.5 (Step 2 for total coliforms and Step 3 for fecal coliforms and/or *E. coli*).

5.3.1.6   Calculations

Follow the general membrane filtration calculations given in Section 5.2.6. The number of colonies per plate should be between 20 and 60.

5.3.2   Membrane Fecal Coliform Method

5.3.2.1   Introduction

Method 9222(D) from *Standard Methods* (APHA et al., 2017) is based on lactose fermentation and the formation of blue colonies at a temperature of $44.5 \pm 0.2\ °C$ in a circulating water bath or dry incubator (if it can maintain the temperature requirement and tolerance) for 24 hours. The blue colonies may range in color, and all should be considered fecal coliforms.

Confirmation can be performed on 10% of blue colonies and atypical colonies monthly by introducing a portion of the colony to tubes of LTB and *E. coli* broth as described in Section 5.1.5 (Step 2 for total coliforms and Step 3 for fecal coliforms and/or *E. coli*).

5.3.2.2   Method-Specific Media Preparation

The m-FC media should be prepared as described in this section. Verify the finished media pH and run a positive and negative control and blank prior to use as described in Section 4.6. The operator–analyst should note that m-FC broth and agar can either be purchased or prepared from purchased media in the laboratory. However, it is strongly recommended that m-FC broth and agar not be prepared from individual ingredients.

- Rosolic acid: Add 100 mL of 0.2N NaOH (0.8 g NaOH/100 mL deionized water) to 1.0 g of rosolic acid to produce a 1% solution. Do not autoclave, as this will break down the rosolic acid. Store stock solution in a refrigerator and discard after 2 weeks or sooner if its color changes from dark red to muddy brown.
- Membrane fecal coliform media:

○ Tryptose or biosate	10.0 g
○ Proteose peptone no. 3 or polypeptone	5.0 g
○ Yeast extract	3.0 g
○ Sodium chloride, NaCl	5.0 g
○ Lactose	12.5 g
○ Bile salts no. 3 or bile salts mixture	1.5 g
○ Aniline blue	0.1 g
○ Agar (optional)	15.0 g

○ Deionized water or distilled water   1 L

1. Add 37.1 g (broth) or 52.1 g (agar) of m-FC media to 1 L of water in a 1.5- to 2-L beaker with a stir bar.

2. Add 10 mL of 1% rosolic acid in 0.2-N NaOH.

3. Heat to near boiling and promptly remove. Cool to 45 °C. Do not autoclave.

4. Final pH should be 7.4 ± 0.2 s.u.

5. Perform a positive and negative control and blank as described in Section 4.6 before using the prepared media for the first time.

5.3.2.3 Procedure

Follow the general membrane filtration method given in Section 5.2.5 using m-FC agar or m-FC broth plates. Incubate at 44.5 ± 0.2 °C for 24 hours ± 2 hours.

5.3.2.4 Interpretation

Count and record various shades of blue. Nonfecal coliform colonies are gray to cream-colored. Normally, few nonfecal coliform colonies will be observed on m-FC medium because of selective action of the elevated temperature and addition of rosolic acid salt reagent.

5.3.2.5 Calculations

Follow the general membrane filtration calculations given in Section 5.2.6. The number of colonies per plate should be between 20 and 60.

5.3.3 Modified Membrane Thermotolerant *Escherichia coli Method*

5.3.3.1 Introduction

U.S. EPA's Method 1603 is a test for *E. coli* only (U.S. EPA, 2002b). The modified medium contains a chromogen (5-bromo-6-chloro-3-indolyl-1-D-glucuronide), which is catabolized to glucuronic acid and a red- or magenta-colored compound by *E. coli* that produces the enzyme 1-D-glucuronidase. After filtration, the membrane containing the bacteria is placed on the modified m-TEC agar and incubated at 35 ± 0.5 °C for 2 hours to resuscitate the injured or stressed bacteria, and then incubated at 44.5 ± 0.2 °C for 22 hours. The target colonies on modified m-TEC agar are red or magenta in color after the incubation period. Confirmation is not required (U.S. EPA, 2002b).

5.3.3.2 Method-Specific Reagents

Prepare the m-TEC media as described in this section. Verify the finished media pH and run a positive and negative control before use, as described in Section 4.6.

The operator–analyst should note the agar can either be purchased or prepared from purchased media in the laboratory. However, it is strongly recommended that it not be prepared from individual ingredients.

- Modified m-TEC agar:
  - Protease peptone #3   5.0 g
  - Yeast extract   3.0 g
  - Lactose   10.0 g
  - NaCl   7.5 g
  - Dipotassium phosphate   3.3 g
  - Monopotassium phosphate   1.0 g
  - Sodium lauryl sulfate   0.2 g
  - Sodium desoxycholate   0.1 g
  - Chromogen (5-bromo-6-chloro-3-indolyl-1-D-glucuronide) 0.5 g
  - Agar reagent   15.0 g
  - Deionized water or distilled water   1 L

    1. Add 45.6 g of dehydrated modified m-TEC agar to 1 L of deionized or distilled water and heat to boiling until the ingredients dissolve.
    2. Autoclave at 121 °C, 103 kPa (15 PSI), for 15 minutes and cool in a 50 °C water bath.
    3. Pour the medium into each 9×50 mm culture dish to a 4- to 5-mm depth (approximately 4 to 6 mL) and allow to solidify.
    4. Final pH should be 7.3 ± 0.2.
    5. Perform a positive and negative control and blank as described in Section 4.6 before using the prepared media for the first time.
    6. Store in a refrigerator.

5.3.3.3 Procedure

Follow the general membrane filtration method given in Section 5.2.5 using m-TEC agar. Incubate at 35 ± 0.5 °C for 2 hours to resuscitate the injured or stressed bacteria, then incubate at 44.5 ± 0.2 °C for 22 hours.

5.3.3.4 Interpretation

Count and record the number of red or magenta colonies with the aid of an illuminated lens with a 2 to 5 times magnification or a stereoscopic microscope.

5.3.3.5 Calculations

Follow the general membrane filtration calculations given in Section 5.2.6. The number of colonies per plate should be between 20 and 80.

5.3.4 Membrane ColiBlue®

The membrane ColiBlue® (m-ColiBlue®) is a membrane filter procedure (U.S. EPA–approved Hach Company Method 10029) for the detection of total coliforms and *E. coli* (Figure 5.12). The formulation contains methylene blue, triphenyl tetrazolium chloride, and 5-bromo-4-chloro-3-indolyl-b-D-glucuronic acid (selective enzyme indicators) for the detection of total coliforms and *E. coli*. A water sample is filtered through a 0.45-μm membrane filter, then placed in a Petri dish containing a filter pad and m-ColiBlue24 nutrient broth and incubated at 35 ± 0.5 °C for 24 hours. Blue colonies are enumerated as *E. coli* and red colonies are enumerated as total coliforms. The operator–analyst should refer to the manufacturer's instructions on the use of this method.

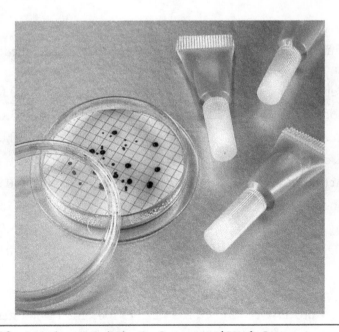

FIGURE 5.12 Membrane ColiBlue®. *Courtesy of Hach Company*

5.3.5   Membrane-Enterococcus Indoxyl 1-D Glucoside Agar

5.3.5.1   Introduction

U.S. EPA's Method 1600 is a membrane filter procedure for the detection and enumeration of enterococci bacteria in water (U.S. EPA, 2002a). The modified medium has a reduced amount of TTC and includes indoxyl 1-D glucoside, a chromogenic cellobiose. In this procedure, 1-glucosidase-positive enterococci produce a blue halo around the colony. The plates are incubated for 24 hours at 41 ± 0.5 °C. All colonies greater than or equal to 0.5 mm in diameter (regardless of color) with a blue halo are recorded as enterococci colonies. A fluorescent lamp with a magnifying lens is used for counting to give maximum visibility of colonies. It is strongly recommended that media or prepared plates be purchased.

5.3.5.2   Method-Specific Media Preparation

The m-EI media should be prepared as described in this section. Verify the finished media pH and run a positive and negative control and blank prior to use, as described in Section 4.6.

The operator–analyst should note that m-EI broth and agar can either be purchased or prepared from purchased media in the laboratory. However, it is strongly recommended that m-EI broth and agar not be prepared from individual ingredients.

Preparation of reagents includes the following:

- Indoxyl 1-D-glucoside   0.75 g
- Membrane enterococcus indoxyl 1-D glucoside agar medium (mEI) agar, Difco 0333

○ Peptone	10.0 g
○ Sodium chloride (NaCl)	15.0 g
○ Yeast extract	30.0 g
○ Esculin	1.0 g
○ Actidione (cycloheximide)	0.05 g
○ Sodium azide	0.15 g
○ Agar	15.0 g
○ Deionized or distilled water	1.0 L

    1. Add reagents to 1 L of reagent-grade water, mix thoroughly, and heat to dissolve completely.
    2. Autoclave at 121 °C, 103 kPa (15 PSI), for 15 minutes and cool in a 50 °C water bath.

3. After sterilization, add 0.24 g of nalidixic acid (sodium salt) and 0.02 g of TTC to the m-EI medium and mix thoroughly.

4. Dispense m-EI agar into 9×50 mm or 15×60 mm Petri dishes to a 4- to 5-mm depth (approximately 4 to 6 mL) and allow to solidify. The final pH of medium should be 7.1 ± 0.2.

5. Run a positive and negative control and blank as described in Section 4.6 before using the prepared media for the first time.

6. Store in a refrigerator.

5.3.5.3   Procedure

Follow the general membrane filtration method given in Section 5.2.5 using m-EI agar or m-EI broth plates. Incubate at 41 ± 0.5 °C for 24 ± 2 hours.

5.3.5.4   Interpretation

Count and record colonies on those membrane filters containing, if practical, 20 to 60 colonies 2:0.5 mm in diameter with a blue halo, regardless of colony color as an enterococci. The operator–analyst should note that when measuring colony size, the halo size should not be included. In addition, the operator–analyst should use magnification for counting and a small fluorescent lamp to give maximum visibility of colonies.

5.3.5.5   Calculations

Follow the general membrane filtration calculations given in Section 5.2.6. The number of colonies per plate should be between 20 and 60.

5.4   Enzymatic Most Probable Number Methods

5.4.1   Colilert® and Colilert®-18

This method is in *Standard Methods* (APHA et al., 2017) Section 9223B.

5.4.1.1   Introduction

The method is based on Defined Substrate Technology® (DST®) (IDEXX Laboratories, Westbrook, Maine). The product uses nutrient indicators that produce color and fluorescence when metabolized by total coliforms and *E. coli*. When the reagent is added to the sample and incubated, it can detect these bacteria at 1 MPN/100 mL within 18 hours for Colilert®-18 and 24 hours with Colilert®, with as many as 2 million heterotrophic bacteria/100 mL. Colilert® and Colilert®-18 can be used with 15-tube serial dilutions. The operator–analyst should consult *Standard Methods* (APHA et al., 2017) for the appropriate dilutions and MPN tables.

For accuracy and counting range, the operator–analyst should use the IDEXX Laboratories Quanti-Tray® system with either the 51-well Quanti-Tray® or the Quanti-Tray® 2000 (97 wells) (IDEXX Laboratories, Westbrook, Maine, www. idexx.com/water). The Quanti-Tray® MPN table should be used for Quanti-Tray® or Quanti-Tray® 2000 (see Tables 5.9 and 5.10). If a dilution is required, sterile deionized or distilled water and not buffered water should be used for making the dilutions. Colilert® or Colilert®-18 should always be added to the diluted sample.

5.4.1.2   Apparatus and Materials

- Colilert-18®
- Quanti-Trays® (51-well or 97-well)
- IDEXX proprietary Quanti-Tray® sealer
- Sterile pipets, graduated cylinders, and assorted laboratory glassware
- Sterile sampling containers
- 6-W, 365-nm wavelength UV black light

5.4.1.3   Reagents

- Colilert® media packs (IDEXX Laboratories)
- Sterile phosphate buffer dilution water

5.4.1.4   Procedure

The operator–analyst should follow the package insert for Colilert® and Colilert®-18 or see *Standard Methods* 9223B (APHA et al., 2017) for additional information. The procedure is as follows:

- **Step 1**—Turn the sealer on to allow it to come to temperature (approximately 10 minutes). The orange light on the top right will light when the sealer power is turned on. When the green light goes on, the sealer is ready for use.
- **Step 2**—Select the appropriate Quanti-Tray®. To remove a sterile Quanti- Tray® from the plastic bag, tear open the plastic bag at the bottom, which has a green line around the bag. Remove the Quanti-Trays® needed and close the bag with the remaining trays using either a clip or tape.

TABLE 5.9 A 51-well Quanti-Tray® MPN table. *Reprinted with permission from IDEXX US, 2022*

IDEXX
51-Well Quanti-Tray®
MPN Table

No. of wells giving positive reaction	MPN per 100 ml sample	95% Confidence Limits Lower	Upper
0	<1.0	0.0	3.7
1	1.0	0.3	5.6
2	2.0	0.6	7.3
3	3.1	1.1	9.0
4	4.2	1.7	10.7
5	5.3	2.3	12.3
6	6.4	3.0	13.9
7	7.5	3.7	15.5
8	8.7	4.5	17.1
9	9.9	5.3	18.8
10	11.1	6.1	20.5
11	12.4	7.0	22.1
12	13.7	7.9	23.9
13	15.0	8.8	25.7
14	16.4	9.8	27.5
15	17.8	10.8	29.4
16	19.2	11.9	31.3
17	20.7	13.0	33.3
18	22.2	14.1	35.2
19	23.8	15.3	37.3
20	25.4	16.5	39.4
21	27.1	17.7	41.6
22	28.8	19.0	43.9
23	30.6	20.4	46.3
24	32.4	21.8	48.7
25	34.4	23.3	51.2
26	36.4	24.7	53.9
27	38.4	26.4	56.6
28	40.6	28.0	59.5
29	42.9	29.7	62.5
30	45.3	31.5	65.6
31	47.8	33.4	69.0
32	50.4	35.4	72.5
33	53.1	37.5	76.2
34	56.0	39.7	80.1
35	59.1	42.0	84.4
36	62.4	44.6	88.8
37	65.9	47.2	93.7
38	69.7	50.0	99.0
39	73.8	53.1	104.8
40	78.2	56.4	111.2
41	83.1	59.9	118.3
42	88.5	63.9	126.2
43	94.5	68.2	135.4
44	101.3	73.1	146.0
45	109.1	78.6	158.7
46	118.4	85.0	174.5
47	129.8	92.7	195.0
48	144.5	102.3	224.1
49	165.2	115.2	272.2
50	200.5	135.8	387.6
51	> 200.5	146.1	infinite

TABLE 5.10A Quanti-Tray®/2000 MPN table. *Reprinted with permission from IDEXX US, 2022*

# Large wells positive	# Small wells positive																								
	0	1	2	3	4	5	6	7	8	9	10	11	12	13	14	15	16	17	18	19	20	21	22	23	24
0	<1	1.0	2.0	3.0	4.0	5.0	6.0	7.0	8.0	9.0	10.0	11.0	12.0	13.0	14.1	15.1	16.1	17.1	18.1	19.1	20.2	21.2	22.2	23.3	24.3
1	1.0	2.0	3.0	4.0	5.0	6.0	7.1	8.1	9.1	10.1	11.1	12.1	13.2	14.2	15.2	16.2	17.3	18.3	19.3	20.4	21.4	22.4	23.5	24.5	25.6
2	2.0	3.0	4.1	5.1	6.1	7.1	8.1	9.2	10.2	11.2	12.2	13.3	14.3	15.4	16.4	17.4	18.5	19.5	20.6	21.6	22.7	23.7	24.8	25.8	26.9
3	3.1	4.1	5.1	6.1	7.2	8.2	9.2	10.3	11.3	12.4	13.4	14.5	15.5	16.5	17.6	18.6	19.7	20.8	21.8	22.9	23.9	25.0	26.1	27.1	28.2
4	4.1	5.2	6.2	7.2	8.3	9.3	10.4	11.4	12.5	13.5	14.6	15.6	16.7	17.8	18.8	19.9	21.0	22.0	23.1	24.2	25.3	26.3	27.4	28.5	29.6
5	5.2	6.3	7.3	8.4	9.4	10.5	11.5	12.6	13.7	14.7	15.8	16.9	17.9	19.0	20.1	21.2	22.2	23.3	24.4	25.5	26.6	27.7	28.8	29.9	31.0
6	6.3	7.4	8.4	9.5	10.6	11.6	12.7	13.8	14.9	16.0	17.0	18.1	19.2	20.3	21.4	22.5	23.6	24.7	25.8	26.9	28.0	29.1	30.2	31.3	32.4
7	7.5	8.5	9.6	10.7	11.8	12.8	13.9	15.0	16.1	17.2	18.3	19.4	20.5	21.6	22.7	23.8	24.9	26.0	27.1	28.3	29.4	30.5	31.6	32.8	33.9
8	8.6	9.7	10.8	11.9	13.0	14.1	15.2	16.3	17.4	18.5	19.6	20.7	21.8	22.9	24.1	25.2	26.3	27.4	28.6	29.7	30.8	32.0	33.1	34.3	35.4
9	9.8	10.9	12.0	13.1	14.2	15.3	16.4	17.6	18.7	19.8	20.9	22.0	23.2	24.3	25.4	26.6	27.7	28.9	30.0	31.2	32.3	33.5	34.6	35.8	37.0
10	11.0	12.1	13.2	14.4	15.5	16.6	17.7	18.9	20.0	21.1	22.3	23.4	24.6	25.7	26.9	28.0	29.2	30.3	31.5	32.7	33.8	35.0	36.2	37.4	38.6
11	12.2	13.4	14.5	15.6	16.8	17.9	19.1	20.2	21.4	22.5	23.7	24.8	26.0	27.2	28.3	29.5	30.7	31.9	33.0	34.2	35.4	36.6	37.8	39.0	40.2
12	13.5	14.6	15.8	16.9	18.1	19.3	20.4	21.6	22.8	23.9	25.1	26.3	27.5	28.6	29.8	31.0	32.2	33.4	34.6	35.8	37.0	38.2	39.5	40.7	41.9
13	14.8	16.0	17.1	18.3	19.5	20.6	21.8	23.0	24.2	25.4	26.6	27.8	29.0	30.2	31.4	32.6	33.8	35.0	36.2	37.5	38.7	39.9	41.2	42.4	43.6
14	16.1	17.3	18.5	19.7	20.9	22.1	23.3	24.5	25.7	26.9	28.1	29.3	30.5	31.7	33.0	34.2	35.4	36.7	37.9	39.1	40.4	41.6	42.9	44.2	45.4
15	17.5	18.7	19.9	21.1	22.3	23.5	24.7	25.9	27.2	28.4	29.6	30.9	32.1	33.3	34.6	35.8	37.1	38.4	39.6	40.9	42.2	43.4	44.7	46.0	47.3
16	18.9	20.1	21.3	22.6	23.8	25.0	26.2	27.5	28.7	30.0	31.2	32.5	33.7	35.0	36.3	37.5	38.8	40.1	41.4	42.7	44.0	45.3	46.6	47.9	49.2
17	20.3	21.6	22.8	24.1	25.3	26.6	27.8	29.1	30.3	31.6	32.9	34.1	35.4	36.7	38.0	39.3	40.6	41.9	43.2	44.5	45.9	47.2	48.5	49.8	51.2
18	21.8	23.1	24.3	25.6	26.9	28.1	29.4	30.7	32.0	33.3	34.6	35.9	37.2	38.5	39.8	41.1	42.4	43.8	45.1	46.5	47.8	49.2	50.5	51.9	53.2
19	23.3	24.6	25.9	27.2	28.5	29.8	31.1	32.4	33.7	35.0	36.3	37.6	39.0	40.3	41.6	43.0	44.3	45.7	47.1	48.4	49.8	51.2	52.6	54.0	55.4
20	24.9	26.2	27.5	28.8	30.1	31.5	32.8	34.1	35.4	36.8	38.1	39.5	40.8	42.2	43.6	44.9	46.3	47.7	49.1	50.5	51.9	53.3	54.7	56.1	57.6
21	26.5	27.9	29.2	30.5	31.8	33.2	34.5	35.9	37.3	38.6	40.0	41.4	42.8	44.1	45.5	46.9	48.4	49.8	51.2	52.6	54.1	55.5	56.9	58.4	59.9
22	28.2	29.5	30.9	32.3	33.6	35.0	36.4	37.7	39.1	40.5	41.9	43.3	44.8	46.2	47.6	49.0	50.5	51.9	53.4	54.8	56.3	57.8	59.3	60.8	62.3
23	29.9	31.3	32.7	34.1	35.5	36.8	38.3	39.7	41.1	42.5	43.9	45.4	46.8	48.3	49.7	51.2	52.7	54.2	55.6	57.1	58.6	60.2	61.7	63.2	64.7
24	31.7	33.1	34.5	35.9	37.3	38.8	40.2	41.7	43.1	44.6	46.0	47.5	49.0	50.5	52.0	53.5	55.0	56.5	58.0	59.5	61.1	62.6	64.2	65.8	67.3
25	33.6	35.0	36.4	37.9	39.3	40.8	42.2	43.7	45.2	46.7	48.2	49.7	51.2	52.7	54.3	55.8	57.3	58.9	60.5	62.0	63.6	65.2	66.8	68.4	70.0
26	35.5	36.9	38.4	39.9	41.4	42.8	44.3	45.9	47.4	48.9	50.4	52.0	53.5	55.1	56.7	58.2	59.8	61.4	63.0	64.7	66.3	67.9	69.6	71.2	72.9
27	37.4	38.9	40.4	42.0	43.5	45.0	46.5	48.1	49.6	51.2	52.8	54.4	56.0	57.6	59.2	60.8	62.4	64.1	65.7	67.4	69.1	70.8	72.5	74.2	75.9
28	39.5	41.0	42.6	44.1	45.7	47.3	48.8	50.4	52.0	53.6	55.2	56.9	58.5	60.2	61.8	63.5	65.2	66.9	68.6	70.3	72.0	73.7	75.5	77.3	79.0
29	41.7	43.2	44.8	46.4	48.0	49.6	51.2	52.8	54.5	56.1	57.8	59.5	61.2	62.9	64.6	66.3	68.0	69.8	71.5	73.3	75.1	76.9	78.7	80.5	82.4
30	43.9	45.5	47.1	48.7	50.4	52.0	53.7	55.4	57.1	58.8	60.5	62.2	64.0	65.7	67.5	69.3	71.0	72.9	74.7	76.5	78.3	80.2	82.1	84.0	85.9

TABLE 5.10A Quanti-Tray®/2000 MPN table. *Reprinted with permission from IDEXX US, 2022*

# Large wells positive	# Small wells positive																								
	0	1	2	3	4	5	6	7	8	9	10	11	12	13	14	15	16	17	18	19	20	21	22	23	24
31	46.2	47.9	49.5	51.2	52.9	54.6	56.3	58.1	59.8	61.6	63.3	65.1	66.9	68.7	70.5	72.4	74.2	76.1	78.0	79.9	81.8	83.7	85.7	87.6	89.6
32	48.7	50.4	52.1	53.8	55.6	57.3	59.1	60.9	62.7	64.5	66.3	68.2	70.0	71.9	73.8	75.7	77.6	79.5	81.5	83.5	85.4	87.5	89.5	91.5	93.6
33	51.2	53.0	54.8	56.5	58.3	60.2	62.0	63.8	65.7	67.6	69.5	71.4	73.3	75.2	77.2	79.2	81.2	83.2	85.2	87.3	89.3	91.4	93.6	95.7	97.8
34	53.9	55.7	57.6	59.4	61.3	63.1	65.0	67.0	68.9	70.8	72.8	74.8	76.8	78.8	80.8	82.9	85.0	87.1	89.2	91.4	93.5	95.7	97.9	100.2	102.4
35	56.8	58.6	60.5	62.4	64.4	66.3	68.3	70.3	72.3	74.3	76.3	78.4	80.5	82.6	84.7	86.9	89.1	91.3	93.5	95.7	98.0	100.3	102.6	105.0	107.3
36	59.8	61.7	63.7	65.7	67.7	69.7	71.7	73.8	75.9	78.0	80.1	82.3	84.5	86.7	88.9	91.2	93.5	95.8	98.1	100.5	102.9	105.3	107.7	110.2	112.7
37	62.9	65.0	67.0	69.1	71.2	73.3	75.4	77.6	79.8	82.0	84.2	86.5	88.8	91.1	93.4	95.8	98.2	100.6	103.1	105.6	108.1	110.7	113.3	115.9	118.6
38	66.3	68.4	70.6	72.7	74.9	77.1	79.4	81.6	83.9	86.2	88.6	91.0	93.4	95.8	98.3	100.8	103.4	105.9	108.6	111.2	113.9	116.6	119.4	122.2	125.0
39	70.0	72.2	74.4	76.7	78.9	81.3	83.6	86.0	88.4	90.9	93.4	95.9	98.4	101.0	103.6	106.3	109.0	111.8	114.6	117.4	120.3	123.2	126.1	129.2	132.2
40	73.8	76.2	78.5	80.9	83.3	85.7	88.2	90.8	93.3	95.9	98.5	101.2	103.9	106.7	109.5	112.4	115.3	118.2	121.2	124.3	127.4	130.5	133.7	137.0	140.3
41	78.0	80.5	83.0	85.5	88.0	90.6	93.3	95.9	98.7	101.4	104.3	107.1	110.0	113.0	116.0	119.1	122.2	125.4	128.7	132.0	135.4	138.8	142.3	145.9	149.5
42	82.6	85.2	87.8	90.5	93.2	96.0	98.8	101.7	104.6	107.6	110.6	113.7	116.9	120.3	123.4	126.7	130.1	133.6	137.2	140.8	144.5	148.3	152.2	156.1	160.2
43	87.6	90.4	93.2	96.0	99.0	101.9	105.0	108.1	111.2	114.5	117.8	121.1	124.6	128.1	131.7	135.4	139.1	143.0	147.0	151.0	155.2	159.4	163.8	168.2	172.8
44	93.1	96.1	99.1	102.2	105.4	108.6	111.9	115.3	118.7	122.3	125.9	129.6	133.4	137.4	141.4	145.5	149.7	154.1	158.5	163.1	167.9	172.7	177.7	182.9	188.2
45	99.3	102.5	105.8	109.2	112.6	116.2	119.8	123.6	124.4	131.4	135.4	139.6	143.9	148.3	152.9	157.6	162.4	167.4	172.6	178.0	183.5	189.2	195.1	201.2	207.5
46	106.3	109.8	113.4	117.2	121.0	125.0	129.1	133.3	137.6	142.1	146.7	151.5	156.5	161.6	167.0	172.5	178.2	184.2	190.4	196.8	203.5	210.5	217.8	225.4	233.3
47	114.3	118.3	122.4	126.6	130.9	135.4	140.1	145.0	150.0	155.3	160.7	166.4	172.3	178.5	185.0	191.8	198.9	206.4	214.2	222.4	231.0	240.0	249.5	259.5	270.0
48	123.9	128.4	133.1	137.9	143.0	148.3	153.9	159.7	165.8	172.2	178.9	186.0	193.5	201.4	209.8	218.7	228.2	238.2	248.8	260.3	272.3	285.1	298.7	313.0	328.2
49	135.5	140.8	146.4	152.3	158.5	165.0	172.0	179.3	187.2	195.6	204.6	214.3	224.7	235.9	248.1	261.3	275.5	290.9	307.6	325.5	344.8	365.4	387.3	410.6	435.2

TABLE 5.10B Quanti-Tray®/2000 MPN table. *Reprinted with permission from IDEXX US, 2022*

# Large wells positive	# Small wells positive																							
	25	26	27	28	29	30	31	32	33	34	35	36	37	38	39	40	41	42	43	44	45	46	47	48
0	25.3	26.4	27.4	28.4	29.5	30.5	31.5	32.6	33.6	34.7	35.7	36.8	37.8	38.9	40.0	41.0	42.1	43.1	44.2	45.3	46.3	47.4	48.5	49.5
1	26.6	27.7	28.7	29.8	30.8	31.9	32.9	34.0	35.0	36.1	37.2	38.2	39.3	40.4	41.4	42.5	43.6	44.7	45.7	46.8	47.9	49.0	50.1	51.2
2	27.9	29.3	30.0	31.1	32.2	33.2	34.3	35.4	36.5	37.5	38.6	39.7	40.8	41.9	43.0	44.0	45.1	46.2	47.3	48.4	49.5	50.6	51.7	52.8
3	29.3	30.4	31.4	32.5	33.6	34.7	35.8	36.8	37.9	39.0	40.1	41.2	42.3	43.4	44.5	45.6	46.7	47.8	48.9	50.0	51.2	52.3	53.4	54.5
4	30.7	31.8	32.8	33.9	35.0	36.1	37.2	38.3	39.4	40.5	41.6	42.8	43.9	45.0	46.1	47.2	48.3	49.5	50.6	51.7	52.9	54.0	55.1	56.3
5	32.1	33.2	34.3	35.4	36.5	37.6	38.7	39.9	41.0	42.1	43.2	44.4	45.5	46.6	47.7	48.9	50.0	51.2	52.3	53.5	54.6	55.8	56.9	58.1
6	33.5	34.7	35.8	36.9	38.0	39.2	40.3	41.4	42.6	43.7	44.8	46.0	47.1	48.3	49.4	50.6	51.7	52.9	54.1	55.2	56.4	57.6	58.7	59.9
7	35.0	36.2	37.3	38.4	39.6	40.7	41.9	43.0	44.2	45.3	46.5	47.7	48.8	50.0	51.2	52.3	53.5	54.7	55.9	57.1	58.3	59.4	60.6	61.8
8	36.6	37.7	38.9	40.0	41.2	42.3	43.5	44.7	45.9	47.0	48.2	49.4	50.6	51.8	53.0	54.1	55.3	56.5	57.7	59.0	60.2	61.4	62.6	63.8
9	38.1	39.3	40.5	41.6	42.8	44.0	45.2	46.4	47.6	48.8	50.0	51.2	52.4	53.6	54.8	56.0	57.2	58.4	59.7	60.9	62.1	63.4	64.6	65.8
10	39.7	40.9	42.1	43.3	44.5	45.7	46.9	48.1	49.3	50.6	51.8	53.0	54.2	55.5	56.7	57.9	59.2	60.4	61.7	62.9	64.2	65.4	66.7	67.9
11	41.4	42.6	43.8	45.0	46.3	47.5	48.7	49.9	51.2	52.4	53.7	54.9	56.1	57.4	58.6	59.9	61.2	62.4	63.7	65.0	66.3	67.5	68.8	70.1
12	43.1	44.3	45.6	46.8	48.1	49.3	50.6	51.8	53.1	54.3	55.6	56.8	58.1	59.4	60.7	62.0	63.2	64.5	65.8	67.1	68.4	69.7	71.0	72.4
13	44.9	46.1	47.4	48.6	49.9	51.2	52.5	53.7	55.0	56.3	57.6	58.9	60.2	61.5	62.8	64.1	65.4	66.7	68.0	69.3	70.7	72.0	73.3	74.7
14	46.7	48.0	49.3	50.5	51.8	53.1	54.4	55.7	57.0	58.3	59.6	60.9	62.3	63.6	64.9	66.3	67.6	68.9	70.3	71.6	73.0	74.4	75.7	77.1
15	48.6	49.9	51.2	52.5	53.8	55.1	56.4	57.8	59.1	60.4	61.8	63.1	64.5	65.8	67.2	68.5	69.9	71.3	72.6	74.0	75.4	76.8	78.2	79.6
16	50.5	51.8	53.2	54.5	55.8	57.2	58.5	59.9	61.2	62.6	64.0	65.3	66.7	68.1	69.5	70.9	72.3	73.7	75.1	76.5	77.9	79.3	80.8	82.2
17	52.5	53.9	55.2	56.6	58.0	59.3	60.7	62.1	63.5	64.9	66.3	67.7	69.1	70.5	71.9	73.3	74.8	76.2	77.6	79.1	80.5	82.0	83.5	84.9
18	54.6	56.0	57.4	58.8	60.2	61.6	63.0	64.4	65.8	67.2	68.6	70.1	71.5	73.0	74.4	75.9	77.3	78.8	80.3	81.8	83.3	84.8	86.3	87.8
19	56.8	58.2	59.6	61.0	62.4	63.9	65.3	66.8	68.2	69.7	71.1	72.6	74.1	75.5	77.0	78.5	80.0	81.5	83.1	84.6	86.1	87.6	89.2	90.7
20	59.0	60.4	61.9	63.3	64.8	66.3	67.7	69.2	70.7	72.2	73.7	75.2	76.7	78.2	79.8	81.3	82.8	84.4	85.9	87.5	89.1	90.7	92.2	93.8
21	61.3	62.8	64.3	65.8	67.3	68.8	70.3	71.8	73.3	74.9	76.4	77.9	79.6	81.1	82.6	84.2	85.8	87.4	89.0	90.6	92.2	93.8	95.4	97.1
22	63.8	65.3	66.8	68.3	69.8	71.4	72.9	74.5	76.1	77.6	79.2	80.8	82.4	84.0	85.6	87.2	88.9	90.5	92.1	93.8	95.5	97.1	98.8	100.5
23	66.3	67.8	69.4	71.0	72.5	74.1	75.7	77.3	78.9	80.5	82.2	83.8	85.4	87.1	88.7	90.4	92.1	93.8	95.5	97.2	98.9	100.6	102.4	104.1
24	68.9	70.5	72.1	73.7	75.3	77.0	78.6	80.3	81.9	83.6	85.2	86.9	88.6	90.3	92.0	93.8	95.5	97.2	99.0	100.7	102.5	104.3	106.1	107.9
25	71.7	73.3	75.0	76.6	78.3	80.0	81.7	83.3	85.1	86.8	88.5	90.2	92.0	93.7	95.5	97.3	99.1	100.9	102.7	104.5	106.3	108.2	110.0	111.9
26	74.6	76.3	78.0	79.7	81.4	83.1	84.8	86.6	88.4	90.1	91.9	93.7	95.5	97.3	99.2	101.0	102.9	104.7	106.6	108.5	110.4	112.3	114.2	116.2
27	77.6	79.4	81.1	82.9	84.6	86.4	88.2	90.0	91.9	93.7	95.5	97.4	99.3	101.2	103.1	105.0	106.9	108.8	110.8	112.7	114.7	116.7	118.7	120.7
28	80.8	82.6	84.4	86.3	88.1	89.9	91.8	93.7	95.6	97.5	99.4	101.3	103.3	105.2	107.2	109.2	111.2	113.2	115.2	117.3	119.3	121.4	123.5	125.6
29	84.2	86.1	87.9	89.8	91.7	93.7	95.6	97.5	99.5	101.5	103.5	105.5	107.5	109.5	111.6	113.7	115.7	117.8	120.0	122.1	124.2	126.4	128.6	130.8
30	87.8	89.7	91.7	93.6	95.6	97.6	99.6	101.6	103.7	105.7	107.8	109.9	112.0	114.2	116.3	118.5	120.6	122.8	125.1	127.3	129.5	131.8	134.1	136.4

TABLE 5.10B Quanti-Tray®/2000 MPN table. *Reprinted with permission from IDEXX US, 2022 (continued)*

# Large wells positive	\# Small wells positive																							
---	25	26	27	28	29	30	31	32	33	34	35	36	37	38	39	40	41	42	43	44	45	46	47	48
31	91.6	93.6	95.6	97.7	99.7	101.8	103.9	106.0	108.2	110.3	112.5	114.7	116.9	119.1	121.4	123.6	125.9	128.2	130.5	132.9	135.3	137.7	140.1	142.5
32	95.7	97.8	99.9	102.0	104.2	106.3	108.5	110.7	113.0	115.2	117.5	119.8	122.1	124.5	126.8	129.2	131.6	134.0	136.5	139.0	141.5	144.0	146.6	149.1
33	100.0	102.2	104.4	106.6	108.9	111.2	113.5	115.8	118.2	120.5	122.9	125.4	127.8	130.3	132.8	135.3	137.8	140.4	143.0	145.6	148.3	150.9	153.7	156.4
34	104.7	107.0	109.3	111.7	114.0	116.4	118.9	121.3	123.8	126.3	128.9	131.4	134.0	136.3	139.2	141.9	144.6	147.4	150.1	152.9	155.7	158.6	161.5	164.4
35	109.7	112.2	114.6	117.1	119.6	122.2	124.7	127.3	129.9	132.6	135.3	138.0	140.8	143.6	146.4	149.2	152.1	155.0	158.0	161.0	164.0	167.1	170.2	173.3
36	115.2	117.8	120.4	123.0	125.7	128.4	131.1	133.9	136.7	139.5	142.4	145.3	148.3	151.3	154.3	157.3	160.5	163.6	166.8	170.0	173.3	176.6	179.9	183.3
37	121.3	124.0	126.8	129.6	132.4	135.3	138.2	141.2	144.2	147.3	150.3	153.5	156.7	159.9	163.1	166.5	169.8	173.2	176.7	180.2	183.7	187.3	191.0	194.7
38	127.9	130.8	133.8	136.8	139.9	143.0	146.2	149.4	152.6	155.9	159.2	162.6	166.1	169.6	173.2	176.8	180.4	184.2	188.0	191.8	195.7	199.7	203.7	207.7
39	135.3	138.5	141.7	145.0	148.3	151.7	155.1	158.6	162.1	165.7	169.4	173.1	176.9	180.7	184.7	188.7	192.7	196.8	201.0	205.3	209.6	214.0	218.5	223.0
40	143.7	147.1	150.6	154.2	157.8	161.5	165.3	169.1	173.0	177.0	181.1	185.2	189.4	193.7	198.1	202.5	207.1	211.7	216.4	221.1	226.0	231.0	236.0	241.1
41	153.2	157.0	160.9	164.8	168.9	173.0	177.2	181.5	185.8	190.3	194.8	199.5	204.2	209.1	214.0	219.1	224.2	229.4	234.8	240.2	245.8	251.5	257.2	263.1
42	164.3	168.6	172.9	177.3	181.9	186.5	191.3	196.1	201.1	206.2	211.4	216.7	222.2	227.7	233.4	239.2	245.2	251.3	257.5	263.8	270.3	276.9	283.6	290.5
43	177.5	182.3	187.3	192.4	197.6	202.9	208.4	214.0	219.8	225.8	231.8	238.1	244.5	251.0	257.7	264.6	271.7	278.9	286.3	293.8	301.5	309.4	317.4	325.7
44	193.6	199.3	205.1	211.0	217.2	223.5	230.0	236.7	243.6	250.8	258.1	265.6	273.3	281.2	289.4	297.8	306.3	315.1	324.1	333.3	342.8	352.4	362.3	372.4
45	214.1	220.9	227.9	235.2	242.7	250.4	258.4	266.7	275.3	284.1	293.3	302.6	312.3	322.3	332.5	343.0	353.8	364.9	376.2	387.9	399.8	412.0	424.5	437.4
46	241.5	250.0	258.9	268.2	277.8	287.8	298.1	308.8	319.9	331.4	343.3	355.5	368.1	381.1	394.5	408.3	422.5	437.1	452.0	467.4	483.3	499.6	516.3	533.5
47	280.9	292.4	304.4	316.9	330.0	343.6	357.8	372.5	387.7	403.4	419.8	436.6	454.1	472.1	490.7	509.9	529.8	550.4	571.7	593.8	616.7	640.5	665.3	691.0
48	344.1	360.9	378.4	396.8	416.0	436.0	456.9	478.6	501.2	524.7	548.3	574.8	601.5	629.4	658.6	689.3	721.5	755.6	791.5	829.7	870.4	913.9	960.6	1011.2
49	461.1	488.4	517.2	547.5	579.4	613.1	648.8	686.7	727.0	770.1	816.4	866.4	920.8	980.4	1046.2	1119.9	1203.3	1299.7	1413.6	1553.1	1732.9	1986.3	2419.6	>2419.6

- **Step 3**—Using a marker, indicate the sample identification and date and time of incubation on the white back of the tray.
- **Step 4**—Dilute samples as needed. Final sample aliquots should be 100 mL each. Samples should be in 120-mL containers.
- **Step 5**—Separate the reagent blister packs (see Figure 5.13). Remove one blister packet and open the packet over the container containing the sample. Allow the media to empty into the container with the sample.
- **Step 6**—Close the cap and mix well to dissolve. Additional samples are processed as indicated previously.
- **Step 7**—When completed, return to the initial sample and mix well one more time prior to adding the mixture to the Quanti-Tray®. All the media should be dissolved before dispensing into the Quanti-Tray®.
- **Step 8**—Open the Quanti-Tray® as per the package insert directions. Remove the cap from the vessel and add the entire contents into the tray (see Figure 5.14).
- **Step 9**—Seal the tray with the Quanti-Tray® sealer (see Figure 5.15). The operator–analyst should note that if using Colilert®-18, the use of antifoam B will reduce foaming of the sample and should be added per the package insert prior to being dispensed into the Quanti-Tray®; if not, let sit until foam has dissipated.
- **Step 10**—Incubate the Quanti-Trays®:
  1. For *E. coli*, incubate the tray(s) at 35 ± 0.5 °C for 18 to 22 hours for Colilert®-18 and 24 to 28 hours for Colilert®. It is a definitive test and no confirmation is required.

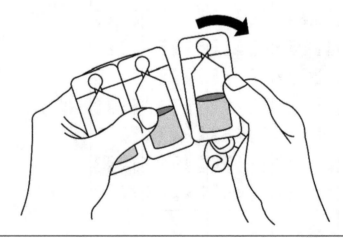

FIGURE 5.13   Separation of blister packs. *Reprinted with permission from IDEXX US, 2022*

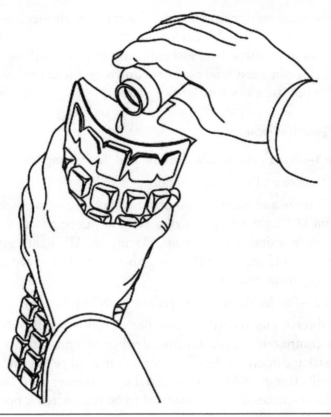

FIGURE 5.14 Pouring sample and reagent into Quanti-Tray®. *Reprinted with permission from IDEXX US, 2022*

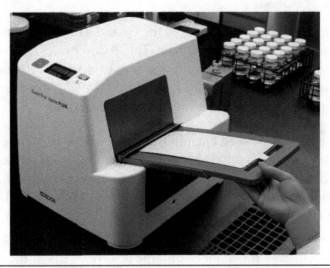

FIGURE 5.15 Placing the filled tray into the Quanti-Tray® sealer to seal and separate wells. *Reprinted with permission from IDEXX US, 2022*

2. For fecal coliforms, incubate the trays in an incubator or water bath at 44.5 ± 0.2 °C. If incubated in the water bath, do not place trays in a plastic bag. Place directly into the bath on the bottom shelf or in a test tube rack held down by a lead ring or equivalent. Do not use a brick or bottled water.

5.4.1.5   Interpretation

- **Step 1**—For *E. coli* results (yellow and fluorescing wells):
1. Observe and record the number of yellow wells.
2. Observe and record the number of fluorescing wells using a 365-nm UV light source to observe the fluorescence. Perform this task in a dark environment, placing the UV light approximately 12.7 cm (5 in.) from the tray. It is important to keep the UV light away from your eyes.
- **Step 2**—For fecal coliforms (yellow wells only):
1. Observe and record the number of yellow wells only. Note that a comparator is recommended when using a Quanti-Tray® (see package insert for detailed instructions). All yellow and fluorescing wells (for *E. coli*) must be equal to or greater than the color and fluorescence of the comparator to be considered a positive result. For fecal coliform, all yellow wells must be equal to or greater than the color of the comparator to be considered positive. See Figures 5.16 and 5.17.

5.4.1.6   Calculations

The operator–analyst should refer to the Quanti-Tray® MPN table for the 51-well (see Table 5.9) or the Quanti-Tray® 2000 (see Tables 5.10a and 5.10b) to obtain the MPN/100 mL (the tray being used). If a dilution was made during sample preparation, multiply the MPN value from the table by the dilution factor. The following are some examples:

- Example 1—a 51-well Quanti-Tray®: 10 yellow and fluorescing wells. For the Quanti-Tray® MPN table first column, look at the number of positive wells column (first column on the left of the MPN table) at the number 10 and read the next column (MPN/100 mL) to obtain the value of 11.1 MPN/100 mL.
- Example 2—97-well Quanti-Tray® 2000: 12 large and 4 small yellow wells. Observe the number of large positive wells in the left vertical column and the number of small positive wells in the top horizontal column. Record the MPN/100 mL value where the two columns meet (18.1/100 mL).

- Example 3—if a 1:10 dilution was made for the sample, multiply the MPN value obtained from the chart by the dilution factor of 10. The MPN value of 11.1/100 mL $\times$ 10 = 111 MPN/100 mL.

5.4.2 Enterolert®

This method is based on Defined Substrate Technology® and is similar to the Colilert® method but analyzes for enterococci rather than fecal coliforms and *E. coli*. The procedures differ in the proprietary media used and the incubation temperature.

When enterococci uses the 1-glucosidase enzyme to metabolize the nutrient indicator, 4-methyl-umbelliferyl-1-D-glucoside, the sample fluoresces under a 6-W, 365-nm UV light source. The method can detect enterococci at 1 MPN/100 mL within 24 hours. Enterolert® can be used with the 15-tube serial dilutions. The operator–analyst should consult Section 9221 of *Standard Methods* (APHA et al., 2017) for 9230D for Enterolert, for the appropriate dilutions and MPN tables. For accuracy and counting range, use the IDEXX Laboratories Quanti-Tray® System with either the 51-well Quanti-Tray® or the Quanti-Tray® 2000. Use the Quanti-Tray® MPN table for Quanti-Tray® or Quanti-Tray® 2000 (see Tables 5.9 or 5.10) or the package insert supplied with the trays. If a dilution is required, use sterile deionized or distilled water rather than buffered water for making the dilutions. Always add Enterolert® to the final 100-mL diluted sample only.

5.4.2.1 Procedure

- **Step 1**—Follow Steps 1 through 9 given for the Colilert® method in Section 5.4.1.4. Be sure to use Enterolert® media in place of Colilert® media.
- **Step 2**—Incubate sealed trays at 41.5 $\pm$ 0.5 °C for 24 to 28 hours.

5.4.2.2 Interpretation

Observe and record the number of fluorescing wells using a 365-nm UV light source to observe the blue fluorescence. Perform this task in a dark environment, placing the UV light approximately 12.7 cm (5 in.) from the tray. Remember to keep the UV light away from your eyes.

5.4.2.3 Calculations

The operator–analyst should refer to the Quanti-Tray® MPN table for the 51-well or the Quanti-Tray® 2000 to obtain MPN/100 mL. If a dilution was used to multiply the MPN value from the table by the dilution factor, the operator–analyst should refer to the calculations for Colilert® in Section 5.4.1.6.

6.0 REFERENCES

American Public Health Association, American Water Works Association, & Water Environment Federation. (2017). *Standard methods for the examination of water and wastewater* (23rd ed.). American Public Health Association.

Federal Register. (2007). Table II Holding Time. *Federal Register 72(47)*, 11236.

Hach Company. Method 10029 m-ColiBlue®24 for coliforms—Total and *E. coli.* Retrieved June 24, 2022, from https://www.hach.com/m-coliblue24-prepared-agar-plates-15-pk/product?id=7640249643&callback=qs

IDEXX Laboratories, Inc. (n.d.) Water testing solutions. Retrieved June 24, 2022, from https://www.idexx.com/en/water/

U.S. Environmental Protection Agency. (1978). *Microbiological methods for monitoring the environment: Water and wastes* (EPA-600/8-78-017). U.S. Environmental Protection Agency, Office of Research and Development, Environmental Monitoring and Support Laboratory.

U.S. Environmental Protection Agency. (2002a). *Method 1600: Entero-cocci in water by membrane filtration using membrane-enterococcus indoxyl-1-D-glucoside agar (m-EI)* (EPA-821/R-02-022). U.S. Environmental Protection Agency, Office of Water.

U.S. Environmental Protection Agency. (2002b). *Method 1603:* Escherichia coli (E. coli) *in water by membrane filtration using modified membrane-thermo-tolerant* Escherichia coli *agar (Modified m-TEC)* (EPA-821/R-02-023). U.S. Environmental Protection Agency, Office of Water.

6

Data Reporting and Monitoring

1.0 DATA RECORDING

Unless facts associated with sample collection, preservation, and handling are recorded carefully, all test results from a particular analysis may be irrelevant. The laboratory has the responsibility to prove that given test results are from the given sample, and that appropriate methods and procedures were used to obtain the results. Therefore, recordkeeping and proper reporting are as important as the actual analysis itself. The first requirement for meaningful data is that a sample be representative and carefully handled before analysis begins. The operator–analyst must then complete proper analysis in the prescribed fashion, record all associated facts and figures, and report the given data in the proper form. The proper way to record and report data is presented in this chapter.

1.1   Benchsheets

Each facility has different needs for collecting and recording data; there are no standard laboratory forms. Most water resource recovery facilities develop worksheets for recording laboratory results; these are typically referred to as *benchsheets*. The benchsheet satisfies the following two requirements: (1) it provides a record of the data and (2) it arranges the information in an orderly manner. Writing laboratory results on scraps of paper may result in important process control information being inadvertently thrown away, misplaced, or otherwise rendered unreadable. Benchsheets for laboratory analysis provide a single uniform location to document work that was performed and make it easy to record, review, and recover data when completing forms such as those from the National Pollutant Discharge Elimination System (NPDES) or State Pollutant Discharge Elimination System (SPDES).

The first laboratory benchsheet that must be maintained is the sample log. This log documents when the sample was collected, from where it was collected, who collected the sample, how it was preserved, the analysis that needs to be run, and a column for recording abnormal conditions. Any departure from standard sampling protocol must be noted in the sample log. An example sample log is provided in Figure 6.1.

XYZ Water Resource Recovery Facility

Date	Time	Sample ID	Sample Location	Sample Collector	Required Testing	Preservatives	Comments
4/7/2022	8:15 AM	0407PCS	Primary Effluent	SCJ	BOD, TSS	none	

FIGURE 6.1   Sample logsheet.

The most common types of benchsheets are those used for analysis documentation of work performed on the samples. At a minimum, this includes

- Analyst's initials
- Date and time of analysis
- Sample date(s)
- Sample name or identification
- Preparation date or identification (if generated) of stock solutions
- Preparation date of identification (if generated) of standards
- Preparation date of buffering solutions
- Method reference
- Instrument and electrode model and serial number
- Location of analysis being performed
- Batch number for reporting results in the data system

Records of all of quality control analyses are kept in daily benchsheets and in a separate quality control logbook. A benchsheet example is shown in Figure 6.2.

XYZ Water Resource Recovery Facility
Ammonia-Nitrogen (NH₃-N)
Standard Methods, 4500-NH₃ D

Instrument Model and Serial #	Thermo-Orion 930A / 56U093856	Analysis Date / Time:	11/29/2021 at 11:17 am
Electrode Model and Serial #	Thermo-Orion 95-12/938475736	Analyst:	JD
		Batch #:	1234567

Intermediate Standard		Calibration Curve	
1st source 10 mg/L intermediate standard prep. Date:	11/15/2021	0.100 mg/L standard prep date:	11/28/2021
2nd source 10 mg/L intermediate standard prep. Date:	11/17/2021	1.00 mg/L standard prep date:	11/28/2021
		10.0 mg/L standard prep date:	11/28/2021
		Calibration slope:	-57.8
NaOH/EDTA buffer preparation date:	11/5/2021	Acceptable range is -54 to -60 mV/dec (may vary per manufacturer)	

QC Prep Date	QC Check		mLs Buffer	Meter Reading	Final Conc.	
	Initial calibration verification					
11/9/2021	(ICV) 0.100 mg/L		1	0.1047	0.105	ICV Recovery = 105% (90 - 110%)
N/A	Method blank		1	0.0002	0	Below MDL
11/9/2021	Laboratory fortified blank (LFB) 1.00 mg/L		1	1.124	1.12	LFB Recovery = 112% (85 - 115%)
Collection Date	Sample ID	Dilution Factor	mLs Buffer	Meter Reading	Final Conc. (mg/L)	Comments / QC Calc.
11/8/2021	Effluent	1x	1	0.2334	0.233	
11/8/2021	Effluent Duplicate	1x	1	0.2222	0.222	Avg = 0.228 %RPD = 4.8%
11/9/2021	Final Clarifier A1	1x	1	1.2530	1.25	
11/9/2021	Final Clarifier A2	1x	2	0.9923	0.990	added additional buffer
11/9/2021	Final Clarifier A3	1x	1	0.6527	0.653	
11/9/2021	Final Clarifier B1	1x	1	0.8740	0.874	
11/9/2021	Final Clarifier B2	1x	1	0.5523	0.552	
11/9/2021	Final Clarifier B3	1x	1	0.6799	0.680	
11/9/2021	Influent A	10x	1	2.3480	23.5	
11/9/2021	Influent A + 2.00 mg/L spike	10x	1	4.4100	44.5	% recovery = 103.1
11/9/2021	Influent B	10x	1	3.2860	32.9	
11/9/2021	Influent C	10x	1	4.5590	45.6	
QC Prep Date	QC Check		mLs Buffer	Meter Reading	Final Conc.	
	Continuing calibration verification					
11/9/2021	(CCV) 0.100 mg/L		1	0.1047	0.105	ICV Recovery = 105% (90 - 110%)

FIGURE 6.2 Sample benchsheet.

1.2   Logbook

Typically, records of all data should be kept for 5 years from the date of the sample, measurement, report, or application. The actual date is set through the regulatory process and documented in a facility's NPDES permit. This information includes

- The date, exact place, and time of sampling or measurements
- The individual(s) who performed the sampling or measurements
- The laboratory that performed the analyses
- The date(s) analyses were performed
- The individual(s) who performed the analyses
- The analytical techniques or methods used, including any modifications
- The results of such analyses, including:
  - Units of measurement
  - Minimum reporting limit for the analysis
  - Results less than the reporting limit but above the method detection limit (MDL)
  - Data qualifiers and a description of the qualifiers
  - Quality control test results (and a written copy of the laboratory quality assurance plan)
  - Dilution factors, if used
  - Sample matrix type
- Electronic data and information regarding influent and effluent flow, pH, and other constituents subject to monitoring or effluent limitations generated by the Supervisory Control and Data Acquisition system

2.0   EXPRESSION OF RESULTS

Generally, the chemical and physical results are expressed in milligrams per liter (mg/L) or parts per million (ppm). If concentrations are less than 0.1 mg/L, it may be more convenient to express results in micrograms per liter (μg/L). If concentrations are more than 10 000 mg/L, results should be expressed as percent (%); 10 000 mg/L equals 1%. The following equations can be used to convert the results from slurries like sludge samples or solids samples to parts per million or percent by mass:

$$\text{Parts per million by weight} = \frac{\text{mg/L}}{\text{specific gravity}} \tag{6.1}$$

$$\text{Percent by mass} = \frac{\text{mg/L}}{(10\,000)(\text{specific gravity})} \tag{6.2}$$

2.1 Definitions

Proper use of significant figures indicates the reliability of the chosen analytical method and eliminates ambiguity in reporting. A number expresses quantity; a figure is any of the characters 0, 1, 2, 3, 4, 5, 6, 7, 8, and 9, which, alone or in combination, express a number. Reported values should contain only significant figures. A value is made up of significant figures when it contains all digits known to be true and one last digit in doubt. For example, if a value is reported as 12.2 mg/L, the "12" must be a firm value while the "0.2" is somewhat uncertain and may be "0.1" or "0.3."

The digit 0 (zero) may or may not be a significant figure as the following instances illustrate:

- Final zeros after a decimal point are always significant figures. For example, 2.34 g to the nearest milligram is reported as 2.300 g.
- Zeros before a decimal point with other preceding digits are significant. If there is no other preceding digit, a zero before the decimal point is not significant.
- If there are no digits preceding a decimal point, the zeros that appear after the decimal point but precede the other digits are not significant. These zeros only indicate the position of the decimal point.
- Final zeros in a whole number may or may not be significant. A good measure of significance of one or more zeros before or after another digit is to determine whether the zeros can be dropped by expressing the number in exponential form. If they can, the zeros are not significant. For example, no zeros can be dropped when expressing a mass of 100.02 g in exponential form; therefore, the zeros are significant. However, a mass of 0.002 g can be expressed in exponential form as 2×10^{-3} g, and the zeros are not significant.

2.2 Rounding Off Numbers

To round off numbers, drop those digits that are not significant. If the digit 6, 7, 8, or 9 is dropped, increase the preceding digit by one unit. For example,

5.48 is rounded off to 5.5. If the digit 0, 1, 2, 3, or 4 is dropped, do not alter the preceding digit. For example, 5.43 is rounded off to 5.4. If the digit 5 is dropped, round off the preceding digit to the nearest even number. For example, 5.35 is rounded off to 5.4, and 5.45 is rounded off to 5.4.

2.3   Rounding Off Single Arithmetic Operations

2.3.1   Addition

When adding a series of numbers, round off to the same number of decimal places as the addend with the smallest number of places. However, the operation is completed with all decimal places intact, and rounding off is performed afterward. As an example, in the following calculation:

$$
\begin{array}{r}
6.123 \\
8.1 \\
+\,11.13 \\
\hline
25.353
\end{array}
\tag{6.3}
$$

The sum is rounded off to 25.4 because the addend 8.1 has the smallest number of decimal places.

2.3.2   Subtraction

Rounding off should be completed before subtracting one number from the other.

2.3.3   Division and Multiplication

When two numbers of unequal digits are to be divided or multiplied, all digits are carried through the operation and then the product is rounded off to as few significant figures as are present in the factor with the fewest significant figures. For example, suppose the following calculation is made to obtain the result of an analysis:

$$
\frac{23 \times 0.0021 \times 53.11}{1.22}
$$

A 10-place calculator yields an answer of 2.10263360656. Round off the number to 2.1 because one of the measurements that entered into the calculation, 23, has only two significant figures.

3.0 COMPLETING THE DISCHARGE MONITORING REPORT

This section explains how to fill out hard copy and electronic discharge monitoring report (DMR) forms. Each regulatory agency is free to adopt its own format. Contact local regulatory agencies for specific information.

3.1 Introduction

Sample collection and analytical results required by the effective permit (U.S. Environmental Protection Agency [U.S. EPA] or state issued) must be reported to the enforcement authority (U.S. EPA or state) through submission of DMRs (U.S. EPA Form 3320-1). An original and one legible copy of the DMRs must be submitted to the enforcement authority by the date specified in the permit. These data are now entered into a national database. It is extremely important that the data reported on the DMR be accurate, timely, and legible to ensure that the facility's compliance status is correctly reflected. The reported data will be compared with current limits contained in the permit or any enforcement order to determine facility compliance. It should be noted that a DMR is required even if the facility did not have a discharge during a reporting period. *It is the responsibility of the permittee to ensure that DMRs are reported to the appropriate agency, either state or federal, and via the appropriate portal.* Completion and submission of the U.S. EPA NetDMR is discussed in Section 3.3.

3.2 Discharge Monitoring Report Hard Copy

The NPDES or SPDES authority typically mails DMRs to permittees for an entire year. These forms can be self-generated; however, they must be submitted for approval by NPDES or SPDES authority before use. A separate DMR form is submitted for each monitoring period and for each discharge number (outfall and pipe) or permitted feature. It is important to verify that you are reporting the correct measurement information on the correct discharge number DMR for the correct monitoring period. A narrative description of the discharge number (outfall and pipe) or permitted feature also appears in the top right corner of the DMR. Figure 6.3 depicts a typical DMR form with numbered areas. The following are instructions for completing the information in each form:

1. Verify or enter the permittee name and address and (and facility name and location, if different).

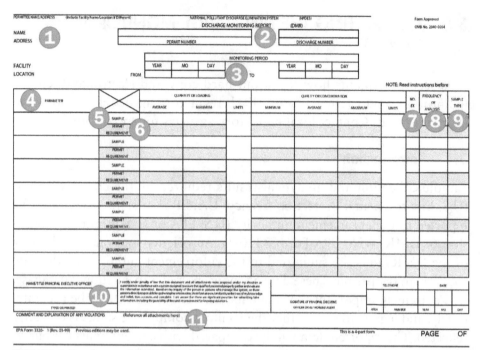

FIGURE 6.3   Parts of the DMR form.

2. Verify or enter the permit number and discharge number where indicated. Note that a separate form is required for each discharge.

3. Verify or enter the monitoring period. All analytical results obtained using the required analytical methodology must be reported even if the analytical frequency exceeds that required by the permit, from the first day of the monitoring period through the last day of the monitoring period. The dates should be displayed as "YEAR," "MO," and "DAY." Applicable monitoring periods will be specified in each permit. Some examples include, but are not limited to:

   ○ Monthly—02 01 01 to 02 01 31

   ○ Quarterly—02 01 01 to 02 03 31

   ○ Semiannual—02 01 01 to 02 06 30

4. Enter each parameter as specified in monitoring requirements of the permit. Each parameter contained in the permit is listed in the far left column titled "parameter." Each box must display the parameter name (on line 1 and, possibly, 2), the 8-digit parameter code number (line 2 or 3), and a narrative description of the sample location (line 3 or 4).

   Seasonal discharge numbers (outfalls and pipes) or permitted features and seasonal parameters are only included on DMRs for the monitoring periods stipulated in the permit. Parameters or entire discharge numbers (outfalls and pipes) or permitted features that must be reported less frequently than monthly (i.e., quarterly, semiannually, etc.) will be included in DMRs for the last month of the reporting period. For example, unless otherwise specified in the permit, reporting for the January through March reporting quarter must be included on the March DMR.

However, sampling for these parameters may be performed any time during the reporting period unless otherwise specified in the permit. Particular attention should be paid to the narrative description of the sampling location that appears on the bottom of the parameter block. Multiple monitoring locations may appear for the same parameter on the same DMR form.

5. Enter sample measurement data for each parameter under "quantity" and "quality" in units specified in the permit. "Average" is typically the arithmetic average (geometric average for bacterial parameters) of all sample measurements for each parameter obtained during the monitoring period; "maximum" and "minimum" are typically extreme high and low measurements. For municipal operations with secondary treatment requirements, enter 30-day averages of sample measurements under "average" and maximum 7-day averages of sample measurements obtained during the monitoring period under "maximum."

6. Enter the permit requirement for each parameter under "quantity" and "quality" as specified in the permit.

7. Enter the number of sample measurements during the monitoring period that exceed maximum (and/or the minimum or 7-day average, as appropriate) permit requirements for each parameter in the column titled "No. Ex." Enter "0" if there are none.

8. The column titled "Frequency of Analysis" applies to both "Sample Measurement" (actual frequency of sampling and analysis used during monitoring period) and "Permit Requirement" specified in the permit (e.g., enter "Cont," for continuous monitoring, "1/7" for one day per week, "1/30" for one day per month, "1/90" for one day per quarter)

9. The column titled "Sample Type" documents the method by which the sample was taken for both sample measurement (actual sample type used during the monitoring period) and permit requirement (e.g., enter "grab" for an individual sample, "24HC" for 24-hour composite, "N/A" for continuous monitoring). If "no discharge" occurs

during the monitoring period, enter "no discharge" across the form in place of data entry.

10. At the bottom of the form, enter the name and title of the principal executive officer, the signature of the principal officer or authorized agent, the telephone number, and date.

11. On the last line of the form, in the section titled "Comment and Explanation of Any Violations," document any clarifying information of permit requirements or reporting instructions. Where violations of permit requirements are reported, attach a brief explanation to describe the cause and corrective actions taken, and reference each violation by date.

The following are some final tips for completing the DMR hard copy successfully:

- Data should be entered in black or blue ink. The person filling out the form should write legibly (e.g., make sure decimals look like decimals). In addition, commas should not be used.

- All data should be reported, as required by the NPDES or SPDES permit, on the preprinted DMR.

- Data should be entered in open boxes only (not shaded boxes or boxes containing asterisks).

- Items appearing on preprinted DMRs should not be changed.

- For monthly reporting requirements, both the maximum and the minimum columns should be filled out (note that if a facility collects a single sample in a given month, the same value must be entered in both boxes, reflecting the one measurement).

- It is important to ensure that the reporting units are the same as those that appear in the permit.

- Values that are less than the detection limit should be reported by entering "<MDL" where MDL is the numeric value of the method detection limit. Different regulatory agencies may have different requirements for reporting values below the detection limit, such as entering BDL when all values for a reporting period are below the detection limit. Be sure to verify the reporting requirements for your regulatory agency when dealing with non-detect sample results.

- Units should not be entered in sample measurement value boxes.

- All pages of the preprinted DMR should be dated and signed before submission.

- A written report should be provided for each instance of noncompliance with a permit requirement.
- For modifications to permits, requests should be made in writing to the regional permit administrator. Requests should not be made on the DMR form.
- Proper copies and attachments should be sent to the appropriate offices.

3.3 Completing U.S. Environmental Protection Agency's NetDMR

On October 22, 2015, the U.S. EPA promulgated the NPDES e-reporting rule that mandates the electronic submittal of all DMRs after December 21, 2016. The goal was to modernize Clean Water Act (CWA) reporting for municipalities, industries, and other facilities, by removing paper-based NPDES reporting requirements. On September 23, 2020, U.S. EPA Administrator Andrew Wheeler signed the final "Phase 2 Extension Rule," which extends the compliance deadline for implementation of Phase 2 of the eRule by 5 years from December 21, 2020, to December 21, 2025. This final rule also provides states with additional flexibility to request additional time as needed. Further, this final rule promulgates clarifying changes to the NPDES eRule and eliminates some duplicative or outdated reporting requirements. Taken together, these changes are designed to save the NPDES authorized programs considerable resources, make reporting easier for NPDES-regulated entities, streamline permit renewals, ensure full exchange of NPDES program data between states and U.S. EPA, enhance public transparency, improve environmental decision-making, and protect human health and the environment.

NetDMR is a national web-based tool for regulated CWA permittees to sign and submit DMRs electronically via the U.S. EPA's Integrated Compliance Information System via the Central Data Exchange (CDX) node on the Environmental Information Exchange Network. All users who will have any sort of access to state DMRs will have to register a login through the CDX NetDMR. Only those with a Signatory role (those who are certified to sign DMRs) will be able to electronically sign a DMR.

Filing through NetDMR allows participants to discontinue mailing in hard-copy forms under 40 CFR 122.41 and 403.12. National Pollutant Discharge Elimination System and SPDES permittees required to submit DMRs may use NetDMR after requesting and receiving permission from their permitting authority. After the state or region has approved the facility's request, permittees may use the NetDMR tool to complete their DMRs via a secure Internet connection. A list of participating states is available at https://usepa.servicenowservices.com/oeca_icis (U.S. EPA, 2022). The map and table (U.S. EPA, 2022) show which states use or plan to use NetDMR, or a state electronic discharge monitoring report (state eDMR) system. There are EPA regional instances of NetDMR and state instances of NetDMR. Where a state eDMR system and NetDMR are available, it is likely that U.S. EPA is also accepting DMRs from permittees using NetDMR where U.S. EPA issued the permit.

The procedure for the operator–analyst to get started is summarized in the following sections.

3.3.1   Creating an Account and Requesting Access

Before submitting data to CDX, you must create an account. Click "Register with CDX" if you are a first-time user (Figure 6.4). You must accept the Terms and Conditions indicating that you are the original registrant and owner of the User ID requested. The first step is to select the program service, NetDMR, from the Active Program Service List (Figure 6.5).

Next, select the appropriate regulatory agency from the list of NetD-MRs (Figure 6.6). Then proceed to select "Role Access" (Figure 6.7). Available options are Data User, Internal User, Permittee (no signature), and Permittee (signature). The NetDMR is role-driven, with each role having specific privileges. It is important to note that the first person creating an account for NetDMR must be a person who will be signing the DMRs (i.e.,

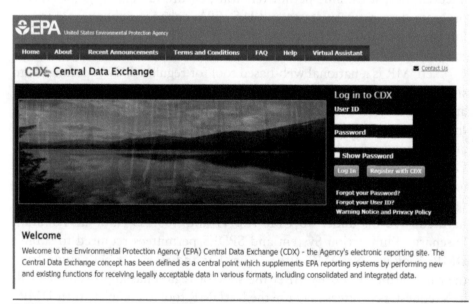

FIGURE 6.4   U.S. EPA Central Data Exchange (CDX). *Courtesy of U.S. EPA*

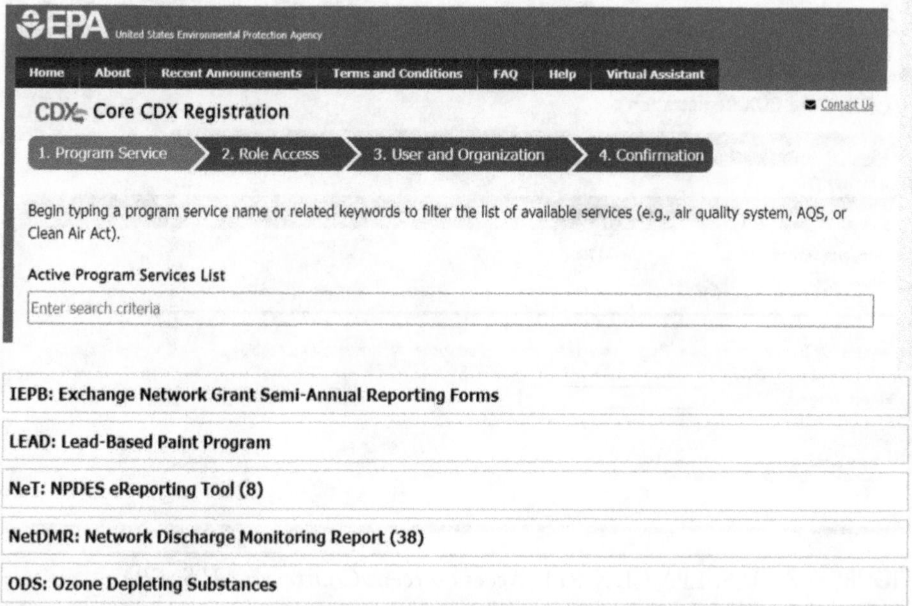

FIGURE 6.5 U.S. EPA CDX Registration Program Services List screen. *Courtesy of U.S. EPA*

signatory role). This person is automatically given the permit administrator role. As others in the company or data providers (such as laboratories) create an account and request permit administrator, edit, or view roles, this person approves access. All other signatory requests must be approved by the regulatory authority. In the following screen, enter user and organization information, security questions, and choose a password (Figure 6.8). Passwords are required to be updated every 3 months in the CDX system.

The screen that follows asks for confirmation that all information has been entered correctly. Once complete, the appropriate regulatory agency is notified to approve access of the new user to the CDX account. After approval has been granted, you will receive an email notifying you that you

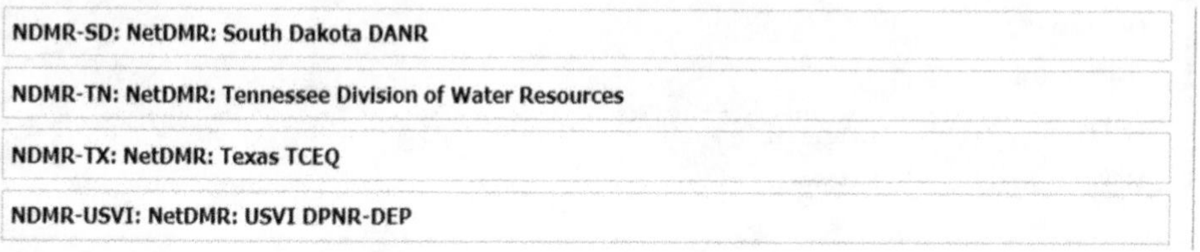

FIGURE 6.6 U.S. EPA CDX NetDMR Agency List screen. *Courtesy of U.S. EPA*

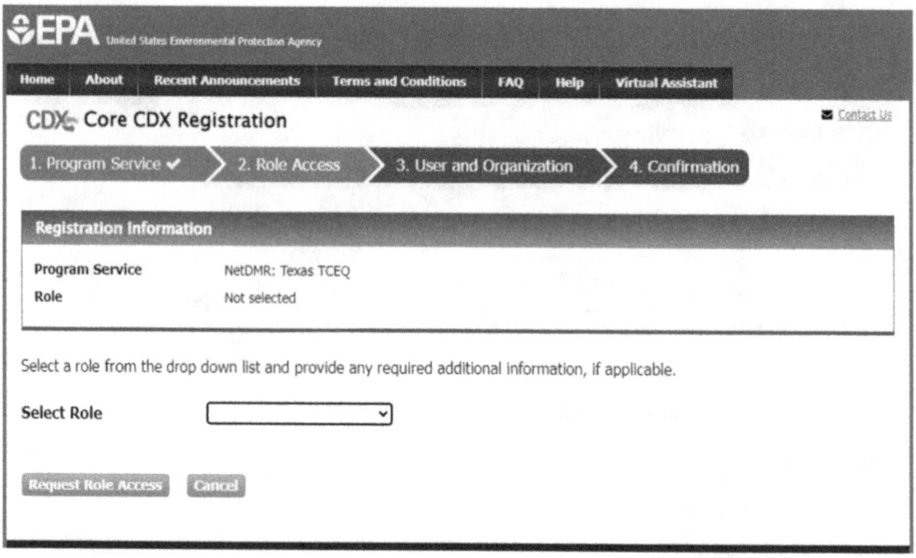

FIGURE 6.7   U.S. EPA CDX Role Access screen. *Courtesy of U.S. EPA*

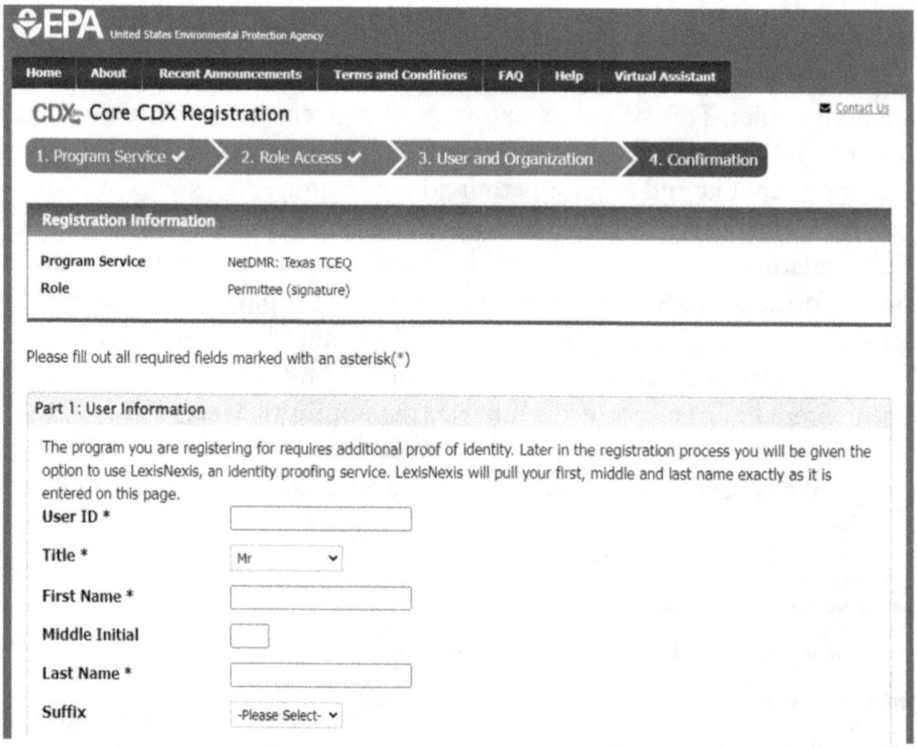

FIGURE 6.8   U.S. EPA CDX User and Organization Information screen. *Courtesy of U.S. EPA*

can now log in to your CDX account using the permit number and password you selected when you requested access (Figure 6.9).

3.3.2 My Account

When you have logged in to CDX, you will see a services screen that lists programs that you have access to (Figure 6.10). The My Profile tab allows the user to access account information, select security questions, and view pending access requests. To access your NetDMRs, click the role link for the NetDMR program. This will direct you to your regulatory agency's NetDMR home screen (Figure 6.11). From here, you can check your permit ID, view your regulatory agency's NetDMR contact info, and view any program news.

Click "Continue to NetDMR." On this screen, you can view your permits IDs, facilities, and the status of signed, upcoming, and in-progress DMRs (Figure 6.12). Select the appropriate DMR status that you wish to view and click "Search." When you find the appropriate DMR, you can view the completed DMR, edit or begin a DMR, and sign completed DMRs. For any questions, contact your regulatory agency's NetDMR administrator.

Information on the account (basic account information, security questions and/or answers, change the password, lock the account, etc.) can be changed by clicking on "Edit Account" (Figure 6.13).

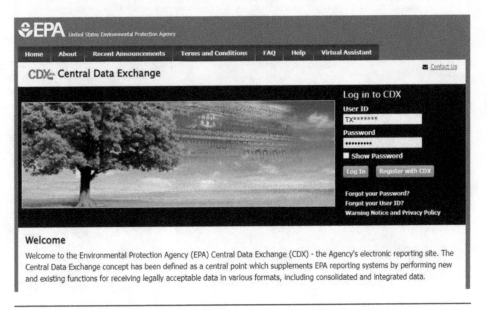

FIGURE 6.9   U.S. EPA CDX log in screen. *Courtesy of U.S. EPA*

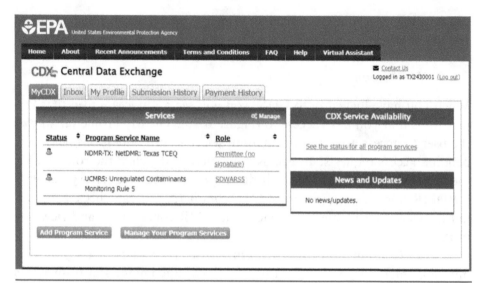

FIGURE 6.10   U.S. EPA User home screen. *Courtesy of U.S. EPA*

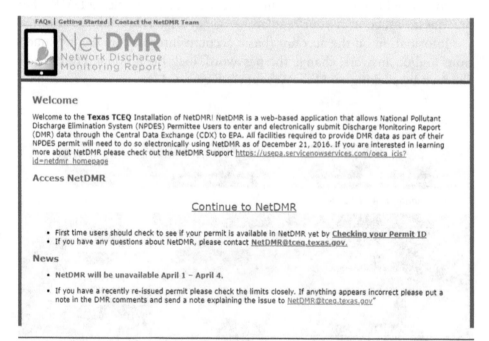

FIGURE 6.11   U.S. EPA NetDMR Welcome screen. *Courtesy of U.S. EPA*

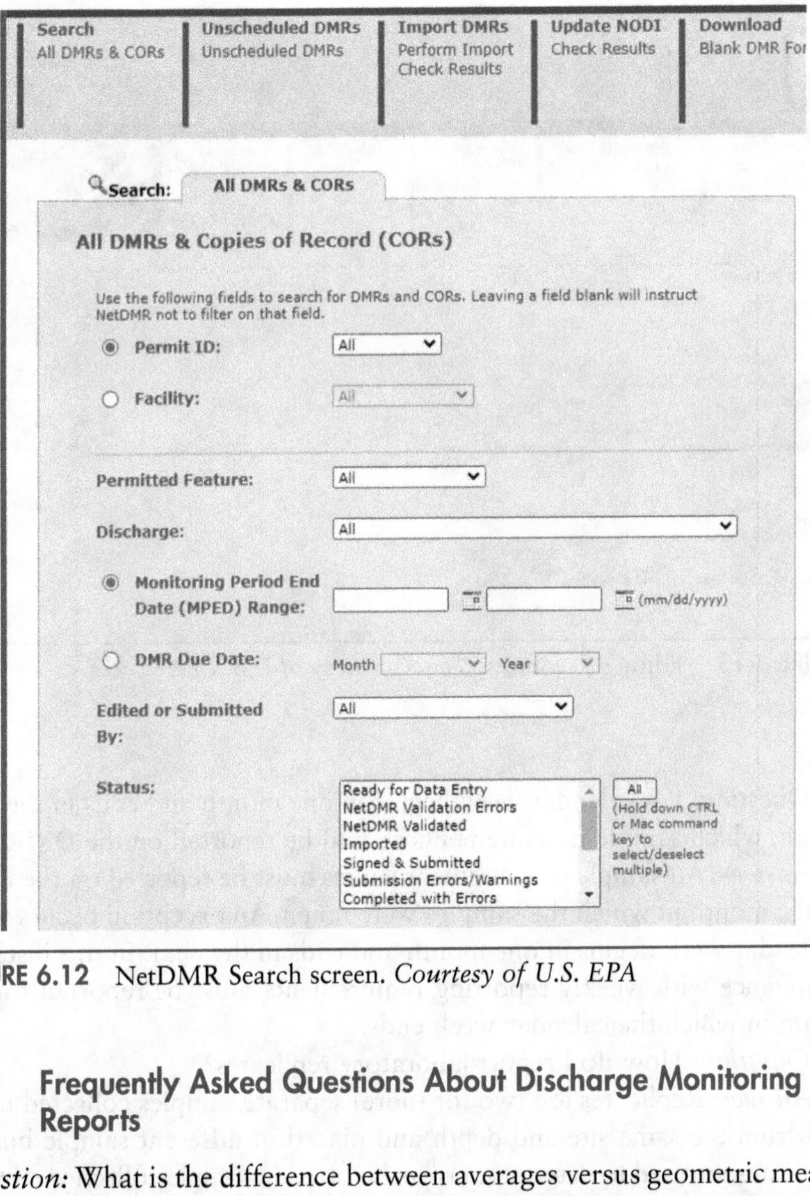

FIGURE 6.12 NetDMR Search screen. *Courtesy of U.S. EPA*

3.4 Frequently Asked Questions About Discharge Monitoring Reports

Question: What is the difference between averages versus geometric means?

Answer: Averages are based on adding numbers up and then dividing by the number of numbers. For example, 23 + 10 + 27 has a total of 60. Now, divide this number by the total number of values (in this example, divide by 3); the average is 20.

A geometric mean is based on multiplying a group of numbers, then finding the root of the numbers. Using the same aforementioned numbers to determine the geometric mean, multiply $23 \times 10 \times 27$; the product is 6210. Because there are 3 values, take the cube root of the product. The geometric mean is 18.

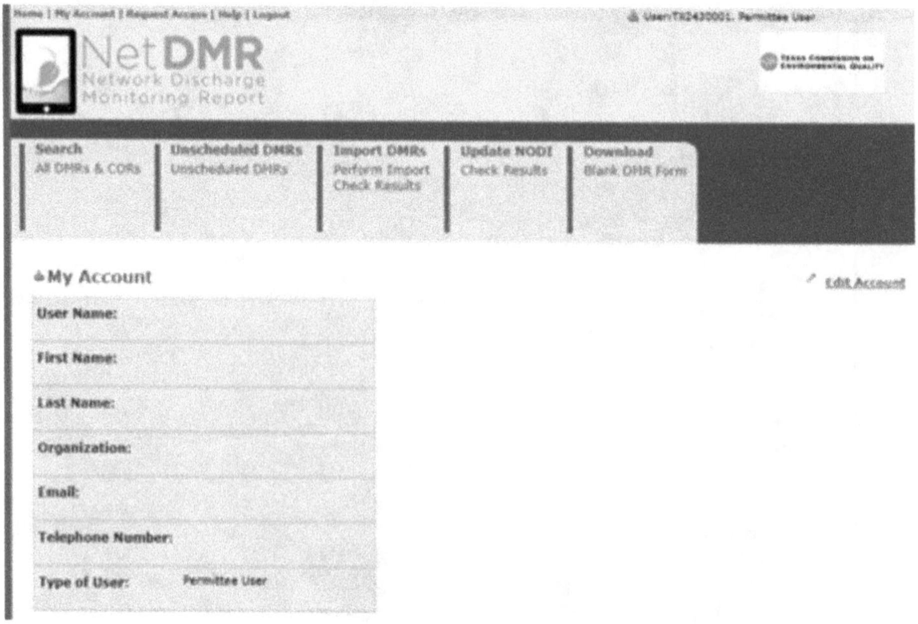

FIGURE 6.13   Editing Account screen. *Courtesy of U.S. EPA*

Question: If a calendar week begins in one month and ends in the next month, which sample measurements should be reported on the DMR?

Answer: All sample measurement values must be reported on the DMR for the month in which the samples were taken. An exception occurs when a calendar week begins in one month and ends in the next. In this instance, compliance with weekly reporting requirements must be reported for the month in which the calendar week ends.

Question: How do I report laboratory replicates?

Answer: Replicates are two (or more) separate samples collected in the field from the same site and depth and placed in different sample bottles. Replicates are used to determine errors involved in sample collection; if there are no errors in collection and analysis, the difference between two replicate analyses indicates the natural variability in the water at that location.

Replicates are part of the quality assurance and quality control (QA/QC) process, the results of which are not required to be reported on the DMR. Nevertheless, QA/QC data be retained by the facility and be available for inspection.

Question: How do I report laboratory duplicates?

Answer: Duplicates are two (or more) laboratory analyses that are performed on the same sample; for example, two separate pipets are used to draw samples from one sample bottle. Duplicates are used to determine the percent difference between two samples to estimate the error involved in analyses. For example, if a phosphate (PO_4^{3-}) number is reported as 3 μM, the error of that measurement must also be reported. If duplicates are within 10%, then the actual value could range from 2.7 to 3.3 μM.

Duplicates are part of the QA/QC process, the results of which are not required to be reported on the DMR. Nevertheless, QA/QC data should be retained by the facility and be available for inspection.

Question: What if my quality control failed?

Answer: If a quality assurance check fails, obviously the first obligation is to determine the cause and make a correction if possible. The test should be repeated. An invalid test is defined as any test that does not satisfy the test acceptability criteria, procedures, and quality assurance requirements specified in the test methods and permit. A repeat test should be conducted within the reporting period of any test determined to be invalid. If the analysis cannot be repeated after correction and if the analyst believes that sample data are in error, the value is reported with an explanation.

3.5   Common Errors When Completing the Discharge Monitoring Report

- Using maximum flow and highest biochemical oxygen demand concentration to calculate peak load
- Using average monthly flow and average monthly load to calculate average load
- Forgetting to sign the form

4.0   REFERENCE

U.S. Environmental Protection Agency. (2022). ICIS, NetDMR, NeT, and ECHO Support Portal web page. Retrieved March 9, 2022, from https:// usepa.servicenowservices.com/oeca_icis

Index

CPSIA information can be obtained
at www.ICGtesting.com
Printed in the USA
JSHW052256200423
40649JS00006B/132